Ultrasonic Nondestructive Evaluation Systems

Pietro Burrascano • Sergio Callegari
Augusto Montisci • Marco Ricci • Mario Versaci
Editors

Ultrasonic Nondestructive Evaluation Systems

Industrial Application Issues

Springer

Editors
Pietro Burrascano
Department of Engineering,
Polo Scientifico Didattico di Terni
University of Perugia
Terni, Italy

Sergio Callegari
Department of Electrical, Electronic
and Information Engineering "G. Marconi"
(DEI) and Advanced Research Center
on Electronic Systems for Information
and Communication Technologies
"E. De Castro" (ARCES)
University of Bologna
Bologna, Italy

Augusto Montisci
Dipartimento di Ingegneria Elettrica
ed (DIEE)
University of Cagliari
Cagliari, Italy

Marco Ricci
Department of Engineering,
Polo Scientifico Didattico di Terni
University of Perugia
Terni, Italy

Mario Versaci
Dipartimento di Ingegneria Civile, dell'
Universitä Mediterranea di Reggio Calabria
Reggio Calabria
Italy

ISBN 978-3-319-10565-9 ISBN 978-3-319-10566-6 (eBook)
DOI 10.1007/978-3-319-10566-6
Springer Cham Heidelberg New York Dordrecht London

Library of Congress Control Number: 2014956873

Printed on acid-free paper

Springer is part of Springer Science+Business Media (www.springer.com)

Quegli che pigliavano per altore altro che la natura, maestra de' maestri, s'affaticavano invano.

Those who took other inspiration than from nature, master of masters, were laboring in vain.

Leonardo da Vinci

Preface

This book originates from a nice discussion among colleagues with different engineering backgrounds and skills. We had so far worked in apparently distant fields, such as ultrasonics, sensing for nondestructive evaluation (NDE) of manufactured goods, signal processing for telecommunications, signal processing for earth exploration and geophysical engineering, and more. We had been involved in these areas in various capacities, for different applications and with different goals. During our exchanges, it clearly emerged that some of the difficulties encountered in the industrial implementation of ultrasonic testing techniques could be better addressed by pursuing a closer interaction among sibling working areas. Specifically, we realized that there were advantages in importing and adapting results or points of view originally developed in areas different from ultrasonic testing, such as telecommunications, radar, or geophysics.

The most immediate result of that conversation was a strong desire to cooperate to translate that intuition into action. This lead to the conception of a research project, targeting "*nondestructive ultrasound test based on pseudoorthogonal sequences for imaging and automatic classification of industrial products.*" The work plan was then funded by the Italian Ministry of Education, University and Research, in the framework of the Italian Research Programs of Relevant National Interest (PRIN).

In the 3-year activity of this research project, a stable network of academic cooperation grew, including working with research groups from different countries and gaining strength from close collaboration with leading industrial groups. In this period, we all gained experience by the opportunity to directly apply our techniques to the production line. We all benefited from a uniquely short path from conception, development through theoretical analysis and discussion, preliminary verification by simple laboratory tests, and real-world on-field validation.

This book aims at disseminating the body of coordinated knowledge developed in this context. With this monograph, our desire is to add a small tile to the extensive and thorough mosaic of scientific literature existing on this subject. Our aim is to offer a contribution based on a strongly interdisciplinary perspective and linked to real-world working examples of ultrasonic nondestructive testing (NDT) techniques for industrial applications. We would like to offer a view on some of the challenging problems that we encountered in our investigation; spark the readers'—and particularly the student readers'—curiosity on them and their many facets; and hint at approaches to tackle them when applying advanced techniques for signal and information processing in a systematic way.

As a consequence of these choices, the book lies at a somehow higher abstraction level than typical texts on the topic. While it certainly cannot ignore some technological aspects, it gives much space to "what happens after the probe," including signal processing and the extraction of useful information out of the collected data. To avoid appearing too abstract, speculative, or theoretic, we have striven to include a large number of practical applications and examples. Furthermore, we hope that this setup can be an opportunity to avoid overlap and enhance complementarity with other publications.

The book is primarily addressed at advanced technicians, engineers, and experienced teachers already involved in ultrasonic NDE and NDT inspection techniques. We expect the reader to be open to the use of advanced signal processing methodologies. For this reason, the book only summarizes the fundamentals of current diagnostic techniques in an introductory chapter. The interested readers who would like more details on fundamental or technological aspects are encouraged to refer to the excellent texts already available in the technical literature, including *Ultrasonic Nondestructive Evaluation Systems, Models and Measurements* from our same publisher.

Although the book is not meant to have a strictly educational purpose, we believe that our approach, always very practical, can have a significant educational value. Student readers can certainly get insight from our attempt at always tying theoretical considerations to implementation and the practice of production systems. That is why all chapters include a significant number of examples. In addition, we have dedicated the entire Part IV to examples of industrial application where the techniques proposed in the previous chapters are put to work.

The book is organized by growing abstraction levels, starting from the physics of ultrasonic NDE, going up to signal processing, information processing, and applications. This is reflected in a structure made of four parts, devoted to:

Part I: A description of the ultrasonic system at the physical level. How to analytically model and numerically simulate the system;

Part II: A description of how to improve system performance by the application of signal processing tools;

Part III: Considerations on how to deal with the collected data by means of high level information processing techniques;

Part IV: Real-world applications of processing techniques in the context of industrial ultrasonic NDE systems.

We chose to include a wide range of application examples to show how the great flexibility of the ultrasonic technique, well known to all who already work in this sector, can further broaden the horizons of this technique by suggesting new applications, thus helping push its limits. We believe that ultrasonics can be applied in industrial evaluation on an even wider scale as long as it sees widespread adoption of advanced signal processing, and information extraction and classification techniques.

We are confident that the research that we carried out with enthusiasm, the intuition of Springer who proposed us to write this text, and our work of publishers and authors can give a small contribution in a field, which is already very important today, but whose full potential remains to be explored.

Terni
July 2014

Pietro Burrascano

Introduction: Generalities on Ultrasonic Nondestructive Testing Systems and Organization of the Book

This volume provides a description of how numerical techniques, signal processing, and information processing can be deployed at different levels in order to improve the effectiveness of ultrasonic nondestructive systems designed to test objects delivered by industrial processes. Before addressing the central issue of the book in its many aspects, it is useful to give a brief description of ultrasound techniques as they are commonly used, underlining their strengths as well as the aspects that may instead show limitations. This introductory chapter provides an overview of the main lines of ultrasonic investigation techniques. Readers interested in deepening their knowledge with respect to these aspects, as well as some technology related issues, can also refer to [4]. The objective here is to describe a standard method of ultrasonic inspection, a reference methodology against which to show the advantages and limitations of the digital signal processing techniques applied to ultrasonic inspection. In the subsequent chapters of this book we will analyze how to address these constraints effectively through digital signal processing techniques.

Background

Ultrasonic testing [2, 3] is a technique in which beams of high-frequency elastic waves are fed into materials for the detection of surface and subsurface discontinuities. The sound waves travel through the material with some associated loss of energy (attenuation) and are partly reflected at interfaces.

Ultrasonic inspection, in its different configurations, detects flaws by monitoring changes in the characteristics of the propagation of elastic waves, and in particular:

- reflections of the propagating waves generated at interfaces consisting of material boundaries or discontinuities within the material itself;
- discontinuities of the time of transit of the propagating wave through the piece under test from the emitting transducer to the receiving transducer;
- attenuation of elastic waves by absorption and scattering within the test piece;
- modifications in the time or spectral response for either a transmitted or a reflected signal.

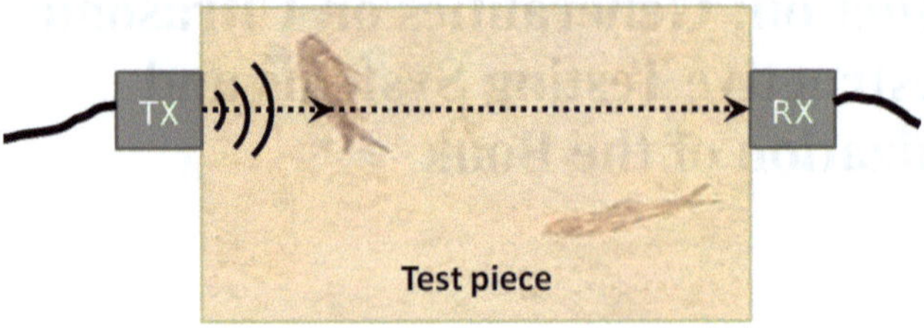

Fig. 1 Through-transmission technique with a transmitter (*TX*) and a receiver (*RX*)

Throughout the chapter, if not otherwise specified, we refer to the propagation of *bulk waves*. Rayleigh (surface) waves or guided waves can also be used for ultrasonic nondestructive testing (NDT) applications, but require different modeling approaches and different transducers from those considered in this book (see, for instance, [4] with particular respect to App. E). Moreover, we assume that the amplitude of vibrations in the propagation media impose stresses well below the elastic limit, thus preventing permanent effects on the parts.

Furthermore, unless explicitly specified, the study of the reflections of the propagating waves will be considered, in the lines of a pulse-echo (PuE) analysis. In this respect, note that the most common techniques for testing a sample of material are *transparency* or through-transmission (TrT), PuE, and *tomography*.

In test methods for transparency or TrT, the emitter and receiving transducers are arranged in corresponding points on two opposite surfaces (Fig. 1). The distance of the defect can be detected, for instance with an oscilloscope, as reflecting flight time or as echo delay.

In PuE techniques, the ultrasonic waves hit the item to be tested, penetrate into it and are then reflected and refracted by both the surfaces bounding the component itself and by the defects (Fig. 2). The reflections (echoes) provide information on the presence, position, and type of defects in the item.

The tomographical ultrasonic technique consists in imaging the object under test by sections (sectioning), in order to produce 2-D and 3-D cross sectional images (Fig. 3).

The focus of this book is on PuE and TrT techniques: in the simple case in which the elastic wave is of the sinusoidal type, we can define the wavelength λ [meter], a period T [second], and a frequency f [hertz]. These quantities are related to each other by the relationships below, where c [meter per second] is the speed of propagation of sound in the material under test.

$$\lambda = \frac{c}{f} = c \cdot T \Rightarrow c = \lambda \cdot f \tag{1}$$

The following definitions apply:

Period T [second]: the time taken for a cycle of expansion and complete contraction of source, or pressure and depression in the medium;

Frequency f [hertz]: the number of cycles (namely, the number of complete oscillations), per second;

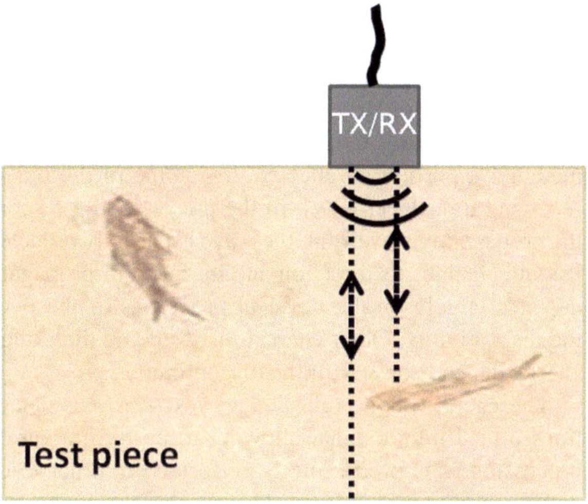

Fig. 2 Pulse-echo technique with a single transducer

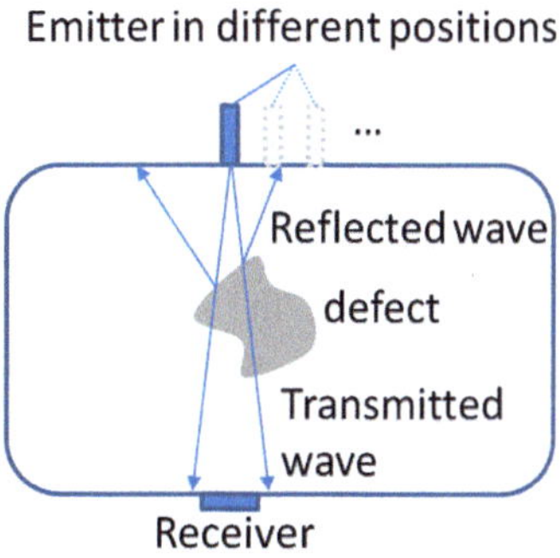

Fig. 3 Tomography principle

Acoustic pressure [pascal]: the variation of the instantaneous pressure with respect to its static component (the atmospheric pressure) present in a given point of the medium;

Wavelength λ [meter]: the distance between two successive wavefronts in the direction of propagation;

Propagation speed c [meter per second]: the propagation speed of the wavefront in the direction of propagation;

Vibration intensity I: the amount of energy that passes through the unitary area in a second. Usually it is measured as the relative magnitude A_1, or intensity I_1, with respect to a reference value A_2 or I_2 in decibel as:

$$I_{\text{dB}} = 20\log\frac{A_1}{A_2} = 20\log\frac{I_1}{I_2} \tag{2}$$

Density ρ [kilogram per cubic meter]: the mass per unit volume of the medium;

Specific acoustic impedance z [kilogram per second-square meter]: for plane waves is equal to the product between density ρ and velocity of propagation c.

An ultrasonic PuE scan system can be compared to a wide-angle searchlight and an observer working besides the light: only if the beam strikes an object, a portion of the emitted light is reflected and returns to the observer to indicate the presence of the object. Similarly, in the ultrasonic scan system, a little of the energy emitted by an transmitting transducer (TX) travels to a discontinuity in the propagating medium, a portion is reflected, and some of the reflected energy is detected by a receiving transducer (RX). In the presence of a discontinuity that affects a part of the propagating wavefront, the wave is partially reflected, and the part that continues its path changes its direction and its speed of propagation. The degree of reflection depends largely on the physical properties of the two elastic media which create the discontinuity. Other characteristics being the same, the amount of the reflected energy depends on *size* of the discontinuity.

The energy of the reflected waves (*echoes*), detected at the receiving end, is transformed into a signal allowing to perform a variety of tasks after appropriate processing. The processing is performed in order to optimize the detection of the defect, in case it is present. This means, for example, determining the position of the defect in the test piece, determining its characteristics, being able to extend the range of investigation exceeding the superimposed background noise, being able to detect defects of small size.

A number of different measurement setups can be adopted: PuE analysis measures the interval required for the acoustic wave to make the trip from the TX to the object and back to the RX that is located at the same position of the TX or quite close to it. Because the speed of propagation in an homogeneous medium is constant, range position, namely the reflection depth, can be determined with an accuracy corresponding to the accuracy in the measurement of the time interval. In this, two different schemes are possible. In the first one, which is by far the most widely used, the TX operates in short bursts and the range is measured in terms of the time interval between the transmission of a pulse and the reception of its echo: this is the traditional PuE setup. In the second scheme, derived from radar techniques [5], the frequency is periodically varied at a constant rate according to a linear rule, and the distance to the reflecting object is determined from the frequency difference between the echo and the wave being emitted at the time the echo returns. This is the principle leading to the frequency-modulated continuous wave (FMCW) ultrasonic radar described in Chap. 12.

Throughout this chapter we will refer to the pulsed system, whose basic operating principle is illustrated by the block diagrams in Fig. 4.

In the traditional case of a pulsed system, the transmitter supplies energy to the TX in the form of pulses of high voltage and brief duration T. In the most classical situation, during this brief period, the waveform has a sinusoidal shape, as indicated below in Fig. 5. Typical frequencies are in the megahertz range. For the rest of the time the transmitter is inactive. The time of operation of the transmitter, denoted as *pulse duration* T is in the order of a few micro second.

The total time between the start of two subsequent pulses is called *repetition period* and is indicated in the following as T_r. Its reciprocal is the *repetition frequency* f_r, i.e., the number of bursts occurring in each second. A typical value of T_r is in the

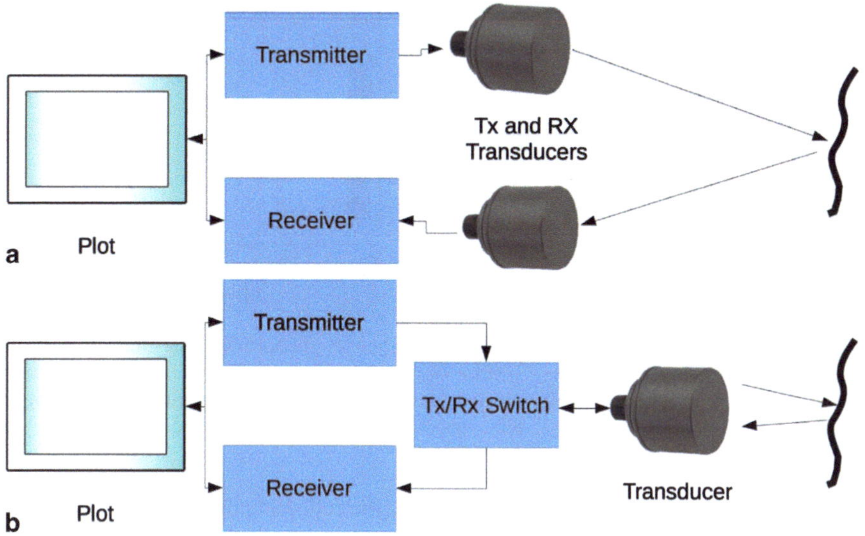

Fig. 4 Schematic representation of the setup for pulse-echo measurements. **a** Setup with separate transducers **b** Setup with a single transducer

order of millisecond with the corresponding value of f_r in the kilohertz range. Note that the transmitted pulse duration T is typically only a very small fraction of the *repetition period* T_r. The ratio between the two quantities is the *duty ratio* of the transmitter. For the typical values introduced above, the duty ratio can be in the order of 10^{-3}. In several cases, the same transducer is used for both transmission and reception, as shown in Fig. 4. This implies that during the first part of each period the transducer is fed by the transmitting channel. After a time interval $T_x > T$, the transducer connection is switched from the transmitting channel to the receiving channel, and this implies that a fraction of the *repetition period* T_r is taken before the switching action is completed. If a reflection occurs before the switching, the system can not detect it. Consequently, the advantage of operating with a single transducer comes at the cost of a dead zone: if there are reflections at a small distance from the transducer, these may lead to reflections that return with just a small delay with respect to the transmitted pulse. If this delay is not enough for the switching action to complete, the system can not detect the echo. In the following, we consider the system as if it operates with two separate transducers for the transmission and for reception, as in Fig. 4.

The transmitter is required to produce a very high level of emission in its brief period of operation. This is needed because of attenuation during propagation, to assure that the tiny portion of it that returns in the echo is still detectable. Because the pulse is very short, a very large transmitter power is required to produce an ultrasonic pulse of sufficient energy. Moreover, because the duty ratio is small, the *average power* output of the transmitter is not large.

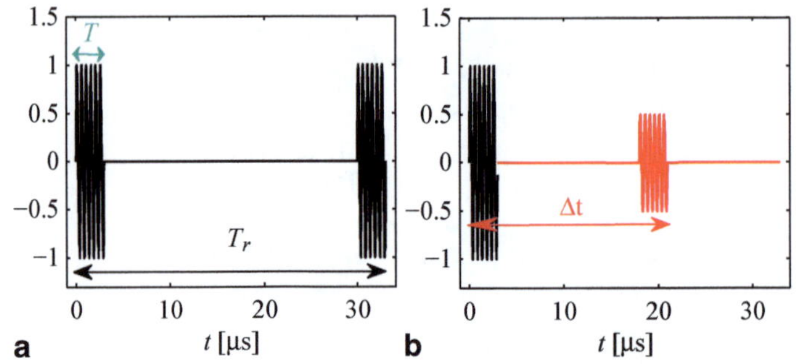

Fig. 5 Schematic representation of transmitted pulses and received echos. **a** The transmitted signal, where the pulse duration T and the repetition period T_r are made evident. **b** A representation of the transmitted and received signals, where the latter is delayed with respect to the former by an interval Δt. For the sake of representation, images are not in scale with respect to amplitude

The elastic waves traveling from the TX to the reflecting discontinuity and back, generate a signal at the receiving end: if only a small part of the wavefront is involved in the phenomenon of reflection (i.e., if the discontinuity is of small size), the received signal takes a small amplitude. In this regard, note that Fig. 5 represents the transmitted and received signals with different scales, for representation purposes. If the scales of the two pulses were the same, in a typical case the received signal amplitude would not even be visible. On a conventional transducer, the amplitude of the emitted signal may be in the order of hundreds of volt, and the received one just as large as some millivolt, with an amplitude ratio that may reach or exceed 10^{-6}.

In Fig. 5, the received pulse has a delay Δt referred to the transmitted pulse. The time interval Δt is the time required by the propagating wave to travel to the reflecting discontinuity and back. If r is the distance from the transducer to the reflecting surface and we assume as known the propagation speed c_{medium}, measured in meter per second, the time required for the round trip is $\Delta t = 2r/c_{\mathrm{medium}}$, and the distance can be estimated as: $r = (c_{\mathrm{medium}}/2)\Delta t$. Table 1 reports typical propagation speed values in some propagation media of interest.

Example 1
If the propagation is in iron, where the propagation speed is $c_{\mathrm{medium}} = 4910$ m/s, measuring a reflection located at 3 m implies a $\Delta t = 2r/c_{\mathrm{medium}} = 2(3/4910)\mathrm{s} = 1.22$ ms. If the propagation was in water, with $c_{\mathrm{medium}} = 1482$ m/s, the same distance would have been associated with a delay of $\Delta t = 2r/c_{\mathrm{medium}} = 2(3/1482)$ s $= 4.05$ ms.

Table 1 Typical propagation speeds of frequently found propagating media

Material	Propagation speed [m/s]
Iron	4910
Beryllium	13,000
Chromium	5940
Mercury	1407
Thallium	818
Concrete	3200–3600
Dry air	344
Water	1482

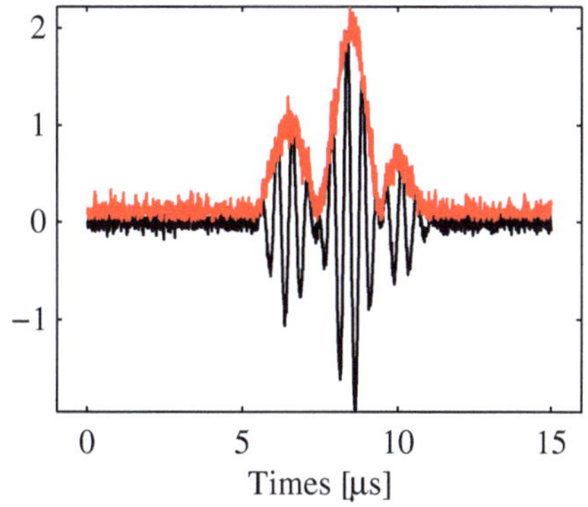

Fig. 6 Typical behavior of the signal at the receiving end, in the presence of more than one reflector. In the figure, pulses are represented by the amplitude of their detected envelope

The simple scheme outlined so far already enables an intuitive appreciation of some of the issues that the designer needs to address, among which are the resolution in range, the effects of superimposed noise, and the characteristics of transducers.

Range Resolution

In the simplest scheme, the behavior of the signal at the receiving end in presence of more than one reflector is as represented in Fig. 6: the horizontal axis represents time and consequently the range, as long as a constant propagation speed can be assumed. The vertical axis represents the amplitude of the envelope of the received signal, where the different *pseudopulses* of different amplitudes and shapes correspond to the echoes from various objects. The duration of the pseudopulses in Fig. 6 can be related to the size, along the range direction, of the objects causing them. This relationship would be exact if the transmitted pulses were of negligible duration. Because the transmitted pulse duration is greater than zero, the apparent duration is increased by an amount depending on the pulse duration itself: the portion of the received signal representing the emitted pulse is convolved with the portion of the signal representing the reflection of the discontinuity obtained after an ideal pulse wave. This relationship will be better clarified in the following chapters. If the distance between two different objects is less than the range equivalent to the pulse duration, the echo pulses run together and the system is not able to show them separately. Following these lines, it is customary to introduce the concept of *range resolution* to indicate the ability to separate objects that are close together along the range direction. Clearly, the range resolution is related to the duration of the transmitted pulse. The need for good range resolution is the principal reason why short pulses are used in ultrasonic PuE systems. However, this means trading resolution for signal-to-noise ratio (SNR). In fact, short pulses cannot have a large energy content (energy being the product of power and duration). Consequently, they also imply a low value of received pulse energy, with all the consequences that this may have, due to the presence of background noise.

Noise

A further aspect that must be considered in this preliminary description of ultrasonic inspection systems is the presence of noise, i.e., of small, randomly fluctuating voltages that unavoidably are present at the input section of the receiver. Noise is unwanted but cannot be avoided in physical systems. Were it not for noise, the maximum range at which a discontinuity can be detected would be extendable almost indefinitely. In absence of noise, even the weakest echo could be amplified to a usable level by a suitable amplifier. However, because of noise, a sensitivity limit is reached when the signal level falls below the noise to such an extent that the useful content is obscured. In this case, an increase of amplification does not help to overcome the problem, because signal and noise would get amplified together. The receivers are designed with an amplification such that large amplitude noise voltages are present at the output when full gain is used: the limit of sensitivity is then set by the noise. The input circuits of the receiver are the most critical parts of the entire electronic system, and this is precisely because of the noise: they must use the small echo signal as effectively as possible, while introducing the least possible amount of noise on top of it. The highest possible SNR is desired so that the signal can be discernible. Subsequent stages are not so critical. In fact, they amplify both the signal and the noise received by the initial stage in the same manner and the noise that they introduce themselves is negligible in comparison to what had already been superimposed.

The noise-induced limitation of the receiver sensitivity is the main reason for the following consideration: *the ease in identifying the presence of an echo pulse depends on the energy of the pulse rather than its peak power.* This is worth underlining: it would be wrong to assume that the peak power is the most important parameter to be taken into account. For example, one may think that halving the duration of the pulse and simultaneously doubling the value of the peak power, while maintaining the energy content unchanged, can increase the pulse amplitude and thus allow the useful signal to come off better from the noise. This would be wrong. To understand why this view is incorrect, one should refer to the concept of signal bandwidth: in order to amplify pulse signals, the receiver must have a bandwidth centered about the nominal frequency of the echo pulse and large enough to accommodate the signal bandwidth. For pulsed signals with very short duration, the bandwidth is inversely proportional to the duration. Thus, halving the pulse duration implies that the receiver bandwidth is doubled. Unfortunately, noise power is proportional to receiver bandwidth, and thus when the duration of the constant energy pulse is halved, not only the power of the pulse but also the power of the noise with which the pulse must be compared must be doubled. Therefore, no improvement is obtained in the ability to identify the presence of an echo. We will return on this subject dealing with *pulse compression techniques* [1] in Part II.

Acoustic Features of Ultrasonic Transducers and Signal Design

Ultrasonic transducers can be applied to estimate a variety of quantities that characterize the physical structures under test; moreover ultrasound measures can be used in a variety of real-world problems. Time of flight (ToF) and wave velocity measures are employed in tasks such as defect detection, thickness measurements, assessment of elastic and mechanical properties, robotics, remote sensing, and surface and internal imaging. Ultrasonic spectroscopy, i.e., measurements that prioritize the frequency characteristics of the response, are applied to the analysis of the microstructure of the system under test (as, for example, grain size in metals or porosity in carbon fiber-reinforced polymers). This variety of measurement possibilities and application fields leads to a variety of parameters needed to define the specific transducer (e.g., inner physical structure, size, ability to operate in hostile environments, coupling characteristics, just to name a few). We will not deal with all these aspects in this chapter, and will only concentrate on the acoustic parameters that characterize the transducer. The main quantities to be considered for our purposes are:

Nominal frequency: typically expressed as the average of the lowest and highest points at a $-6[dB]$ level on the amplitude response vs frequency diagram;

Bandwidth: i.e., the difference between the highest and lowest frequency points introduced above. Note that the bandwidth is in some cases expressed as a fraction of the nominal frequency;

Sensitivity: typically expressed with reference to a given transmission medium, defined as $-20 \log(V_x/V_0)$, where V_0 is the excitation pulse in V, and V_x is the received signal, also measured in volt;

Signal-to-noise ratio (SNR): defined as $20 \log(V_x/V_n)$, where V_x is the received signal amplitude in volt, and V_n is the noise floor, also in volt.

As a thumb rule, the maximum spike excitation voltage is related to transducer thickness and thus, in turn, to the operating frequency. For instance, transducers operating at relatively low frequencies (5 MHz and lower) can be driven up to 500–600 V. Doubling the frequency (e.g., with 10 MHz transducers) leads to a much thinner piezoelectric layer, and thus the driving voltage should be halved. Moreover, it can be noted that piezoelectric transducers are designed for short-burst excitation:

continuous wave or long-tone bursts, i.e., bursts active for a relatively large fraction of the repetition time, can be used, but overheating the transducer must be avoided.

In designing the signal to be transmitted by the ultrasonic system, the bandwidth of the transducer plays a fundamental role: if the nominal frequency of the transducer is f_0, the simplest signal to drive the transducer is a sinusoidal pulse of amplitude A and frequency f_0, time limited by a rectangular window of duration T. This is an amplitude-modulated signal, for which a complex representation can be adopted. Hence, the excitation $s(t)$ can be written as:

$$s(t) = \begin{cases} Ae^{2i\pi f_0 t} & \text{for } 0 < t \le T \\ 0 & \text{otherwise} \end{cases} \tag{3}$$

Let us assume, in a first approximation, that the return signal $r(t)$ is just a time-shifted, attenuated and delayed replica of the emitted signal with superimposed white Gaussian noise $n(t)$. With this, $r(t)$ takes on the form:

$$r(t) = \begin{cases} KAe^{2i\pi f_0(t-\Delta t)} + n(t) & \text{for } \quad 0 < t \le T \\ n(t) & \quad\quad \text{otherwise} \end{cases}$$

where $K < 1$ represents the attenuation and Δt is the delay.

At the receiving end, a filter is present in order to minimize the effect of environmental noise: from the theory of matched filters we know that, in order to maximize the SNR, the optimum filter to detect the received signal in additive white Gaussian noise (AWGN) is the one whose impulse response is a reversed replica of the sent signal. For the filter output $y(t)$, we have:

$$y(t) = \int_0^t s(\tau)r(t-\tau)\mathrm{d}\tau \tag{4}$$

In this simple case, $y(t)$ can be expressed in closed form as:

$$y(t) = KA^2 \Lambda\left(\frac{t-\Delta t}{T}\right) e^{2i\pi f_0(t-\Delta t)} + \hat{n}(t) \tag{5}$$

where $\hat{n}(t)$ derives from the convolution between the noise and the impulse response of the matched filter, and $\Lambda(\cdot)$ is the triangle function, generated by convolving two rectangular signals; $\Lambda(\cdot)$ is different from zero only when its argument is in $[-T/2, T/2]$.

If more than a single reflection occur after the received signal has been processed by the matched filter, the echoes produce separated pulses only if the reflecting objects are sufficiently separated. Figure 7 shows two different situations: the two pulses must be separated by a time interval at least equal to the pulse duration T so that the maxima of both pulses can be separated. This situation is shown in plots Fig. 7 a, b. If this condition is not satisfied, the two triangular waveforms get mixed,

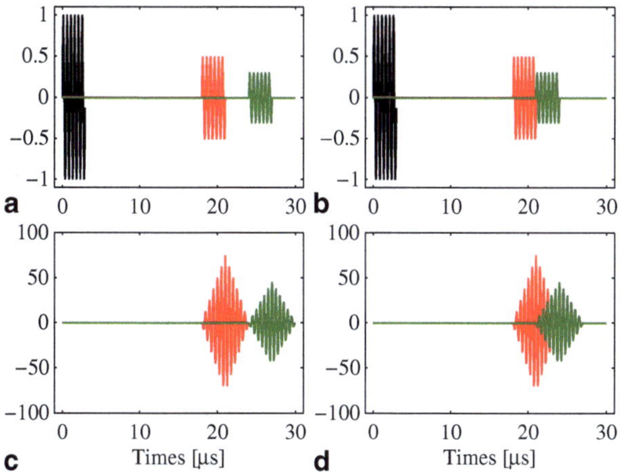

Fig. 7 Sample responses obtained when a double reflection occurs. **a** The two echo pulses received by the transducer are separated by a time interval larger than T. **b** The time separation between the two echos is less than T. **c** The response obtained from case **a** is illustrated. After the matched filter, the maxima corresponding to the two pulses are clearly separable. **d** The response obtained from case **b** is shown. Here, the responses corresponding to the two signals overlap and the two maxima cannot be separated

and therefore, become impossible to separate. In terms of time, T is the minimum distance allowing the two pulses to be distinguishable. Since in a time interval T the elastic wave travels a distance equal to $c_{\text{medium}}T$, having to consider the round-trip propagation, we have that the range resolution with a sinusoidal pulse of duration T is $c_{\text{medium}}T/2$.

Apparently, to increase the range resolution, one needs to reduce the length T of the pulse. However, as already noted, a reduction in T, operated while keeping the same pulse amplitude, reduces the SNR. In fact, the instantaneous power of the received pulse is $P_r(t) = |r(t)|^2$. The energy contained in a period of the signal is:

$$E = \int_0^T P_r(t)\mathrm{d}t = K^2 \frac{A^2}{2} T \tag{6}$$

Indicating as σ the standard deviation of the noise, the SNR at the receiver is: $(K^2A^2T)/(2\sigma^2)$ which shows that, on equal basis for the other parameters, the SNR is proportional to the pulse duration T. In these conditions, it is evident that an increase of range resolution conflicts with an increase of the SNR.

Organization of the Book

The book builds upon the previous general concepts, discussing many related aspects. These include means to address the inherent limitations of the most basic system and to extract useful information out of the sensed data. To this aim, the book is structured in levels, progressively moving from the physical one to the application. In better detail, the organization is the following:

Part I provides a description of the ultrasonic system at the *physical level*. Here, Chap. 1 is dedicated to *waves propagation* and describes the physical phenomena and the governing equations. Furthermore, the chapter addresses the different kind of waves that may arise, described from both the physical and analytic point of view, as well as the effects caused by a density change in the medium where the waves propagate;

Chapter 2 is concerned with the *numerical simulation of wave propagation* and gives an overview on the finite element method (FEM). Furthermore, it describes how to use FEM for an accurate analysis of waves propagation;

Chapter 3 considers *modeling ultrasounds for NDT applications* and discusses how FEM techniques can be used to estimate the effects of the presence of defects on the propagation of waves. To this aim, the chapter considers how to describe the behavior of waves in passing through an interface and the problem of modeling the transmitting and receiving transducers.

Part II deals with the *signal processing level* and proposes how to improve the ultrasonic system by means of suitably designed excitations and effective operations on signals. In detail, Chap. 4 deals with aspects related to *excitation and deconvolution in ultrasonic NDT systems*. After a quick review of PuE systems, looking at their strengths and limitations, a minimal channel model is defined to motivate the introduction of pulse-compression (PuC) systems and to discuss how to evaluate the performance of a PuC architecture by means of appropriate merit factors. The chapter also discusses how *linear chirps* (LChirps) and *nonlinear chirps* (NLChirps) excitation signals can be used in PuC systems and how they compare. It is illustrated how NLChirps can help matching the hardware characteristics. Particularly, the possibility of adjusting the characteristics of the nonlinear chirp in order to adapt it to the probe features is focused.

Chapter 5 looks at *excitations and signal processing for multiprobe setups*. Systems where multiple transducers are available are considered, reviewing different ways of using them. After hinting at the classical approach of coordinating the transducers in arrays of nearby elements in order to obtain focused beams (phased arrays), the problem of using spatially distributed probes to get different views at the material under test is considered. In this arrangement, if the TXs and RXs are all to be used simultaneously, one gets a multiple input, multiple output (MIMO) system. To deal with this case, the minimal channel is extended and so are the merit factors to evaluate system performance. Intuitively, to avoid interference, the excitations must be made orthogonal to each other. The problem of generating *families* of excitation signals with good orthogonality, while preserving other properties indispensible for PuC is investigated.

Part III is devoted to *information processing* and considers how to use the data collected from the probes in order to gather high-level information, such as defect classification. This part of the book is mainly focused on how to use *soft computing* techniques for these goals. Specifically:

Chapter 6 shows how to take into account the residual uncertainties and vagueness inherent in any measurement process and how this applies to ultrasonic inspection for NDT/nondestructive evaluation (NDE) of industrial artifacts;

Chapter 7 illustrates how the classification of defects in ultrasonic NDT/NDE can be mapped onto the concept of fuzzy similarity (FS), so that data can be processed by considering appropriate labels in a fuzzy set of points put together for similarity.

Part IV provides the reader with many examples of *real world applications of ultrasonic NDE systems*, including *industrial application issues*. Furthermore, this part offers an example of how the many techniques described in the previous parts can be deployed together for educational purposes. This part consists of five chapters:

Chapter 8 covers the application of *ultrasonics for the imaging of concrete*;

Chapter 9 considers the *ultrasonic inspection of large forgings*;

Chapter 10 deals with the use of *fuzzy geometrical techniques for characterizing defects in ultrasonic NDE*;

Chapter 11 illustrates some *industrial applications of non-contact ultrasonic techniques*;

Chapter 12 is about the use of the *ultrasonic frequency-modulated continuous wave* technique *for range estimation*.

Bibliography

1. Cook CE (1960) Pulse compression—key to more efficient radar transmission. Proceedings of the IRE 48(3):310–316. doi:10.1109/JRPROC.1960.287599
2. Kundu T (2012) Ultrasonic and electromagnetic NDE for structure and material characterization: engineering and biomedical applications. CRC, Boca Raton
3. Mix PE (2005) Introduction to nondestructive testing: a training guide. Wiley, New York
4. Schmerr L, Song JS (2007) Ultrasonic nondestructive evaluation systems: models and measurements. Springer, Berlin
5. Skolnik MI (1970) Radar handbook. McGraw-Hill handbooks, vol 1. McGraw-Hill, New York

Acknowledgements

Part of the material included in this book is the result of research activity conducted with the financial support of the Italian Ministry for Education, University and Research in the framework of Research Programs of Relevant National Interest (PRIN). Specifically, cooperation between the authors was fostered via a grant for the PRIN 2009 Project "Non-destructive ultrasound test based on pseudo-orthogonal sequences for imaging and automatic classification of industrial products."

Contents

Contributors

David Arthur Hutchins School of Engineering, University of Warwick, Coventry, UK

Luigi Battaglini Department of Engineering, Polo Scientifico Didattico di Terni, University of Perugia, Perugia, Italy

Pietro Burrascano Department of Engineering, Polo Scientifico Didattico di Terni, University of Perugia, Perugia, Italy

Matteo Cacciola DICEAM Department, Mediterranea University, Reggio Calabria, Italy

Salvatore Calcagno DICEAM Department, Mediterranea University, Reggio Calabria, Italy

Sergio Callegari Department of Electrical, Electronic and Information Engineering "G. Marconi" (DEI) and Advanced Research Center on Electronic Systems for Information and Communication Technologies "E. De Castro" (ARCES), University of Bologna, Bologna, Italy

Salvatore Caporale Advanced Research Center on Electronic Systems for Information and Communication Technologies "E. De Castro" (ARCES), University of Bologna, Bologna, Italy

Sara Carcangiu Dipartimento di Ingegneria Elettrica ed (DIEE), University of Cagliari, Cagliari, Italy

Lee Andrew John Davis School of Engineering, University of Warwick, Coventry, UK

Renato Forcinetti Dipartimento di Ingegneria Elettrica ed (DIEE), University of Cagliari, Cagliari, Italy

Stefano Laureti Dipartimento di Ingegneria, Polo Scientifico Didattico di Terni, University of Perugia, Terni, Italy

Augusto Montisci Dipartimento di Ingegneria Elettrica ed (DIEE), University of Cagliari, Cagliari, Italy

Francesco Carlo Morabito DICEAM Department, Mediterranea University, Reggio Calabria, Italy

Isabella Palamara DICEAM Department, Mediterranea University, Reggio Calabria, Italy

Diego Pellicanò DICEAM Department, Mediterranea University, Reggio Calabria, Italy

Marco Ricci Department of Engineering, Polo Scientifico Didattico di Terni, University of Perugia, Perugia, Italy

Luca Senni Department of Engineering, Polo Scientifico Didattico di Terni, University of Perugia, Perugia, Italy

Mariangela Usai Dipartimento di Ingegneria Elettrica ed (DIEE), University of Cagliari, Cagliari, Italy

Mario Versaci DICEAM Department, Mediterranea University, Reggio Calabria, Italy

Acronyms

ADC	Analog to digital converter
AI	Artificial intelligence
ANFIS	Adaptive neuro-fuzzy inference system
APC	Acyclic pulse compression
AVG	Abstand verstärkung grosse
AWG	Arbitrary waveform generator
AWGN	Additive white gaussian noise
BP	Band pass
CAD	Computer-aided design
CC	Crosscorrelation
CFRP	Carbon fiber-reinforced plastics
CPC	Cyclic pulse compression
CW	Computing with words
CZT	Chirp z-transform
DAC	Distance amplitude correction
DGS	Distance gain size
DTT	Direct transmission technique
DFT	Discrete Fourier transform
EBP	Energy bandwidth product
ED	Explanatory database
EMAT	Electromagnetic–acoustic transducer
ESD	Energy spectral density
FDM	Finite difference method
FE	Fuzzy entropy
FEM	Finite element method
FFT	Fast Fourier transform
FIR	Finite impulse response
FIS	Fuzzy inference system
FL	Fuzzy logic
FM	Frequency modulation
FMCW	Frequency-modulated continuous wave
FMF	Fuzzy membership function

FR	Fuzzy relation
FS	Fuzzy similarity
FS	Fuzzy similarities
FSL	Far sidelobe level
FSO	Fuzzy subsethood operator
GCS	Golay complementary sequence
GENFIS	Generate fuzzy inference system
GI	Generalized index
IDS	Initial data set
IFFT	Inverse fast Fourier transform
INT	Parabolic analytical interpolation
IRS	Inverse repeated sequences
LChirp	Linear chirp
LFSR	Linear feedback shift register
LI	Linear index
LMS	Least mean square
LNA	Low noise amplifier
LTI	Linear time invariant
MDD	Minimum detectable defect
MFEDM	Minimal fuzzy entropy decisional model
MIMO	Multiple input, multiple output
MISO	Multiple input, single output
MLS	Maximum length sequence
MLW	Main lobe width
MMM	Minimum detectable defect
NDE	Nondestructive evaluation
NDT	Nondestructive testing
NFS	Neuro-fuzzy system
NL	Natural language
NLChirp	Nonlinear chirp
NSL	Near sidelobe level
PAM	Pulse amplitude modulation
PDE	Partial differential equation
PDF	Probability density function
PN	Pseudonoise
PSD	Power spectral density
PZT	Lead zirconate titanate
PuC	Pulse-compression
PuE	Pulse-echo
RR	Range resolution
RX	Receiving transducer
SC	Self-correlation
SE	Shannon entropy
SEF	Shannon entropy fuzzification
SI	International system

SISO	Single input, single output
STP	Standard temperature and pressure
SUT	Sample under test
TBP	Time bandwidth product
TDS	Terminal data set
TF	Time frequency
ToF	Time-of-flight
TrT	Through-transmission
TX	Transmitting transducer
UT	Ultrasonic transducer
VCO	Voltage-controlled oscillator
VGA	Variable gain amplifier

Part I
Describing the Ultrasonic System at a Physical Level: Modeling and Simulation

In this part, we discuss the analytical and numerical modeling of the ultrasonic nondestructive testing (NDT) systems. The motivation of this part is to provide a methodology that in many practical cases is mandatory to define a suitable NDT system. Indeed, designing these systems requires defining a great number of parameters related to transducers, stimuli signals, postprocessing of acquired signals and so on, which strongly affect the performance of the diagnostic systems. The assessment of these parameters cannot be done on the real system because it would be too expensive and most cases of interest cannot be created artificially. On the contrary, the numerical simulation of the system makes it possible to test automatically a great number of scenarios, which represents a basis of knowledge that allows one to properly design the diagnostic system. For this reason, the most important characteristic of a numerical model is the correspondence with the real system. Finite element method (FEM) analysis represents a valid option in this sense. Indeed, it makes it possible to deal with nonlinearities, discontinuities, and anisotropies, by properly adapting the degrees of freedom of the model in the different regions of the domain of interest. This property is particularly important in the case of ultrasound NDT, where the objective is to detect anomalies into a mean where a wave propagates. Furthermore, the presence of small irregularities and the boundaries of the domain contributes to complicate the study.

The software that apply the FEM use a general mathematical formulation that needs to be adapted to the specific physical laws involved in the problem at hand. Usually, they are organized in modules where the equations which describe a physics are implemented, so that to define the analytical model which describes the system under study, in general, it is sufficient in selecting a proper module. After that, the specific problem has to be described. Most part of software gives the possibility to draw simply geometries, in particular when the domain can be represented into a 2D space, but in the case of complex geometries, the most suitable solution consists of importing a computer-aided design (CAD) model into the FEM software. A great number of standard procedures usually are available in order to perform all the steps of the FEM analysis, namely, choice of the physics, definition of the geometry, fixing the boundary conditions, optimizing the mesh of the domain, and finally, postprocessing the results of the simulation.

The three chapters of this part aim to enlighten the FEM analysis in the case of ultrasonic NDT. Chapter 1 describes the phenomenon of the ultrasonic wave propagation and the fundamental equations are given. Chapter 2 gives an introduction to the FEM and it shows how it can be exploited in developing NDT systems. Chapter 3 shows some examples of numerical models of nondestructive diagnostic systems.

Chapter 1
Waves Propagation

Sara Carcangiu, Augusto Montisci and Mariangela Usai

Abstract Wave phenomena are used to evaluate material properties nondestructively as well as to locate and measure defects in critical structures. In fact, physical acoustics is directly related to ultrasound since the ultrasonic waves are employed to determine the properties of materials. The objective of this chapter is to present a brief overview of the physics involved in the propagation of the sound through a material. We start with a description of the physics of the acoustic wave considering the governing equations and the parameters that characterize the means in order to obtain the wave equation in the time domain, which results in the Helmholtz equation in the frequency domain. Successively, various phenomena of wave propagation are described. There is much information available in the acoustic wave that is transmitted and received through a material. In particular, the reflection of elastic waves at a free boundary and reflection and refraction at an interface can be used in order to detect an anomaly in the material tested.

1.1 Introduction

Wave propagation concerns the effects of a local perturbation which spreads in a medium. When the physical quantity which propagates is the variation of the pressure, we in general speak of "sound" or "ultrasound" depending on the range of frequency we are working on. There are several applications of the phenomenon of propagation in materials, for example, testing quickly and cheaply the integrity of a material. This chapter gives the basic mathematical instruments to deal with the phenomenon of wave propagation. At first the propagation of pressure waves in solids

S. Carcangiu (✉) · A. Montisci · M. Usai
Dipartimento di Ingegneria Elettrica ed (DIEE), University of Cagliari, Cagliari, Italy
e-mail: s.carcangiu@diee.unica.it

A. Montisci
e-mail: amontisci@diee.unica.it

M. Usai
e-mail: musai@diee.unica.it

P. Burrascano et al. (eds.), *Ultrasonic Nondestructive Evaluation Systems*,
DOI 10.1007/978-3-319-10566-6_1

and fluids is described, then the parameters of the medium affecting the phenomenon are considered, and finally the equations which describe it are presented.

The sound waves are classified depending on frequency and direction of particles oscillation with respect to the propagation direction. After that, we will analyze how the interfaces and the discontinuities affect the propagation of waves, deflecting, reflecting, and attenuating them. Experimental studies showed that to properly describe a wave propagation phenomenon, modeling the interference effects at high frequency is mandatory. This is accomplished by means of acoustic fields, which are described by partial differential equations (PDEs) as a function of time and space. These equations describe the relationship between field quantities such as instantaneous density, instantaneous pressure, and mass velocity. In general, integrating the PDEs is a hard task, even when dealing with homogeneous materials. Furthermore the hypothesis of homogeneity holds valid only for a very limited real cases, and phenomena of distortion and attenuation, which are always present in nature, cannot be modeled unless the inhomogeneities are properly taken into account. For this reason, the validity of the models is also strongly linked to the possibility of making preliminary tests to define the parameters of the tested materials (as Poisson's ratio, attenuation coefficient, density, acoustic impedance) with precise instrumentation and accurate measurement procedures.

In the experimental tests, acoustic waves are produced by means of transducers. In this chapter we refer to piezoelectric transducers which are the most commonly used.

Particular attention is devoted to how the waves propagate in the material taking into account the distance from the transducer and the frequency of the wave. We will show that this frequency is a crucial parameter for applying ultrasounds to nondestructive testing (NDT). Some general criteria for optimizing test parameters will be provided.

1.2 Physical Phenomena and Equations

The sound propagates in elastic bodies through the vibration of atoms and molecules of the material, at a velocity dependent on the mechanical characteristics of the crossed material. The presence of imperfections or inhomogeneities in the body causes the occurrence of scattering phenomena such as spurious echoes, reverberations, and in general, distortion and attenuation of the sound wave [8, 12].

Wave equations are hyperbolic PDEs which describe the propagation of various types of waves, such as acoustic, elastic, and electromagnetic waves. The usual assumption, for acoustic problems, is that acoustic waves are perturbations of the fluid, which can be either a liquid or a gas medium, with (at any point) instantaneous density $\rho(r,t)$ (International System (SI) unit: kilogram per cubic meter), instantaneous pressure $p(r,t)$ (SI unit: pascal) and velocity mass $\mathbf{v}(r,t)$ (SI unit: meter per second), where t (SI unit: second) is the time. The wave equation in an ideal fluid can be derived from hydrodynamics and the adiabatic relation between pressure and density [5].

Standard acoustic problems involve small acoustic pressure variation over the stationary background pressure. This allows for the linearization of the problem, namely for assuming a first-order approximation as a model of the system. Considering an initial spatially uniform state (stationary fluid ($v_0 = 0$) of density ρ_0 and at pressure p_0), a small parameter expansion is performed on such that:

$$\begin{aligned} p &= p_0 + p' \quad \text{with } p' \ll p_0 \\ \rho &= \rho_0 + \rho' \quad \text{with } \rho' \ll \rho_0 \\ v &= 0 + v' \end{aligned} \tag{1.1}$$

where the primed variables indicate the small variations.

In case of compressible lossless fluids, the system is governed by Euler's and continuity equations[1]:

$$\begin{aligned} \frac{\partial \mathbf{v}}{\partial t} + (\mathbf{v} \cdot \nabla)\mathbf{v} &= -\frac{1}{\rho}\nabla p \\ \frac{\partial \rho}{\partial t} + \nabla \cdot (\rho v) &= 0. \end{aligned} \tag{1.2}$$

In classical pressure acoustics all thermodynamic processes are assumed reversible and adiabatic.

Inserting the linear approximation (1.1) into Eq. (1.2) yields:

$$\begin{aligned} \frac{\partial v'}{\partial t} &= -\frac{1}{\rho_0}\nabla p' \\ \frac{\partial \rho'}{\partial t} + \rho_0(\nabla \cdot v') &= 0. \end{aligned} \tag{1.3}$$

The state equation, which relates pressure and density $p = p(\rho)$ completes the set of equations.

The state equation is exploited in order to eliminate density. The relation is approximated by its first-order Taylor series expansion:

$$p = p(\rho_0) + \left.\frac{\mathrm{d}p}{\mathrm{d}\rho}\right|_S (\rho - \rho_0) + O(\rho - \rho_0)^2 \tag{1.4}$$

Let us define $\left.\frac{\mathrm{d}p}{\mathrm{d}\rho}\right|_S = c_S^2$ where c_S is the speed of sound. The subscript S indicates that the derivative is calculated at constant entropy.

Neglecting the second-order terms, considering $p(\rho_0) = p_0$ and using the relations (1.1), we obtain

$$p = p_0 + c_S^2(\rho - \rho_0) \Longrightarrow p' = c_S^2 \rho'. \tag{1.5}$$

[1] This equation is known also as the equation of mass conservation, describing that the mass is neither created nor destroyed within a volume element.

Differentiating the latter equation of the system (1.3) with respect to time t and replacing the former equation, we obtain

$$\frac{\partial^2 \rho'}{\partial t^2} + \nabla \cdot \left(\rho_0 \frac{\partial v'}{\partial t} \right) = 0 \Longrightarrow \frac{\partial^2 \rho'}{\partial t^2} + \nabla \cdot \left(\rho_0 \left(-\frac{1}{\rho_0} \nabla p' \right) \right) = 0 \Longrightarrow \frac{\partial^2 \rho'}{\partial t^2} = \nabla^2 p'. \tag{1.6}$$

Finally, using the relation (1.5) and omitting the primes for pressure and density perturbations, the wave equation for pressure is:

$$\nabla^2 p - \frac{1}{c_S^2} \frac{\partial^2 p}{\partial t^2} = 0. \tag{1.7}$$

A similar equation can be obtained in terms of density.

If the density is not constant in space, Eq. (1.7) becomes:

$$\frac{1}{\rho c^2} \frac{\partial^2 p}{\partial t^2} + \nabla \cdot \left[-\frac{1}{\rho} (\nabla p) \right] = 0. \tag{1.8}$$

The speed of sound depends on the compressibility of the fluid. The term ρc^2 is called bulk modulus K (SI unit: newton per square meter).

Equation (1.7) does not contain source terms. These can be introduced as boundary conditions when the source is external to the region of interest, otherwise a source term has to be included into the equation.

There are two types of acoustic energy sources: the dipole source and the monopole source [10].

A monopole source q_d (SI unit: per square second) is a closed surface that changes volume.

A dipole source Q_m (SI unit: newton per cubic meter) is a body oscillating back and forth without any change in volume.

The internal sources are taken into account by means of an inhomogeneous term:

$$\frac{1}{\rho c^2} \frac{\partial^2 p}{\partial t^2} + \nabla \cdot \left[-\frac{1}{\rho} (\nabla p - q_d) \right] = Q_m. \tag{1.9}$$

while an additional term has to be introduced in the case of lossy media, to take into account the attenuation effects:

$$\frac{1}{\rho c^2} \frac{\partial^2 p}{\partial t^2} - d_a \frac{\partial p}{\partial t} + \nabla \cdot \left[-\frac{1}{\rho} (\nabla p - q_d) \right] = Q_m \tag{1.10}$$

where d_a is the damping coefficient. Attenuation terms have to be considered even in lossless means, due to boundary effects.

The wave Eq. (1.9) is linear and has particular solutions that are periodic in time. Then it can be described in the frequency domain by applying a Fourier transformation of the acoustic pressure $p(\boldsymbol{r}, t)$. A special case is represented by the time-harmonic wave, where the pressure is described by:

$$p(\boldsymbol{r}, t) = p(\boldsymbol{r}) e^{i\omega t} \tag{1.11}$$

where $i = \sqrt{-1}$, $\omega = 2\pi f$ (SI unit: radian per second) is the angular frequency and f (SI unit: hertz) is the frequency.

In this case the direct and inverse Fourier's transforms of the acoustic pressure with respect to the time are given as

$$P(\boldsymbol{r},\omega) = \mathcal{F}_t\{p(\boldsymbol{r},t)\} = \int_{-\infty}^{+\infty} p(\boldsymbol{r},t)\mathrm{e}^{-\mathrm{i}\omega t}\mathrm{d}t$$
$$p(\boldsymbol{r},t) = \mathcal{F}_t^{-1}\{P(\boldsymbol{r},\omega)\} = \frac{1}{2\pi}\int_{-\infty}^{+\infty} P(\boldsymbol{r},\omega)\mathrm{e}^{\mathrm{i}\omega t}\mathrm{d}t. \tag{1.12}$$

Substituting $P(\boldsymbol{r},\omega)$ in Eq. (1.7) and applying the differentiation theorem of the Fourier transform, the following time-harmonic *Helmholtz equation* is obtained:

$$\nabla^2 P(\boldsymbol{r},\omega) + \frac{\omega^2}{c_S^2} P(\boldsymbol{r},\omega) = 0 \Longrightarrow \nabla^2 P(\boldsymbol{r},\omega) + k^2 P(\boldsymbol{r},\omega) = 0. \tag{1.13}$$

Considering the *wavelength* $\lambda = 2\pi/k = c/f$ (SI unit: meter), the constant k (called *wavenumber*) denotes the number of wave per 2π units of length. This equation can be treated as an eigenvalue PDE to solve for eigenmodes and eigenfrequencies [6].

When the source terms have an harmonic time dependence, Helmholtz's equation can be expressed in the frequency domain:

$$\nabla \cdot \left[-\frac{1}{\rho}(\nabla P - q)\right] - \frac{k^2 P}{\rho} = Q. \tag{1.14}$$

When both q and Q source terms are zero, the solution of Eq. (1.14) gives the plane wave:

$$p = P_0 \mathrm{e}^{\mathrm{i}(\omega t - \boldsymbol{k}\cdot\boldsymbol{r})} \tag{1.15}$$

where P_0 is the wave amplitude and it is moving in the $\boldsymbol{k}$ direction with angular frequency ω and wave number $k = |\boldsymbol{k}|$.

1.3 Wave Propagation

Sound waves propagate due to the compressibility of a medium ($\nabla \cdot \boldsymbol{v} \neq 0$). Depending on frequency one can distinguish (Fig. 1.1):

- Infrasound waves—below 20 Hz
- Acoustic waves—from 20 Hz to 20 kHz
- Ultrasonic waves—above 20 kHz

According to the direction of particle oscillation with respect to the direction of propagation, one can distinguish among longitudinal waves, shear waves, and surface waves [13].

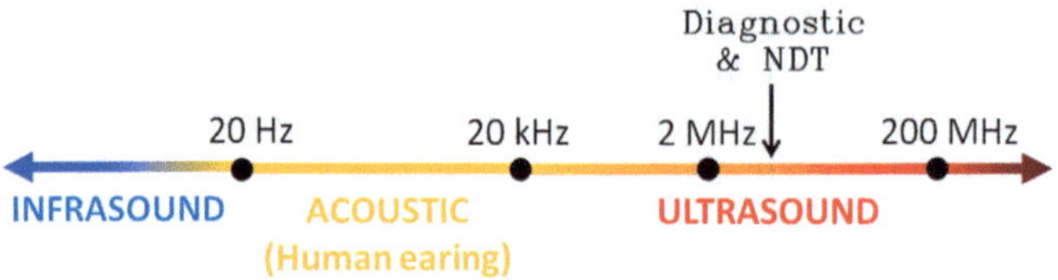

Fig. 1.1 Frequency ranges of sound

Fig. 1.2 Longitudinal waves

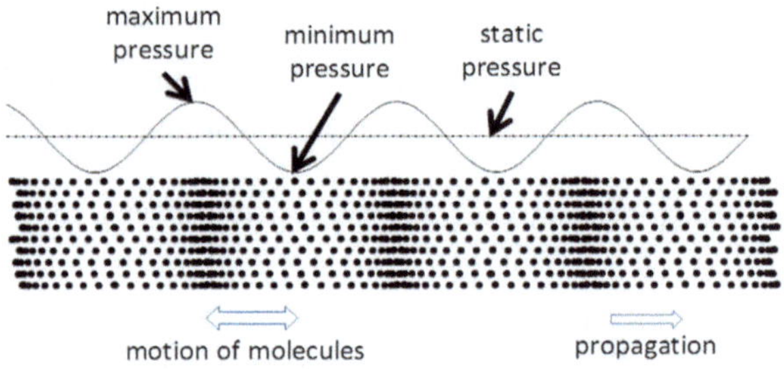

Longitudinal and shear waves are the two modes of propagation most widely used in ultrasonic testing. With the *longitudinal waves*, also called pressure traction waves or P-waves, the particles oscillate in the same direction of wave propagation (Fig. 1.2). Longitudinal waves can be generated in liquids, as well as in solids and gaseous materials, because the energy travels through the atomic structure by means of a series of compressions and expansion (rarefaction) movements. The speed propagation of P-wave is given by:

$$v_P = \sqrt{\frac{E_d}{\rho}\frac{1-\nu_d}{(1+\nu_d)(1-2\nu_d)}}\ [\mathrm{m/s}] \tag{1.16}$$

where E_d (SI unit: megapascal) is the dynamic Young's modulus, ρ (SI unit: kilogram per cubic meter) is the mass per unit of volume of the material, and ν_d is the dynamic Poisson's ratio.

In *shear waves*, or S-waves, the particles oscillate at a right angle to the direction of propagation (Fig. 1.3). Shear waves require an acoustically solid material for effective propagation, and therefore, they are not effectively propagated in materials such as liquids or gases. Shear waves are relatively weak when compared to longitudinal waves. In fact, shear waves are usually generated in materials using part of the energy of longitudinal waves. The speed propagation of S-waves is given by:

$$v_S = \sqrt{\frac{E_d}{\rho}\frac{1}{2(1+\nu_d)}}\ [\mathrm{m/s}]. \tag{1.17}$$

The surface waves propagate by affecting surface layers. Their use is very special and limited compared to P-waves and S-waves. They are slightly slower than S-waves ($v_{sf} = 0.92v_s$) (Fig. 1.4).

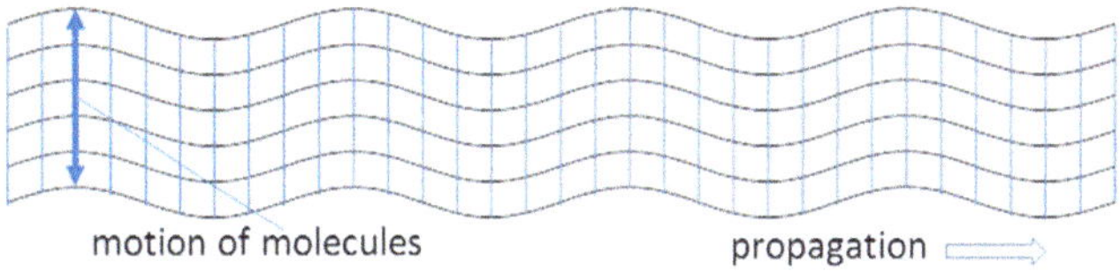

Fig. 1.3 Shear or S-waves

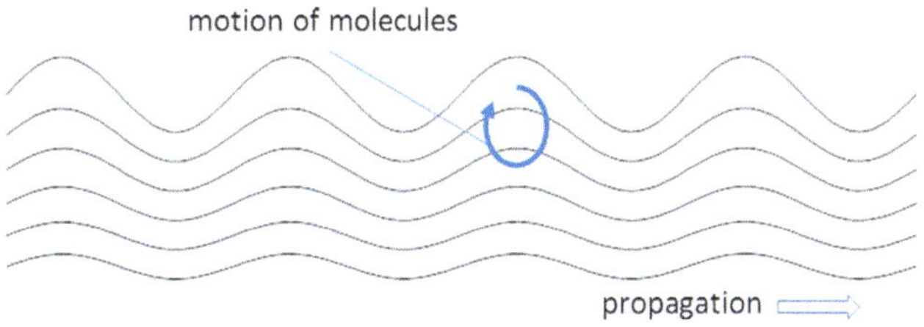

Fig. 1.4 Surface or Rayleigh waves

Using the Bergmann's formula [3], the speed propagation of surface waves is given by

$$v_p = \frac{0.87 + 1.12\nu_d}{1 + \nu_d}\sqrt{\frac{E_d}{\rho}\frac{1}{2(1 + \nu_d)}}\ [\mathrm{m/s}]. \tag{1.18}$$

1.4 Effects of Interfaces and Discontinuities

When an ultrasonic wave hits the surface separating two media a longitudinal wave component is partly reflected and partly transmitted. In turn, the reflected wave and the transmitted wave are split into a longitudinal wave (P-type) and a shear wave (S-type). The incident wave and the reflected one are inclined by the same angle, according to the Snell's law. The angles that the incident and the transmitted waves form with the normal to the surface, are bonded according to the law of Snellius–Descartes (Fig. 1.5).

These angles depend on the values of the impedances and thus by the transmission speed of the two means:

$$\frac{v_{P1}}{\sin \alpha_1} = \frac{v_{S1}}{\sin \beta_1} = \frac{v_{P2}}{\sin \alpha_2} = \frac{v_{S2}}{\sin \beta_2}. \tag{1.19}$$

The v_i are the propagation speeds of the P-waves (longitudinal) and S-type (shear) in the materials 1 and 2 respectively, the α_i are the angles that longitudinal waves form with the surface separating the two media, while the β_i are the angles of the shear waves.

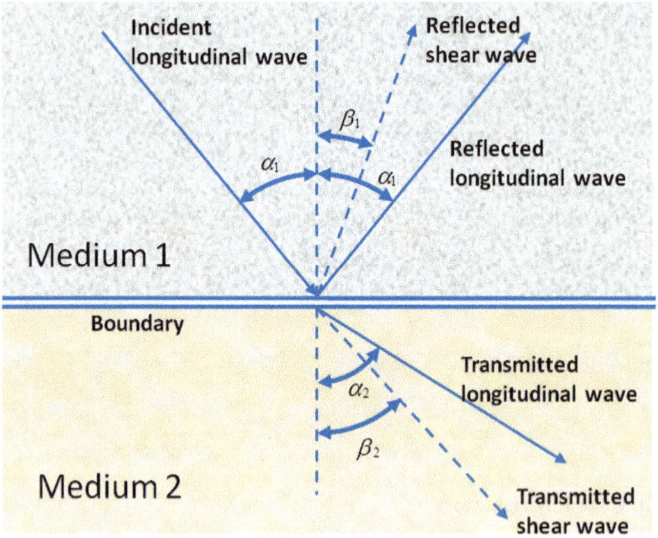

Fig. 1.5 Law of Snellius–Descartes

The higher is the variation of the acoustic impedances between two media, the greater is the intensity of the reflected wave and the lower is the transmitted wave. Since longitudinal waves travel more faster than shear waves (the ratio is about 2:1), the angles of reflection and transmission β_i are in general smaller than those of α_i, as shown in Fig. 1.5.

The fundamental relationship that governs the reflection of an ultrasonic wave that hits on the separation surface between two media of different acoustic impedance Z_1 and Z_2 was formulated in the nineteenth century by Poisson and the reflection coefficient is expressed mathematically as follows:

$$R = \frac{Z_2 - Z_1}{Z_2 + Z_1}. \tag{1.20}$$

A similar relationship can be written regarding the transmission coefficient T

$$T = \frac{2Z_2}{Z_2 + Z_1}. \tag{1.21}$$

Tables of values of transmission speed for different materials are reported in the manuals [9]. The acoustic impedance is very low in the gas (about four orders of magnitude less than that of the solids), and this results in very high values of the reflection coefficient at the solid–gas interface (for example steel–air, which is a frequent case in practice).

This results in the inability to propagate the ultrasonic waves in the air after they have gone through a material solid and this also explains the need to insert a layer of an appropriate substance, liquid or viscose, between the transducer and the external surface of the tested object, to prevent the transmitted signal from being too attenuated and distorted.

The ultrasonic beam emitted by a transducer, as will be better seen below, is divergent, and this leads to a progressive reduction of the energy density along the direction of propagation, as well as an increase of the illuminated area.

In addition, crossing the material, the ultrasonic waves are subject to phenomena of attenuation caused by four different mechanisms: heat conduction, viscous friction, elastic hysteresis, and scattering. It is difficult to determine by simple experimentation which of these mechanisms are of major importance. According to the classical theory, sound absorption is caused only by thermal conduction and internal or viscous friction. Sound waves cause adiabatic compression and rarefactions as they propagate through material. There is a momentary rise of the temperature in the compressions and a momentary fall of temperature in the rarefactions. Thermal energy is released during the compressions and restored during the rarefaction, which decreases the energy of the wave and causes attenuation. In addition, there is a loss of energy for reflection and dispersion (or scattering) due to the elastic discontinuity at the grains boundaries. Experimental results demonstrate that the attenuation increases with grain size [2]. These phenomena lead to a distortion and an attenuation of the ultrasonic waves. In general, the reduction of intensity for a length path z in a given material can be expressed using the relationship:

$$I_z = I_0 e^{-\mu z} \tag{1.22}$$

where I_z is the intensity of sound after traveling a distance z, I_0 is the intensity of sound, μ is the attenuation constant or absorption coefficient, and z is the path length.

The attenuation constant can be ideally divided into the sum of two contributions: one related to the "real" absorption, which depends on the frequency of the incident wave and which is related to the energy dissipation due to the friction among molecules and one derived from scattering, which is a function mainly of the size of the particles of the medium:

$$\mu = \mu_\tau + \mu_s. \tag{1.23}$$

During the test, acoustic waves are produced by means of transducers. The initial deformation of the transducer is obtained by exploiting the piezoelectricity phenomenon. This phenomenon that expresses the combination of two events (mechanical and electrical origin) refers to the occurrence of one of these events:

- If an electric voltage is applied to two opposite surfaces of a block of piezoelectric material, this material expands (or contracts).
- Conversely, if the transducer element undergoes a stress capable of deforming it, between the opposite surfaces an electric voltage is created.

The sound field of a transducer is divided into two zones: the *near field* and the *far field*.

The *near field* is the region directly in front of the transducer where the sound pressure is irregularly distributed on the wave fronts (region of turbulence). The *far*

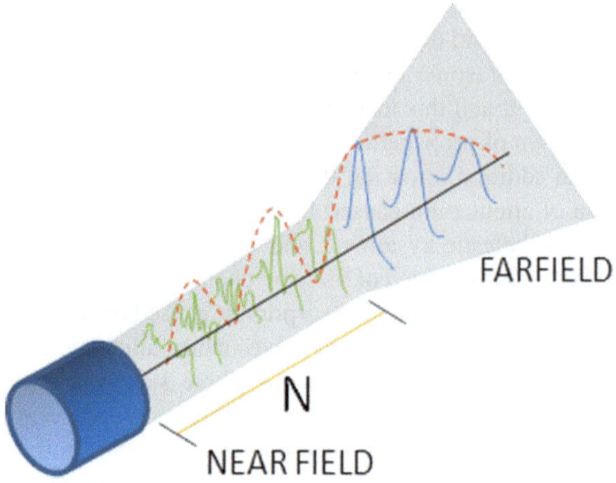

Fig. 1.6 Near field and far field of the transmitted wave

field is the area beyond N (see Fig. 1.6), where the sound field pressure gradually drops to zero. Only in the far field the sound pressure is distributed more regularly. For these reasons, for a given sample having assigned size and material, the frequency of the signal has to be properly chosen in order to work in the far field, because the variation of the sound pressure in the far field provides a sufficiently approximate representation of the variation of the acoustic field with the distance.

The near field distance N is given by the following formula [11]:

$$N = \frac{D^2}{4\lambda} \tag{1.24}$$

where D is the diameter of the transducer source and $\lambda = v/f$ is the wavelength of the wave (Fig. 1.7).

Due to dissipative effect, elastic waves are strongly attenuated; therefore, the emission cone of the signal source has to be as less divergent as possible. Let us define the *beam spread* and the *half angle* α:

- Half angle α is the angle of divergence of the beam.
- Beam spread is equal to $\sin(\alpha)$.

The beam spread depends on wavelength and on the diameter of the transducer

$$\sin\alpha = \frac{1.2\lambda}{D}. \tag{1.25}$$

If $D \gg \lambda$ the divergence of the wave is contained. As λ is in inverse proportion to the frequency f, high frequency signals enable waves to be highly directional. On the other hand, if λ approaches the diameter of the transducer, the waves are sent out in all directions. The attenuation can be evaluated by means of the distance amplitude correction (DAC) curves (see Fig. 1.8), which provide a reference level sensitivity as a function of the sweep distance [4].

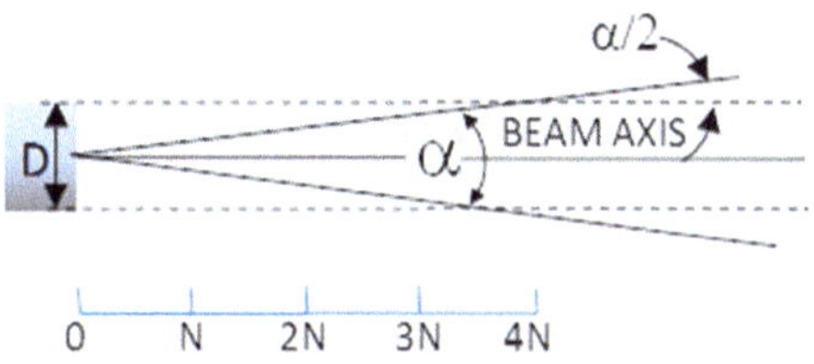

Fig. 1.7 Half angle of the emission cone

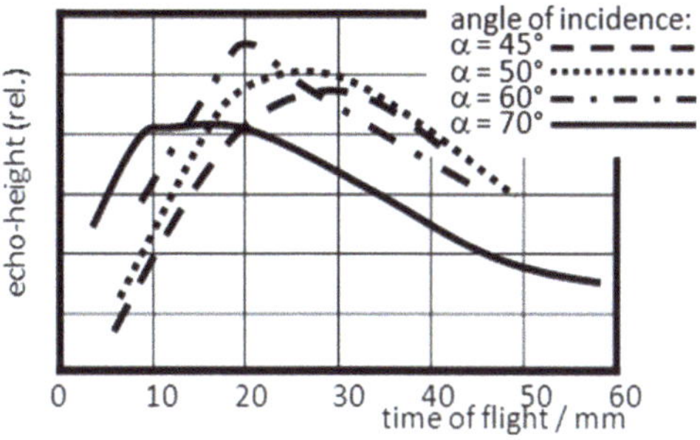

Fig. 1.8 Dependency of the amplitude versus the time of flight for a number of angle of incidence

1.5 Effects of the Presence of Defects

The characteristics of the defects that affect the amplitude of the reflected wave are [7]:

- the conformation (geometry contour)
- the orientation (90° reflectors visible, at 0° practically invisible)
- the roughness (high roughness involves scattering)
- the size (area)

The characteristics of the test sample that affect the amplitude of the signal reflection are related to its degree of roughness, which depends on the transmission of ultrasonic beam through the interface transducer material and its "transparency" (absorption characteristics).

The discontinuity is an obstacle to the propagation of ultrasonic waves in the material. Depending on the ratio between the size of the beam produced by the transducer and the size of the defect or discontinuity, there are three main cases of interest:

1. $a \gg \lambda$: The size of the defect is much larger than the wavelength of the vibration incident. It clearly reflects back the wave according to the rules geometrical acoustics, creating a well-defined reflected beam, and of course over the obstacle, an equally well-defined shadow zone in the transmission.
2. $a = \lambda$: The dimensions of the defect are of the same order of the wavelength of the vibration incident. The diffraction broadens the reflected wave in a divergent cone. Both the reflected beam and the shadow become blurred and not well defined.

3. $a \ll \lambda$: The size of the defect is very much smaller than the wavelength of the vibration of the incident. The obstacle reflects in all directions in the half space, and it becomes completely invisible.

In conclusion, to detect a defect it is necessary that the wavelength vibration emitted is much smaller of the geometric dimension of the defect [1]. In particular, this recommendation applies to the intrinsically inhomogeneous media (concrete, aggregates), which are continuously inhomogeneous.

1.6 Conclusion

In this chapter, the fundamentals of the study of waves propagation in a mean are given. Firstly, the analytical formulation is derived referring to the case of indefinite homogeneous mean, taking into account both the cases of source external and internal to the domain of interest. Starting from the simplest case, a number of terms need to be introduced in order to take into account boundary conditions, inhomogeneities, and discontinuities. On the other hand, these ones characterize most NDT problems, where tests are performed on limited objects; often the propagating mean is not continuous, and finally discontinuities—represented by interfaces with transducers, boundaries, and defects—are always present in practice. For this reason, the formulation of the analytical model of the acoustic problem needs several terms which can describe phenomena related to the nonideality of the system at hand. Solving the PDE analytically in a realistic case is often a prohibitive task, so that resorting to a numerical solution is almost always mandatory. Nevertheless, the equations which describe the problems have to take into account all the elements in order to obtain meaningful results by the numerical simulation. In this chapter, all these aspects have been described, indicating how the nonidealities affect the formulation of the problem. Finally, the phenomena of interest for the NDT, such as the propagation at interfaces and dumping effects have been analyzed.

References

1. Alleyne D, Cawley P (1992) The interaction of lamb waves with defects. IEEE Trans Ultrason Ferroelectr Freq Control 39(3):381–397
2. Badidi Bouda A, Lebaili S, Benchaala A (2003) Grain size influence on ultrasonic velocities and attenuation. NDT&E Int 36(1):1–5
3. Bergmann L (1957) Der Ultraschall und seine Anwendung in Wissenschaft und Technik — Nachtrag zum Literaturverzeichnis der 1954 erschienenen, 6th edn. S. Hirzel, Stuttgart
4. Erhard A, Schenk G, Möhrle W, Montag HJ (2000) Ultrasonic phased array technique for austenitic weld inspection. In: 15th World conference on non-destructive testing, Rome, pp 674–677
5. Jensen FB, Kuperman WA, Porter MB, Schmidt H (2011) Computational ocean acoustics, 2nd edn. Springer, New York

6. Karageorghis A (2001) The method of fundamental solutions for the calculation of the eigenvalues of the Helmholtz equation. Appl Math Lett 14(7):837–842
7. Kinra VK, Vu BQ (1986) Diffraction of Rayleigh waves in a half-space. II. Inclined edge crack. J Acoust Soc Am 79(6):1688–1692
8. Kinsler LE, Frey AR, Coppens AB, Sanders JV (1999) Fundamentals of acoustics, 4th edn. Wiley, New York
9. Lide DR (2004) CRC handbook of chemistry and physics. CRC Press, Boca Raton
10. Nelson PA (2007) Sound sources. In: Crocker MJ (ed) Handbook of noise and vibration control. Wiley, Hoboken, pp 43–51
11. Olympus (2007) Ultrasonic transducers technical notes. Olympus NDT. www.olympus-ims.com/data/File/panametrics/UT-technotes.en.pdf. Accessed June 2014
12. Pierce AD (1981) Acoustics: an introduction to its physical principles and applications. McGraw-Hill, New York
13. Szabo TL, Wu J (2000) A model for longitudinal and shear wave propagation in viscoelastic media. J Acoust Soc Am 107(5):2437–2446

Chapter 2
Numerical Simulation of Wave Propagation

Sara Carcangiu, Augusto Montisci and Renato Forcinetti

Abstract Wave equations are hyperbolic partial differential equations (PDEs) which describe the propagation of various types of waves, such as acoustic, elastic, and electromagnetic waves. In order to solve PDEs, the finite element method (FEM) can be used. After a brief introduction to the mathematical method used by FEM to evaluate the solution in nodes, where the polynomial curve that interpolates the differential equation has to be solved, we will describe the methodology used to solve the wave propagation problem described by the Helmholtz's equation. The wave propagation problem is analyzed by following specific steps: construction of the geometry to study, application of the boundary conditions, and meshing of the domain to be solved. The same procedure is used to simulate the behavior of piezoelectric transducers and the problem of wave propagation in medium with defects. Finally, the procedure followed for the simulation of acoustic problems using a specific software, i.e., COMSOL Multiphysics, is illustrated.

2.1 Introduction

The numerical simulation is commonly considered a mandatory passage in synthesizing a nondestructive testing (NDT) diagnostic system. Indeed, a number of design parameters need to be set on the basis of the response of the diagnostic system in different scenarios, which would be hardly reproducible in real physical setup. Whichever numerical tool is used to simulate the real system, it has to guarantee reliability, convergence, and possibly, a limited computational cost. The problem can be approached by using several different methods, such as analytical, finite difference, moment, finite element, and Monte Carlo, each having its strong and weak points. Among these, finite element method (FEM) has the advantage of combining good

---

S. Carcangiu (✉) · A. Montisci · R. Forcinetti
Dipartimento di Ingegneria Elettrica ed (DIEE), University of Cagliari, Cagliari, Italy
e-mail: s.carcangiu@diee.unica.it

A. Montisci
e-mail: amontisci@diee.unica.it

R. Forcinetti
e-mail: renato.forcinetti@diee.unica.it

P. Burrascano et al. (eds.), *Ultrasonic Nondestructive Evaluation Systems*,
DOI 10.1007/978-3-319-10566-6_2

model precision with a typical easiness of use. Furthermore, it allows for tuning the degree of precision on the simulation domain on the basis of the specific requirements. This aspect is crucial in our application, because in general, the object to detect consists of a discontinuity of the propagating mean.

The FEM, such as the finite difference method (FDM), is a numerical technique designed to seek approximate solutions of problems described by a system of partial differential equations (PDEs), reducing them to a system of algebraic equations [4]. With this method, it is possible to solve problems whose analytical models described by a system of differential equations cannot be solved analytically. To this aim, the problem is discretized by subdividing the domain in a proper number of elements (from which the name of the method), then the solution is found only in specific points of those elements (nodes), and finally, a continuous solution is obtained by interpolating the solution on the nodes. The possibility to entrust the solution of the problem to a computer represents a great advantage. The more the discretization is fine, namely the more the elements are small, the more the FEM model is precise but also computationally costly. Therefore, dealing with such method requires to find the best compromise between such two conflicting issues, by taking into account the resources on hand.

This chapter is organized as follows. First, a general formulation of the FEM is given. Then the method is described in more detail for the case of a wave propagation problem. A specific paragraph is devoted to the study of the behavior of transducers, whose function is the conversion of the electrical signal into pressure waves and vice versa. After that, the problem of detecting defects into a mean by means of ultrasonic waves is described in depth. Appendix reports a tutorial for the development of a model of NDT ultrasonic system.

2.2 The Finite Elements Method

The first thing one has to define in using the FEM is the domain of the problem. After that, the state variables of the problem have to be defined. The domain is delimited by a boundary, except for the cases where it could be useful considering an unlimited domain; in such cases, particular elements are adopted, but this is not a case of interest for the scope of this book.

As for the continuous differential problems, the boundary conditions have to be defined in order to solve the problem. The boundary is subdivided into two parts that are characterized by a specific boundary condition. The nature of this condition depends on the physics of the problem, but in general, we can distinguish between boundary conditions where the value of the state variable is given (Dirichlet conditions) and that where the normal derivative of the state variable is given (Neumann conditions). Corresponding to such distinction, we can subdivide the whole boundary of the domain into a Dirichlet and a Neumann boundary, respectively.

The problem at hand in the FEM is first expressed in the so-called strong form and then it is translated in the weak or variational form to be solved. The first one can be formalized in general as follows:

$$\begin{aligned} L\Phi &= g \quad \text{in} \quad \Omega \\ \Phi &= g_0 \quad \text{in} \quad \Gamma_D \\ \partial_n \Phi &= g_1 \quad \text{in} \quad \Gamma_N \end{aligned} \tag{2.1}$$

where L is a differential (integral) operator, Φ is the state variable, Ω is the domain, g, g_0, and g_1 are known functions, Γ is the boundary of the domain, which is subdivided into the Dirichlet boundary Γ_D and Neumann boundary Γ_N, and finally, ∂_n denotes the derivative of the state variable in the direction normal to Γ.

It is often possible to replace the problem of integrating the equations in (2.1) by an equivalent minimization problem. Problems like these are called variational problems and the methods which allow one to reduce the original problem in variational form are called variational methods. Such methods represent a common basis for different kind of analyses, among which, the FEM is included [12, 13, 22].

The first step to apply the variational methods consists of finding the functional of the problem, which is the function having a minimum in correspondence of the solution of the original problem [9]. This function is commonly called energy function W due to the fact that, like in nature, the systems spontaneously evolve toward states of minimal energy, but in most cases where the problem at hand is physical, the functional is actually an energy function, so that searching the solution of the system just corresponds to find a state of minimal energy.

Generally speaking, the application of the FEM involves the following steps:

1. Discretize the domain of application of the differential equation as function of physical magnitude of field quantity V in a finite number of elements.
2. Define the governing algebraic equations as a function of field quantity V for a generic element.
3. Assembling of all elements in the solution region and determination of the total energy W, associated with the assembly of the elements expressed as a function of the values that the field quantity V assumes in each of the n nodes of the mesh: $W = f(V_1, V_2, V_3, \ldots, V_n)$.
4. Solving the system of linear equations resulting from the application of the variational principle, by imposing the condition of minimum energy stored, equivalent to the equilibrium condition of the system. Thus, we require that the partial derivatives of the energy function W with respect to each nodal value of the potential are zero i.e.,: $\partial W / \partial V_k = 0; \quad k = 1, 2, \ldots, n$.

So we get a system of n algebraic equations whose unknowns are values that the field quantity V assumes in the nodes of the mesh, except for the nodes on the boundary of the domain, in which the magnitude (*Dirichlet conditions*) or the normal derivative (*Neumann conditions*) of the field quantity are known.

Therefore, the FEM is based on the discretization of the domain into elements. Without loss of generality, we can assume that the domain at hand is 2D and that the elements we have chosen are triangular. The value of the field quantity at any point inside of the generic triangular element will be determined by interpolating the values of the field quantity in the nodes of the corresponding element (Fig. 2.1).

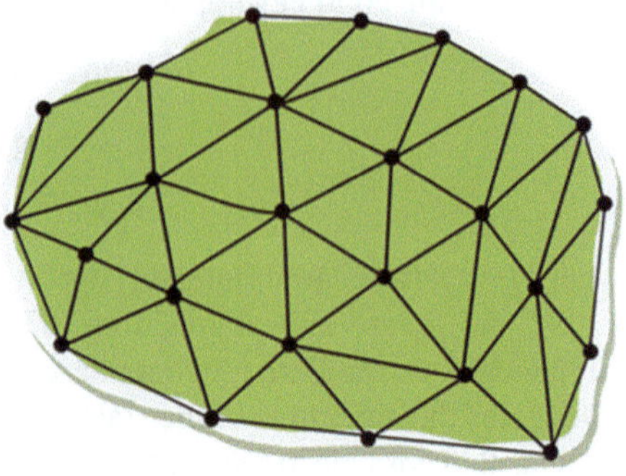

Fig. 2.1 Discretization of the domain by means of triangular elements

The triangular elements are those that fill better the surfaces of the most varied forms. Actually, the elements with the minimum number of facets (simplex: segments in R^1, triangles in R^2, and tetrahedrons in R^3) are the most commonly used as default-element for the mesh by commercial softwares. The elements have a certain number of nodes, which in the simplest case, coincide with the vertices of the element. The FEM solves Eq. (2.1) only on the nodes of the mesh rather than on the whole domain, while the solution within the element is expressed in function of such nodal values. The relationship among the values of the nodes and the other points of the element is fixed a priori, so that in general, it does not fit exactly the actual values of the function. This relationship is called shape function and in general can be fixed arbitrarily, but for simplicity, it is often preferred to have a great number of elements and for each element a limited number of degrees of freedom. This is the reason why the most common choice is to assume linear shape functions.

We seek an approximation of the function $V(x, y)$ in the whole domain as the sum of the function within the elements $V_e(x, y)$:

$$V(x, y) \cong \sum_{e=1}^{N} V_e(x, y) \tag{2.2}$$

where N is the number of elements and

$$V_e(x, y) = a + bx + cy \tag{2.3}$$

is an interpolating linear function which gives the approximation within the element. More in general, the number of nodes could be much greater so that the interpolating function could be different.

For the convergence of the solution, the approximating polynomial has to satisfy the compatibility and completeness criteria [11]. This means that considering, for example, 2D elements:

- the functions (2.3) have to be continuous at element interfaces, i.e., along element interfaces, we must have continuity of V as well as continuity of the derivative of V normal to the interface;

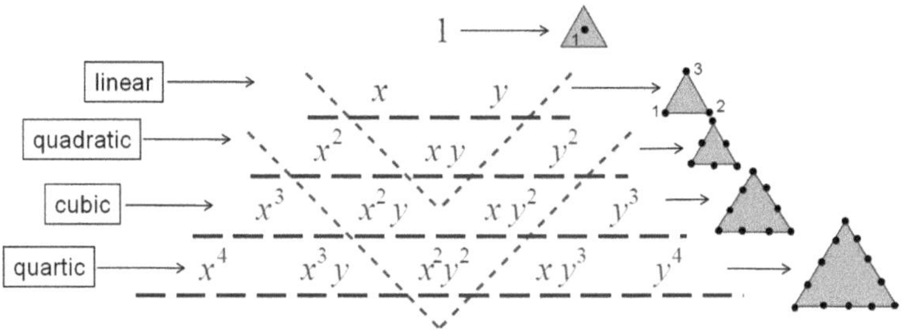

Fig. 2.2 Pascal's triangle (2D case)

- if the size of the elements tends to zero, the function (2.3) and its gradient must be constant within the element.

Furthermore, the approximating polynomial should be isotropic, i.e., the approximating polynomials should remain unchanged under a linear transformation from one Cartesian coordinate system to another. This is achieved if the components of the approximating polynomial are complete or, if incomplete, they are symmetric with respect to the independent variables.

In order to correctly select the terms of the polynomials, we can use the Pascal's triangle (see Fig. 2.2).

Assuming a linear interpolation within the elements implies that a proper size of elements (*mesh*) has to be considered in order to avoid introducing too coarse approximations.

The three coefficients in (2.3) are related to the value of the V_e in the nodes:

$$\begin{bmatrix} a \\ b \\ c \end{bmatrix} = \begin{bmatrix} 1 & x_1 & y_1 \\ 1 & x_2 & y_2 \\ 1 & x_3 & y_3 \end{bmatrix}^{-1} \begin{bmatrix} V_{e1} \\ V_{e2} \\ V_{e3} \end{bmatrix}. \tag{2.4}$$

The coefficients obtained by Eq. (2.4) can be substituted in Eq. (2.3), obtaining at the end the linear function within the element, given by:

$$V_e(x, y) = \sum_{i=1}^{3} \alpha_i(x, y) V_{ei} \tag{2.5}$$

that is, it is given by a linear combination of three functions, one for each node, where the weights are values of the V_e in correspondence of the nodes. The functions $\alpha_i(x, y)$ are called shape functions and, due to the initial choice, are linear.

As said before, a functional has to be defined to solve the problem [9]. The precise form of such functional depends on the form of the equation which describes the problem.

To fix the ideas, let us consider the case of a problem described by the Laplace's equation:

$$\nabla^2 V = 0. \tag{2.6}$$

The corresponding functional evaluated for the single element has the following form:

$$W_e = \frac{1}{2} \int |\nabla V_e|^2 \mathrm{d}S. \tag{2.7}$$

The gradient of V can be expressed by means of Eq. (2.8):

$$\nabla V_e = \sum_{i=1}^{3} V_{ei} \cdot \alpha_i \tag{2.8}$$

which can be substituted in Eq. (2.7), to give

$$W_e = \frac{1}{2} \sum_{i=1}^{3} \sum_{j=1}^{3} V_{ei} \left[\int \nabla \alpha_i \nabla \alpha_j \mathrm{d}S \right] V_{ej}. \tag{2.9}$$

Let us indicate the term in brackets with the notation C_{ij}. With this, Eq. (2.9) can be rewritten in matrix form:

$$W_e = \frac{1}{2} V_e^T [C_e] V_e. \tag{2.10}$$

The terms C_{ij} are obtained by combining gradients of linear functions, so they are constant and depend only on the coordinates of the elements nodes.

The complete functional to minimize is obtained by summing Eq. (2.10) for all the elements, as

$$W = \sum_{e=1}^{N} W_e = \frac{1}{2} V^T [C] V. \tag{2.11}$$

In Eq. (2.11), the variable vector V comprises the function calculated in all the nodes of the domain. As the same node belongs to more than one element, the matrix $[C]$ will not be concentrated around the diagonal, but in general, any entry could be different from zero. The definitive distribution of the elements of $[C]$ will depend on the order according to which they are considered and it is arbitrary, but the computational cost of minimization depends on such distribution. The minimum of Eq. (2.11) can be found by imposing the gradient to be equal to zero. Not all the values of V at nodes are variable. Indeed, the boundary conditions have to be fulfilled according to the formulation of the problem. As a consequence, the equations' system that gives us the minimum will be a nonhomogeneous system. Indeed, the terms of W that are the product between a variable and a fixed value in the derivation give rise to constant terms.

The minimum of the solution can be found by solving a linear equations' system:

$$[C_{\text{free}}]V_{\text{free}} = [C_{\text{bound}}]V_{\text{bound}} \tag{2.12}$$

where V_{free} is the vector of the variable values of the function and C_{free} is the corresponding coefficient matrix, while V_{bound} is the vector of values at the boundary of the domain and C_{bound} is the corresponding matrix of coefficients.

The number of nodes needs to be higher in the regions where the field variable has strong gradients. In these regions, in order to apply the method preserving the required accuracy, it may be necessary to thicken the nodes. The FEM allows to do that. Solving the system (2.12) is in general troublesome due to the high order of coefficients matrix, so that a number of procedures have to be used in order to reduce the computational cost of minimization. Anyway, analyzing these methods is beyond the scope of this book. It is sufficient for the reader to know that this is a task of the software that he/she uses to apply the FEM, and all the information about the implemented minimization algorithms is usually described in the documentation of the software.

2.3 Use of FEM for the Analysis of Waves Propagation

Typically, the wave propagation phenomena are described by the Helmholtz's equation:

$$\nabla^2 \Phi + k^2 \Phi = g \tag{2.13}$$

where Φ is field variable and g is the source function. It is possible to demonstrate that the functional corresponding to Eq. (2.13) is equal to:

$$I(\Phi) = \frac{1}{2} \iint \left[|\nabla \Phi|^2 - k^2 \Phi^2 + 2\Phi g \right] \mathrm{d}S. \tag{2.14}$$

We need to express both the field variable and the source function in terms of shape functions over a triangular element:

$$\Phi_e(x, y) = \sum_{i=1}^{3} \alpha_i \Phi_{ei}$$

$$g_e(x, y) = \sum_{i=1}^{3} \alpha_i g_{ei}.$$

The functional in Eq. (2.14), calculated for the single element, becomes:

$$I(\Phi_e) = \frac{1}{2} \sum_{i=1}^{3} \sum_{j=1}^{3} \Phi_{ei} \Phi_{ej} \iint \nabla \alpha_i \nabla \alpha_j \mathrm{d}S -$$

$$\frac{k^2}{2}\sum_{i=1}^{3}\sum_{j=1}^{3}\Phi_{ei}\Phi_{ej}\iint \alpha_i\alpha_j \mathrm{d}S +$$
$$\sum_{i=1}^{3}\sum_{j=1}^{3}\Phi_{ei}g_{ej}\iint \alpha_i\alpha_j \mathrm{d}S. \tag{2.15}$$

Let us define $C_{ij} = \iint \nabla\alpha_i \cdot \nabla\alpha_j \mathrm{d}S$ and $T_{ij} = \iint \alpha_i \cdot \alpha_j \mathrm{d}S$.
The Eq. (2.15) can be then rewritten in matrix form:

$$I(\Phi_e) = \frac{1}{2}\Phi_e^T[C_e]\Phi_e - \frac{k^2}{2}\Phi_e^T[T_e]\Phi_e + \Phi_e^T[T_e]G_e. \tag{2.16}$$

By assembling such functional for all the elements, one gets

$$I(\Phi) = \frac{1}{2}\Phi^T[C]\Phi - \frac{k^2}{2}\Phi^T[T]\Phi + \Phi^T[T]G. \tag{2.17}$$

Once again, in order to find the solution of the wave problem, the derivative of Eq. (2.17) has to be taken with respect to all the free nodes. By splitting the vector Φ into a free and a prescribed subsets Φ_f and Φ_p, also the matrices $[C]$ and $[T]$ are consequently subdivided according to the fact that the entries multiply only free components, only prescribed ones or both in Eq. (2.17). An important class of problems is represented by the cases where the source term is null. In such cases, Eq. (2.17) can be rewritten as:

$$I(\Phi) = \frac{1}{2}[\Phi_f^T\Phi_p^T]\begin{bmatrix} C_{ff} & C_{fp} \\ C_{pf} & C_{pp} \end{bmatrix}\begin{bmatrix} \Phi_f \\ \Phi_p \end{bmatrix} - \frac{k^2}{2}[\Phi_f^T\Phi_p^T]\begin{bmatrix} T_{ff} & T_{fp} \\ T_{pf} & T_{pp} \end{bmatrix}\begin{bmatrix} \Phi_f \\ \Phi_p \end{bmatrix}. \tag{2.18}$$

In order to minimize this functional, the derivatives with respect to the free nodes have to be set equal to 0. We obtain:

$$\nabla I(\Phi) = [C_{ff}C_{fp}]\begin{bmatrix} \Phi_f \\ \Phi_p \end{bmatrix} - k^2[T_{ff}T_{fp}]\begin{bmatrix} \Phi_f \\ \Phi_p \end{bmatrix} = 0 \tag{2.19}$$

which is a system of linear equations. We can further specialize Eq. (2.19) by assuming that the prescribed values are all equal to 0. In this case, we obtain:

$$\nabla I(\Phi) = [[C_{ff}] - k^2[T_{ff}]]\Phi_f = 0 \tag{2.20}$$

which is a standard eigenproblem [2, 10].

2.4 Developing the FEM Model for a Wave Propagation Problem

In order to study a wave propagation problem using FEM, the steps to be performed will be illustrated below [7, 17]. These steps reflect the procedure commonly followed by the commercial packages, such as described in the Appendix for the software COMSOL™.

First of all we need to define a coordinates system. More specifically, we need to choose how many dimensions have to be considered. Because the 3D geometry is usually a large computational problem, it would be better to avoid it if possible. One needs to exploit symmetries in order to create 2D or 2D axisymmetric geometries that involve a lower computational cost.

After that we need to define the relevant physics involved in the problem. In general, the software gives in hand a collection of specific modules for each physics, so in general, one needs just to select the proper module/modules.

At this point, the PDE and the number of spatial dimensions have been chosen. For acoustic pressure, the PDE is the Helmholtz's equation (1.13) in Chap. 1, that is obtained by considering a constant density value.

This equation governs the spatial dependency of p, which permits us to know p completely, since the temporal part is already known. The goal is then to solve Eq. (1.14), defined in Chap. 1, for the frequencies of interest.

For axisymmetric geometries, the axis of symmetry is $r = 0$. For the acoustic pressure in 2D axisymmetric geometries, the wave equation becomes:

$$\frac{\partial}{\partial r}\left(-\frac{r}{\rho}\frac{\partial p}{\partial r}\right)+r\frac{\partial}{\partial r}\left(-\frac{1}{\rho}\frac{\partial p}{\partial z}\right)-\left[\left(\frac{\omega}{c}\right)^2-\left(\frac{m}{r}\right)^2\right]\frac{rp}{\rho}=0 \qquad (2.21)$$

where m denotes the circumferential wave number. In this case, k_z is the out-of-plane wave number. In 2D axisymmetric geometries, the independent variables are the radial coordinate r and the axial coordinate z.

It is a very good approach to parameterize all the dimensions and link them together in order to easily change the geometry lately. The other way is to design the device using specialized computer-aided design (CAD) software and then import the geometry in the FEM sofware.

The various regions must match the materials they represent. The material properties represent all the constants that appear in the PDE. The type of material can be chosen among those already present in the database, or one can create his/her own materials, specifying the proprieties that have to be used during the analysis.

For acoustic pressure, the principal properties of the medium are density and speed of sound.[1]

[1] If the problem needs to be coupled, others properties are to be entered. For example, if a coupled thermal problem is studied, the density and the speed of sound could be functions of temperature.

Initial conditions are used in conjunction with time studies and this condition adds initial values for the sound pressure p and the pressure time derivative $\mathrm{d}p/\mathrm{d}t$ that can serve as an initial guess for a nonlinear solver.

Boundary conditions define the nature of the boundaries of the computational domain. Some define real physical obstacles like a sound hard wall or a moving interface. Others, called artificial boundary conditions, are used to truncate the domain. The artificial boundary conditions are, for example, used to simulate an open boundary where no sound is reflected. For acoustic pressure mode, it is possible to choose among the following boundary conditions [6]:

- Neumann Condition
  The normal component of particle velocity disappears. This condition is equivalent to the normal acceleration equal to 0:

$$n \cdot \left(\frac{1}{\rho_0} (\nabla p) - q \right) = 0.$$

  If the dipole source q is null, the normal derivative of the pressure is equal to 0:

$$\frac{\mathrm{d}p}{\mathrm{d}n} = 0 \quad \text{(normal velocity = 0).}$$

  This condition is used to model rigid surfaces.

- Dirichlet Condition
  This condition nullifies in the boundary the acoustic pressure value:

$$p = 0.$$

  It is an appropriate approximation for a liquid–gas interface.
- Pressure (Dirichlet Condition) $p = p_0$
  The pressure condition specifies in the boundary, the acoustic pressure amplitude (oscillating as $\mathrm{e}^{\mathrm{i}\omega t}$) in the frequency analysis. In the transient analysis, the time dependence of the pressure source has to be explicated.
- Normal Acceleration (Neumann Condition)
  This condition sets the inward normal acceleration amplitude at the boundary:

$$n \cdot \left(\frac{1}{\rho_0} (\nabla p) \right) = a_n \qquad (2.22)$$

  where a_n represents an external source term.
  Using this condition, it is possible to obtain also an acoustic source: a sound can be produced through the vibration of the boundary (e.g., a baffled piston or a desired simple source strength if a small boundary is selected).
  This condition is used to couple acoustic domains to adjacent mechanical domains through a function that sets the acceleration on the boundary of the latter as the condition (see Sect. 2.5).

- Impedance Condition
  Considering the acoustic input impedance of the external domain Z (ratio between pressure and normal particle velocity), this condition is expressed in the time domain as

$$n \cdot \left(\frac{1}{\rho_0} (\nabla p - q) \right) + \frac{1}{Z} \frac{\partial p}{\partial t} = 0$$

  whereas in the frequency domain it is

$$n \cdot \left(\frac{1}{\rho_0} (\nabla p - q) \right) + \frac{\mathrm{i}\omega p}{Z} = 0.$$

  This condition is an approximation for a locally reacting surface (e.g., an absorbing panel).
- Radiation Condition
  This condition permits to approximate an infinite space. In fact, it sets a boundary that will not reflect normally incident waves and allow an outgoing wave to leave the domain. Such kind of waves can be respectively:
  - Plane waves: used for both far-field boundaries and ports (e.g., waveguide structures);
  - Cylindrical waves;
  - Spherical waves: used to allow a radiated or scattered wave (emanating from an object) to leave the domain without reflections.

  Furthermore, using this condition, an acoustic source can be obtained. In fact, sending a plane wave from a plane wave radiation boundary condition, or an incoming spherical wave from a spherical radiation condition, it is possible to simulate a source at infinity.
- Sources
  Another way to create an acoustic source is to add a point or an edge source. An acoustic source is also a monopole distribution that is applied over a volume in order to radiate a uniform sound field in all directions, or a dipole source in order to radiate a sound field that is typically stronger in two opposite directions.

When the geometry is complete, it has to be divided into finite elements.

In order to get a convergent and accurate solution, the mesh in acoustic computations should be fine enough to both resolve the geometric features and the wavelength of the problem. If an insufficient number of elements is used, the wave will not be well modeled (see Fig. 2.3). The “rule of thumb” in the literature is that the maximal mesh size should be less or equal to λ/N, where N is between 5 and 10 [6, 14].

Increasing k, the number of elements required to maintain accuracy grows at least linearly with respect to k in 1D, with the cost growing at a faster rate in higher dimensions, and this leads to prohibitive computational cost for large values of k. For this reason, it is necessary to model the system to be studied trying to simplify as much as possible the geometry taking advantage of symmetries.

In most part of the simulations that involve fluids, the internal dumping effects can be neglected. Indeed, they could affect meaningfully the behavior of the system

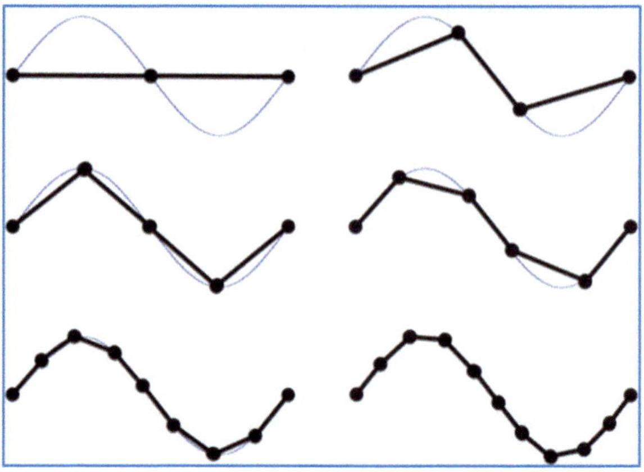

Fig. 2.3 Modeling a single wave with $N = 2, 3, 4, 5, 8$, and 10 elements

only at very high frequencies or when the dynamic viscosity of the fluid is much different from that of both air and water. This is not the case of solids. There are two effects occurring in the solids which give rise to dumping effects: absorption and scattering [6, 15]. The first one consists in a conversion of mechanical energy into heat, due to viscosity that breaks the oscillation of the particles. Furthermore, due to its viscous nature, it rises with the frequency. The scattering is due to the nonperfect homogeneity of the medium where the wave propagates. The abrupt variation of acoustic impedance in correspondence of irregular and randomly oriented surfaces gives rise to a diffusion of the wave energy, and as a consequence, an attenuation of the measured value. Both contributions of attenuation depend on the frequency, even if for different reasons. From a numerical point of view, we can take into account the attenuation by means of an exponential term:

$$p = p_0 \cdot \mathrm{e}^{\alpha d} \tag{2.23}$$

where the amplitude p is the reduced amplitude after the wave has traveled a distance d from the initial location and the attenuation coefficient α has to be defined for the material and in general, it will be a function of the frequency.

The frequency domain modeling appears to be the most convenient way to study damping materials. A material is characterized by two complex functions which are the wave number k and the impedance Z. If these two parameters are known, it is possible to define the complex speed of sound $C_c = \omega/k$ and the complex density $\rho_c = kZ/\omega$, and the problem can be solved in terms of equivalent fluid model. Regarding the mesh, the dumping materials require an unstructured mesh, as it is impossible to establish a priori which direction needs a more detailed analysis.

If the linearity hypothesis holds valid, the time evolution can be treated as the superimposition of sinusoidal regimes, each one having specific amplitude, frequency, and phase. This allows one to eliminate the time variable in the solution of the problem. In order to retrieve the solution in the time domain, all one needs is to apply the inverse Fourier transform to the frequency domain solution.

Even if solving the problem in the time domain is always possible, it should be avoided when possible because of the much higher computational complexity, like for example, the presence of nonlinear terms.

Finally, it is possible to calculate the solution to PDE.

We need to establish which kind of study has to be performed. In general, we have three possibilities: steady state, transient, and frequency domain analysis. The first one is obviously the less demanding one in terms of calculation, but it cannot be used in all that cases, like the one at hand, where the dynamics of the phenomenon is a fundamental aspect. The frequency domain analysis can be performed both for permanent periodical regime and for the transient analysis. The use of this kind of study in general requires much less time than the time domain analysis.

It is possible to change the type of solver under various steps. There are a variety of direct and iterative solvers. Direct solvers tend to be more robust (they are most likely to converge) but require more memory.

In acoustic pressure studies, the parametric sweeps are useful to solve the same geometry for a range of frequencies.

For transient acoustic problems, several new time scales are introduced. One is given by the frequency contents of the signal and by the desired maximal frequency resolution: $T = 1/f_{\max}$. The other one is given by the size of the time step Δt used by the numerical solver. In fact, the distance traveled by the fastest wave in the model should be smaller than the characteristic element size in the mesh. A condition on the so-called *CFL* number [8] dictates the relation between the time step size and the minimal mesh size l_e as:

$$CFL = \frac{c \cdot \Delta t}{l_e} \tag{2.24}$$

where c is the speed of sound of the medium.

Once the solution has converged, the results can be plotted. These are all derived from the solution for the pressure, which is the unknown quantity of the PDE. Obtaining the pressure, one can compute velocity, intensity, etc.

2.5 Modeling Transmitting and Receiving Transducers

An ultrasonic measurement process involves the generation of ultrasound by the transducer, propagation of ultrasonic waves into the propagating medium, and reception of these waves through the transducer again. A complete measurement system consists of a pulser, a transducer, and a receiver. The pulser sends the electrical pulse via a cable to an ultrasonic transducer. The piezoelectric effect implies the conversion of the electrical pulse into an acoustical pulse in the generation system and an acoustical into an electrical signal in the reception system. In fact, certain materials will generate an electric charge when subjected to a mechanical stress and change its dimensions when an electric field is applied across the material (Fig. 2.4). These are called respectively, the direct and inverse piezoelectric effect [20].

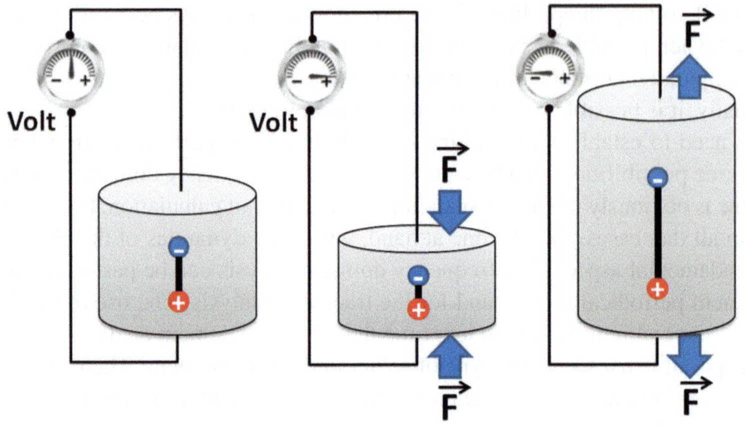

Fig. 2.4 The piezoelectric effect

This effect is observed in a variety of materials, such as quartz, dry bone, polyvinylidene fluoride, and lead zirconate titanate (PZT). The latter of these is a man-made ceramic with a perovskite structure and is one of the most commonly used piezoelectric ceramics today.

A piezoelectric material is described by both the laws of mechanics and electromagnetics [1, 3, 21].

In COMSOL Multiphysics, the modeling of transmitting and receiving transducers is done with the piezoelectric devices interface that is a combination of the solid mechanics and electrostatics interfaces [6]. The electric potential V and the three components of the displacement, x, y, and z are the dependent variables.

The mechanical properties of a piezoelectric material can be described as a linear-elastic material. Consequently, the relation between stress[2] tensor T and strain[3] tensor S in a material is linear and it follows the generalized Hooke's law [18]:

$$T = c \cdot S \text{ or } S = s \cdot T \tag{2.25}$$

where c is the (elastic) stiffness matrix of the material and s is the compliance matrix of the material, which is the inverse of the stiffness matrix.

Electrically, a piezoelectric material is a polarizable dielectric that follows the electrical equation

$$D = \epsilon_0 E + P \tag{2.26}$$

[2] Stress is defined as the force per unit area in direction i acting on the surface of the unit cube whose normal is direction j. The components where $i = j$, are the normal or longitudinal stresses, while those where $i \neq j$, are the shear stresses.

[3] Strain is a quantification of the deformation in direction i of the unit cube whose surface deformed is indicated with j. The strain tensor is expressed as $S = \frac{1}{2}(\nabla u + \nabla u^t)$, where u is the mechanical displacement of a piezoelectric material.

where D is the electric displacement field [C/m^2], ϵ_0 is the vacuum permittivity, E is the electric field [V/m], and P is the polarization density.

Since the material is a dielectric, the divergence of the electric displacement field is null:

$$\nabla \cdot D = 0. \tag{2.27}$$

Piezoelectric materials combine these two constitutive equations into one coupled equation. Then, the constitutive relations of piezoelectric devices are expressed in the subsequent strain–charge form as:

$$\begin{aligned} S &= s_E T + d^t E \\ D &= dT + \epsilon_T E \end{aligned} \tag{2.28}$$

where s_E is the compliance matrix whose coefficients were measured while the electric field across the material was zero or constant (subscript E) and d is the coupling matrix[4]. The permittivity matrix ϵ has been introduced to replace the vacuum permittivity and the polarization vector, and the permittivity data were measured under at least a constant, and preferably a zero stress field (subscript T).

The electric field is connected to the voltage by:

$$E = -\nabla V. \tag{2.29}$$

In the same way as in the acoustic interface (see Sect. 2.4), piezoelectric material model must be applied to all domains governed by the piezoelectric interface. Also, here the temperature and absolute pressure are set with the same standard values as in the acoustics. A feature in this model is the possibility to align the model with another coordinate system than the global Cartesian system. This is often necessary, as piezoelectric materials are generally defined to be poled in the three directions, which coincide with the z-direction in the global system. A 2D model is however set in the x–y plane, causing the piezoelectric to be poled in the out-of-plane direction [16].

All features and conditions that are applied to this interface are described in the following list [6]:

- Free
  This is the standard mechanical boundary condition, and therefore initially applied to all boundaries in domains governed by the piezoelectric interface. It defines the boundary as free to move in any direction and without any loads acting on it.
- Zero charge
  This is the default electrostatic boundary condition. It defines, as the name implies, that there is no electric charge on the boundary.
- Initial values

[4] d is the matrix for the direct piezoelectric effect and transposed matrix d^t is the matrix for the inverse piezoelectric effect.

This option allows to set an initial displacement field, velocity field, electric potential, or time derivative of said potential in any of the domains governed by the piezoelectric interface.

- Axial symmetry
  This option is available only in the axisymmetrical models, where it defines the axis of symmetry. It is set as a standard condition on all boundaries that lie along the line where $r = 0$.
- Electrical potential
  This option sets the electrical potential to a given value on the boundary the condition is applied to.
- Ground
  This option sets the electrical potential to zero at the boundary it applies to.
- Symmetry
  This indicates that the model is a part of a larger structure that is mirror symmetrical around the boundary where this condition is applied, and thus, overrides the free boundary condition.
- Roller
  This condition suppresses the standard mechanical condition, and defines that there can be no displacement perpendicular to the boundary, but tangential displacement is still allowed.
- Damping and loss
  This option allows the inclusion of dielectric, mechanical, and piezoelectric damping in various domains. The loss factors can be set in this feature, and thus, apply to all domains with loss. Alternatively they can be set in each material as a property that is only included when required.
- Boundary load
  This condition replaces the standard free condition. It states that a given mechanical load is applied on the boundary. This can be used to couple a mechanical domain to a bordering acoustic domain by setting the pressure in the acoustic domain as a force per unit area on the boundary of the mechanical domain.

2.6 Dealing with Interfaces and Defects

As seen in the introduction of this book, the ultrasonic direct transmission technique (DTT) [19] is employed to obtain useful, rapid, and relatively low-cost information on the studied structure and it permits to measure the propagation speed.

This technique uses a beam of ultrasonic waves emitted by a transducer and received by another one placed on the opposite surface of the structure. The waves pass through the tested material and they are reflected and refracted by incidental discontinuities encountered along the path. In fact, as can be noted in Fig. 2.5, the sound vibration can be subjected to several conditions. If the path between transmitter **T** and receiver **R** is a medium free of defects, the flight time will be characteristic of that medium. If the path crosses an area with a lack of structural homogeneity,

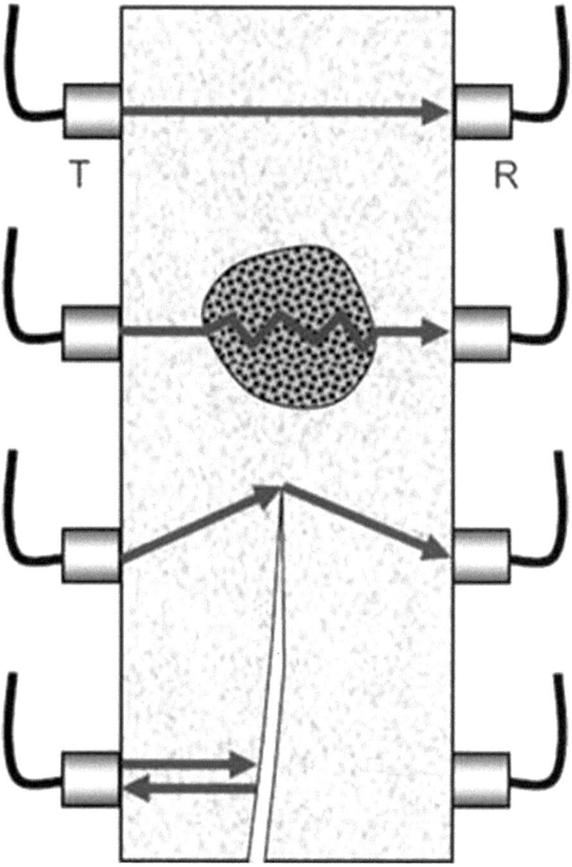

Fig. 2.5 Possible paths of transmitted signal

the signal will be dispersed and both, the path followed and the flight time, will be greater and the speed will be reduced.

If instead, the two transducers are positioned in such a way that the direct path passes close to the edge of a crack, the signal does not travel through the solid–air interface, but there will be a diffraction signal at the edge of the crack resulting in a path greater than the distance between the two transducers. Therefore, it will have a lower speed than that of the sound characteristic of the medium tested. Finally, in the case in which the wave crosses a cavity, the wave will be largely reflected and the time of flight will be hardly measurable.

The inner conformation of the structure, i.e., the possible presence of discontinuities and defects, introduces reflections at the interfaces between the different materials, and alters the amplitude, direction, and frequency content of the signal.

In order to show how the transmitted wave behaves in the presence of interfaces and defects, in the following, a model of a structure with variation of density within the medium, will be analyzed as an example [5].

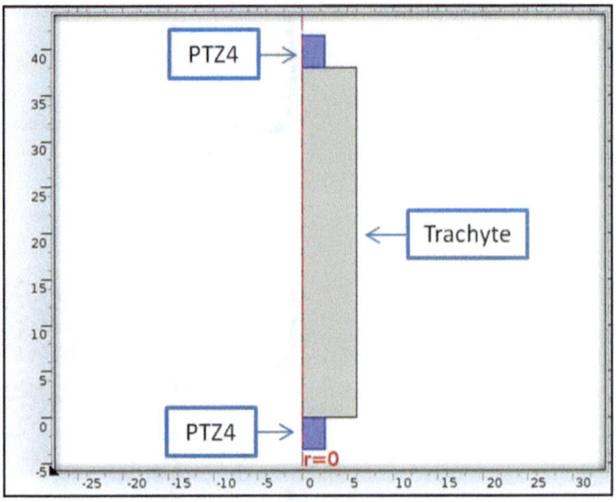

Fig. 2.6 First model: block of a single material

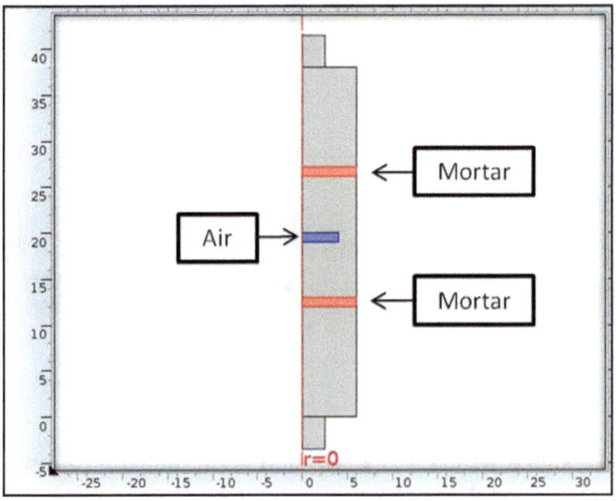

Fig. 2.7 Second model: geometry with insertion of interspaces of mortar and air

Let us consider two axisymmetric bidimensional models with a simple geometry in order to reduce the calculation time, exploiting the cylindrical geometry of the transducers and, with a certain approximation, the volume of material crossed by the ultrasonic wave. The first one consists of a rectangle $6 \times 38\ \text{cm}^2$, which represents the section of a trachite block. Two other rectangles of size $2.5 \times 3.5\ \text{cm}^2$ are inserted at the ends of the block in order to simulate the transducers. In this way, a path of the acoustic wave through a block of a single material (in this example, trachite) is simulated (Fig. 2.6).

The second model is similar to the first one with the insertion of interspaces of mortar and air, transversal to the propagation direction (Fig. 2.7).

The commercial software usually has the interface which allows one to study the behavior of piezoelectric devices. As an example, we can consider the COMSOL

Table 2.1 Materials properties

Material	Sound velocity c_s (m/s)	Density ρ (kg/m^3)
Trachite	1968	1750
Mortar	1768	1800
Air (20°)	340	1.2
PZT4	–	7500

environment. In this software, there is the "transient acoustic piezoelectric interaction" interface that permits to simulate the transient performance of an acoustic wave in a structural element of known geometry and average density variable that interacts with two piezoelectric transducers [6]. This interface combines in a single multiphysics module, the features of "transient pressure acoustics," "solid mechanics," "electrostatics," and "piezoelectric devices" interfaces. The unknowns are the acoustic pressure p, the vector of displacement u, and the electric potential V. The equations that describe the phenomenon are the same as described previously in the Sects. 2.4 and 2.5.

The materials used in the model are listed in Table 2.1.

The conditions applied to the models are (see Sects. 2.4 and 2.5 for details):

- For the transducers:
  - Free
  - Ground
  - Roller
  - Electric potential.
- For the structure:
  - Sound hard boundary wall.

The excitation wave is an acceleration signal obtained by the piezoelectric transducer (made of the crystal PZT4) by applying the impulsive voltage signal shown in Fig. 2.8. This electric pulse with maximum amplitude of 500 V and duration of 9.3 μs is applied to the upper part of the emitter transducer, while the bottom part of both the transducers is grounded.

In order to obtain a fairly accurate analysis, it is useful to adopt elements that are particularly small near the notches or other discontinuities, and then the models are meshed in a different way. The first model, which is done by a homogeneous material, will have a uniform mesh, with thickening at the interface with the transducers (Fig. 2.9).

The model with insertion of interspaces of mortar and air is inhomogeneous and will have mesh thickened at each separation surface between the different materials (Fig. 2.10).

The vibration produced by the emitter arrives at the receiver, which transforms the acoustic vibration into an electric voltage.

The trends of the electrical potential at the center of the receiver transducer have been analyzed. In Fig. 2.11, the trend of the electrical potential for the homogeneous

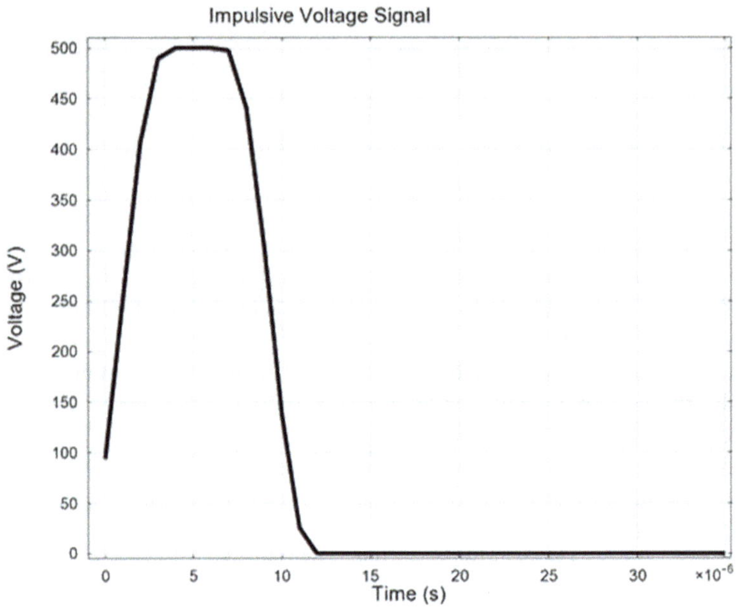

Fig. 2.8 500 V excitation pulse modeled by COMSOL

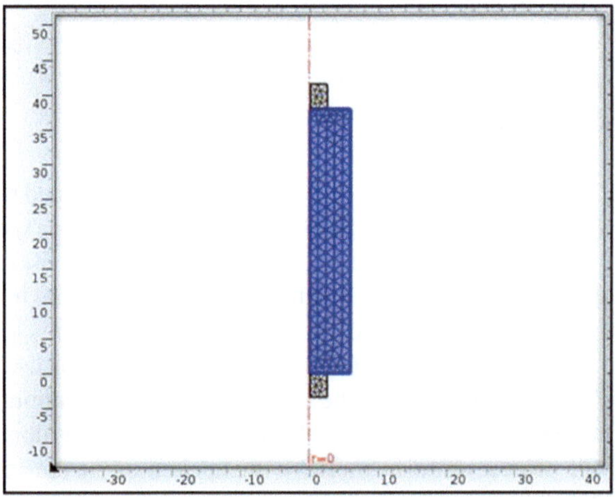

Fig. 2.9 Mesh of the homogeneous model

trachite block is shown, and in Fig. 2.12, we can see the trend of the electric potential for the model with different density (structure with trachite, mortar, and air).

As can be seen in Figs. 2.11 and 2.12, by applying a load in the wall in contact with the transmitter, the received signals are detected with some delay on the opposite side of the wall.

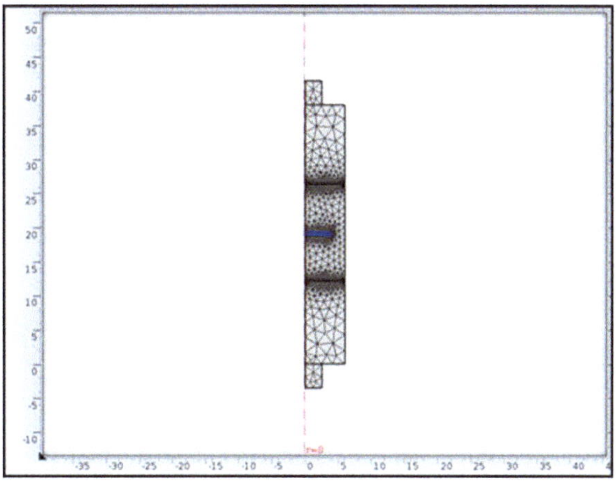

Fig. 2.10 Mesh of the model with different density

The time between the start time of the signal from the emitter and the time of arrival at the receiver, is the *flight time* of the signal and it provides useful information for the characterization of the path followed by the wave.

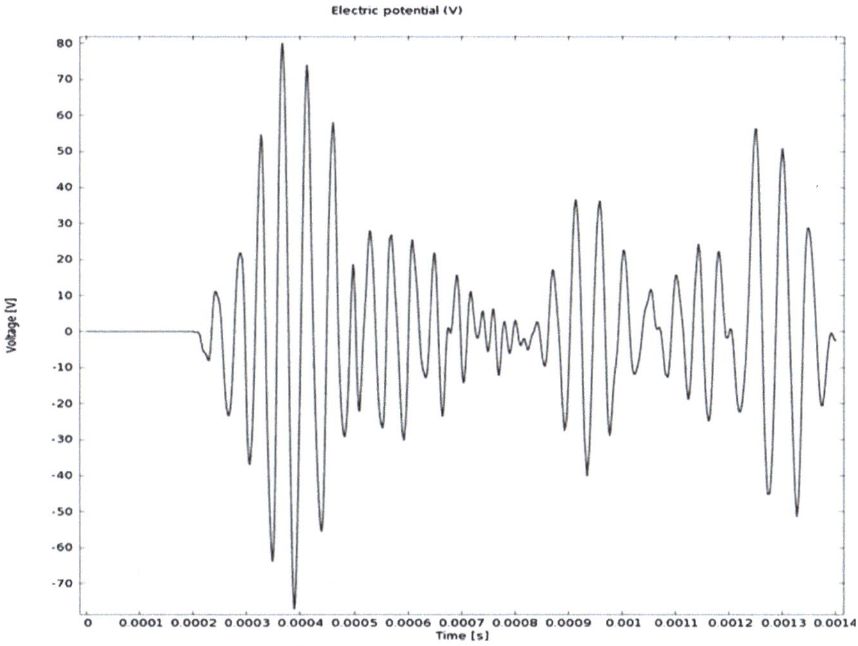

Fig. 2.11 Trend of the electric potential for the first model (homogeneous trachite block)

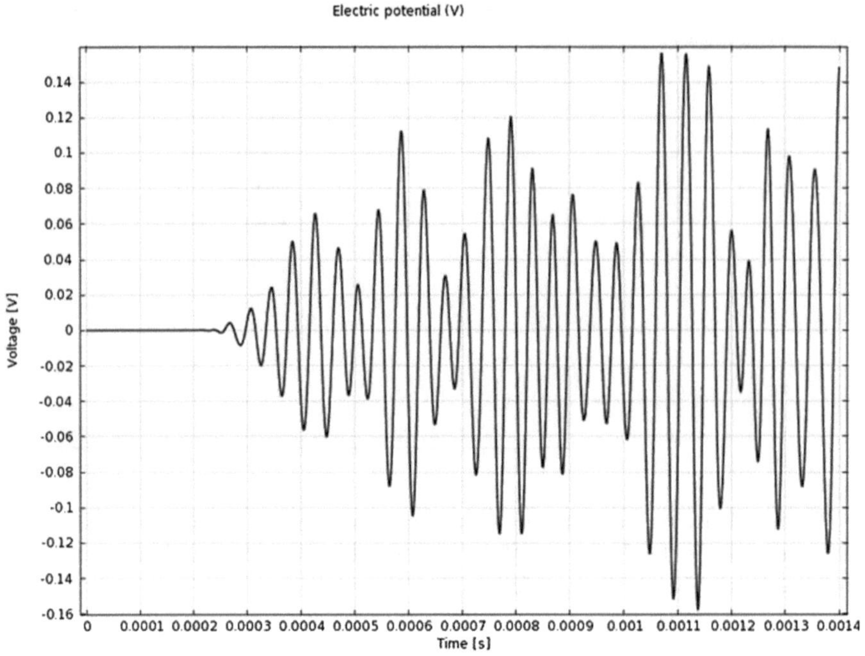

Fig. 2.12 Trend of the electric potential for the second model (structure with trachite, mortar, and air)

As can be observed in Fig. 2.11, considering the length of the path (0.38 cm) and the sound velocity of the material (in this case, trachite) equals 1968 m/s, a flight time equal to 0.193 ms is expected and obtained.

The presence of the mortar and air introduces a delay in the flight time (see Fig. 2.12) and the signal is strongly attenuated.

The inner conformation of the structure, i.e., the presence of discontinuities and defects introduces reflections at the interfaces between different materials, and then, the amplitude, direction, and the frequency content of the signal will be altered.

2.7 Conclusion

In this chapter, the FEM formulation is described in general terms, and then the specific case of waves' propagation problems is analyzed in details. On these bases, the problem of developing a model of an ultrasonic NDT system is described, taking into account three different aspects of the problem, namely, propagation of ultrasounds into a mean, modeling the transducers, and detecting the presence of defects in the analyzed system. In the following appendix, the procedure is shown for a particular case by using COMSOL^{TM} software.

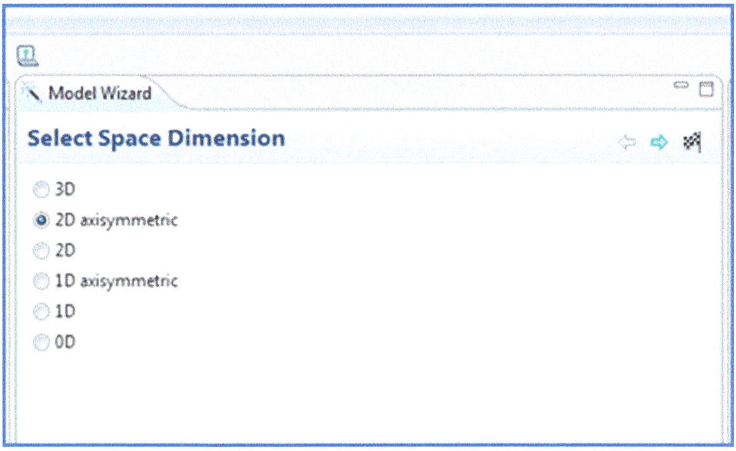

Fig. 2.13 Selection of space dimension

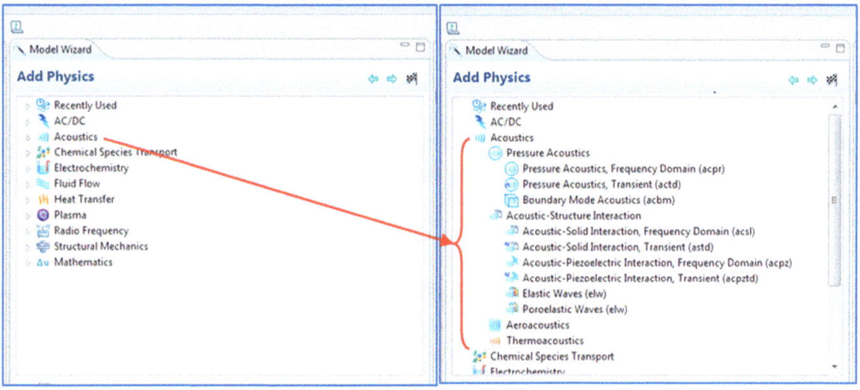

Fig. 2.14 The acoustics module physics interfaces

Appendix: Developing the Model with COMSOL

In this book, all the simulations are done using the FEM software package called "COMSOL Multiphysics" [7]. This software is used for various physics and engineering applications and it is focused especially in coupling different physics together (e.g., acoustics and solid mechanics). COMSOL Multiphysics offers an extensive interface to MATLAB and its toolboxes for a large variety of programming and preprocessing and postprocessing possibilities. In comparison to conventional physics-based user interfaces, COMSOL Multiphysics is highly flexible and it allows to program in your own PDEs if they are not already implemented.

In order to study a wave propagation problem using COMSOL Multiphysics, the analysis of the model can be divided in different fundamental steps that will be

Fig. 2.15 Selection of study type

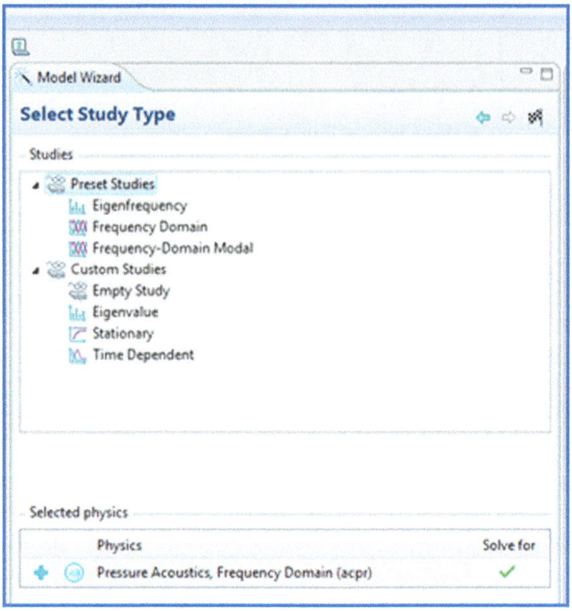

illustrated in the following focusing on acoustics as application [6]. Every step is managed with specific tools provided by the software. Of course, the steps are similar for other physics.

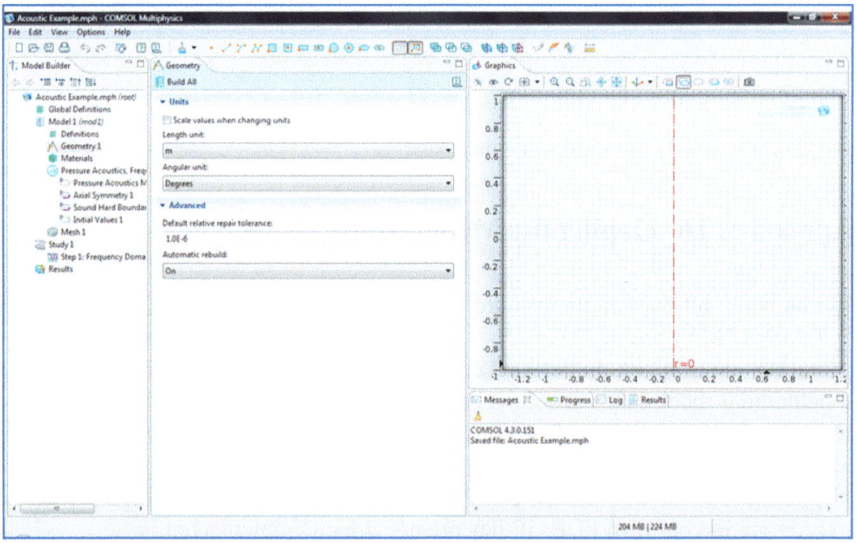

Fig. 2.16 Graphical user interface

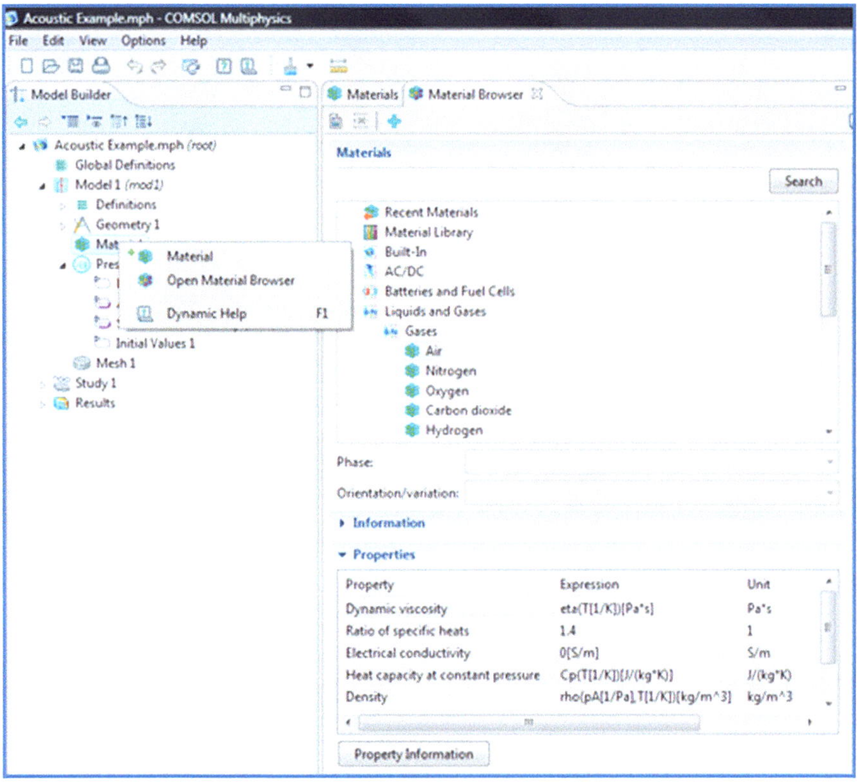

Fig. 2.17 Material properties

1. Start COMSOL Multiphysics by clicking the icon.
2. When COMSOL starts, the Model Wizard will open automatically. It will require you to select:
   - the coordinate system for the model, i.e., choose how many dimensions to work in (Fig. 2.13);
   - the relevant physics to the problem taking into account that multiple physics can be added to a single model if you want coupling (Fig. 2.14);
   - the type of study you wish to perform: time domain, frequency domain, stationary, eigenvalue, ... (Fig. 2.15).
3. Construction of the geometry. The geometry could be drawn directly inside the finite element software, using the internal CAD tool: this is the easiest and fastest way, if and only if the geometry is not very complicated.

   Default units are *mks* units (International System (SI) units). It is possible to change units by selecting the root object in the model tree. For axisymmetric geometries, the axis of symmetry is drawn as a line in the graphics window (see Fig. 2.16).

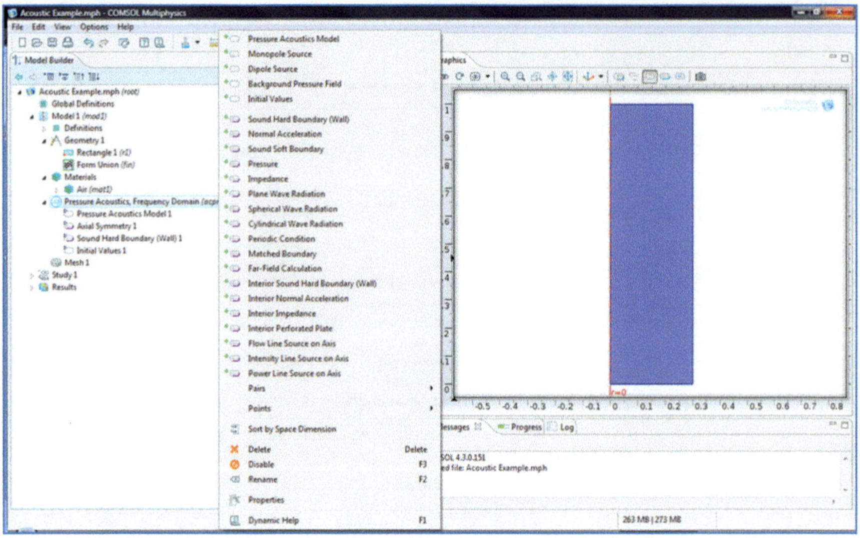

Fig. 2.18 Boundary conditions

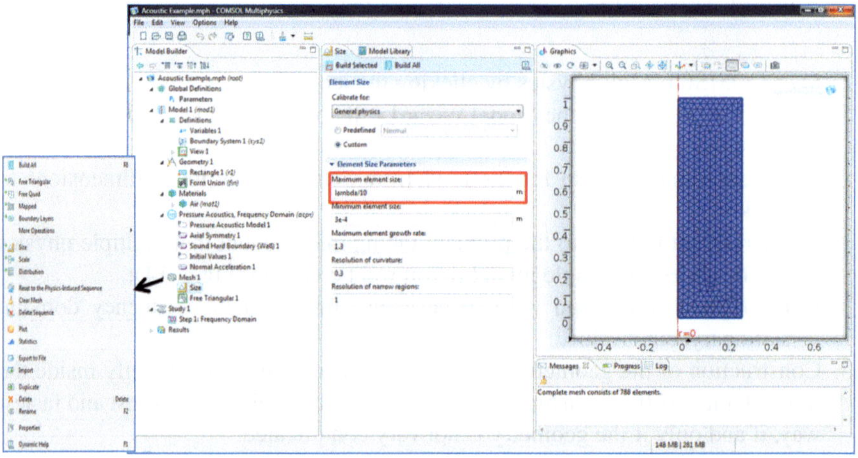

Fig. 2.19 Mesh settings

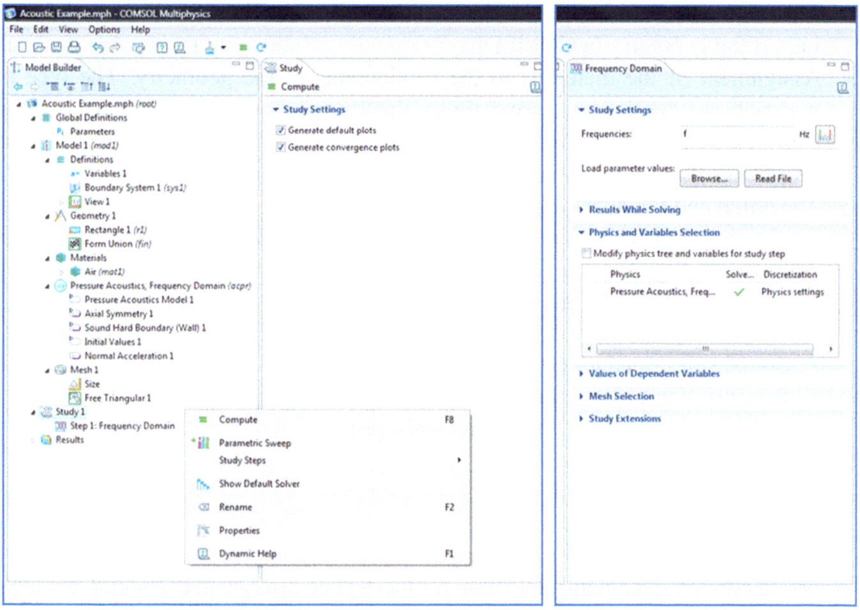

Fig. 2.20 Solving options

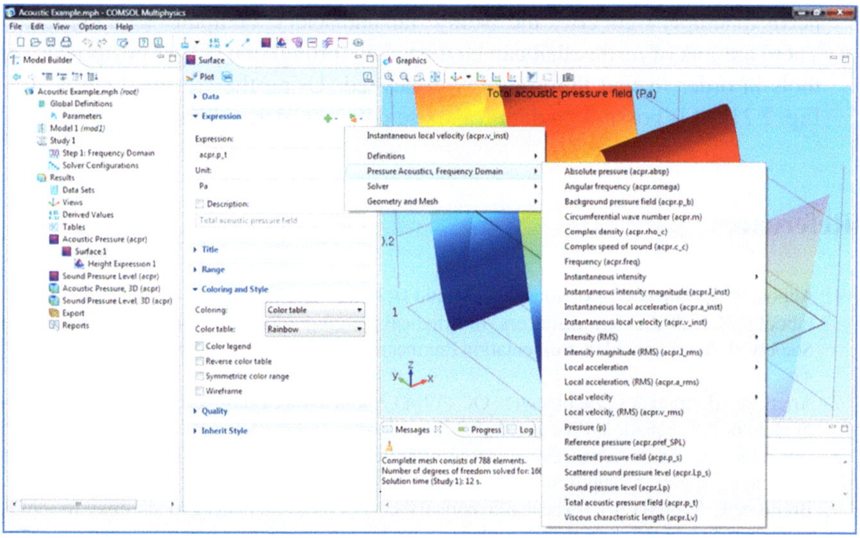

Fig. 2.21 Postprocess the results

The other way is to design the device using specialized CAD software and then import the geometry in COMSOL.

4. Assignment of the materials to the subdomains of the geometry. The type of material can be chosen among those already present in the database, or one can create his/her own materials with the option "+Material," specifying the proprieties that have to be used during the analysis (see Fig. 2.17).
5. The initial and boundary conditions. The default initial conditions are $p = 0[\mathrm{Pa}]$, for the sound pressure, and $\mathrm{d}p/\mathrm{d}t = 0[\mathrm{Pa/s}]$, for the pressure time derivative. As shown in Fig. 2.18 for acoustic pressure mode, it is possible to choose among many boundary conditions.
   - Sound hard boundary (Neumann condition)
   - Sound soft boundary (Dirichlet condition)
   - Pressure (Dirichlet condition) $p = p_0$
   - Normal acceleration (Neumann condition)
   - Impedance condition
   - Radiation condition
   - Sources, etc.
6. Discretizations of the domain through the use of the finite elements. The software has an efficient automatic algorithm for mesh generation with triangular elements (Lagrange second order) or rectangular ones.
   In Fig. 2.19, the mesh options are shown. Mesh sizes and distribution can be parameterized in order to modify the mesh varying the frequency of analysis.
7. Solving and postprocessing of results. Finally, in order to solve the problem, one needs to simply right click the "study" tab and select "compute" (see Fig. 2.20). There are lots of results that can be displayed through the proper postprocessing tools in order to find the information desired. Using the "results branch" (see Fig. 2.21), it is possible to define and use data sets, derived values, and tables.

References

1. Abboud NN, Wojcik GL, Vaughan DK, Mould JJ, Powell DJ, Nikodym L (1998) Finite element modeling for ultrasonic transducers. In: Medical imaging: ultrasonic transducer engineering, Society of Photo-optical Instrumentation Engineers (SPIE) Conference Series, vol 3341, pp 19–42
2. Arlett PL, Bahrani AK, Zienkiewicz OC (1968) Application of finite elements to the solution of Helmholtz's equation. Proc IEE 115:1762–1766
3. Benjeddou A (2000) Advances in piezoelectric finite element modeling of adaptive structural elements: a survey. Comput Struct 76(1):347–363
4. Brenner SC, Scott R (2010) The mathematical theory of finite element methods. Texts in applied mathematics, vol. 15. Springer, New York
5. Cannas B, Carcangiu S, Fanni A, Forcinetti R, Montisci A, Sias G, Usai M, Concu G (2012) Frequency analysis of ultrasonic signals for non-destructive diagnosis of masonry structures. In: Chang S, Bahar S, Zhao J (eds) Advances in civil engineering and building materials, vol 831, Taylor & Francis, pp 807–811
6. Comsol (2011) Acoustics Module User's Guide for COMSOL 4.2. Comsol
7. Comsol (2011) Multiphysics Reference Guide for COMSOL 4.2. Comsol

8. Courant R, Friedrichs K, Lewy H (1967) On the partial difference equations of mathematical physics. IBM J Res Dev 11(2):215–234
9. Finlayson BA (2013) The method of weighted residuals and variational principles. Classics in applied mathematics. SIAM, Philadelphia
10. Harari I (2006) A survey of finite element methods for time-harmonic acoustics. Comput Methods Appl Mech Eng 195:1594–1607
11. Hutton DV (2003) Fundamentals of finite element analysis. Engineering series. McGraw-Hill, New York
12. Kardestuncer H, Norrie DH, Brezzi F (1987) Finite element handbook. McGraw-Hill reference books of interest: handbooks. McGraw-Hill, New York
13. Liu GR, Quek SS (2013) The finite element method: a practical course. Elsevier, Oxford
14. Marburg S (2002) Six boundary elements per wavelength: is that enough? J Comput Acoust 10(01):25–51
15. Morse PMC (1948) Vibration and sound. International series in pure and applied physics. McGraw-Hill, New York
16. Nygren MW (2011) Finite element modeling of piezoelectric ultrasonic transducers. Master's thesis, Norwegian University of Science and Technology, Department of Electronics and Telecommunications
17. Pryor R (2009) Multiphysics modeling using COMSOL®: a first principles approach. Jones & Bartlett, Sudburg
18. Sokolnikoff I (1956) Mathematical theory of elasticity. McGraw-Hill, New York
19. UNI ISO (2004) Natural stone test methods—determination of sound speed propagation. Published standard EN 14579:2004, UNI ISO
20. Vives AA (2008) Piezoelectric transducers and applications. Springer, New York
21. Yang J (2005) An introduction to the theory of piezoelectricity. Advances in mechanics and mathematics. Springer, New York
22. Zienkiewicz OC, Taylor RL, Zhu JZ (2005) The finite element method: its basis and fundamentals. Butterworth-Heinemann, Oxford

Chapter 3
Modeling Ultrasounds for Nondestructive Testing Applications

Sara Carcangiu, Augusto Montisci and Mariangela Usai

Abstract The ultrasonic techniques are today one of the most used methods for nondestructive testing (NDT) and for the characterization of the materials. Often it is necessary to simulate the system under test before to apply this technique to the real case. In fact, during the test in general a great number of parameters have to be calibrated in order to maximize the information captured by means of the measurements. In this chapter we will give some examples of ultrasonic NDT applications. The first one is a tutorial example and it considers the simulation of ultrasonic waves within a fish tank. The next two examples are concerned with the application of ultrasonic testing to masonry and concrete structures, respectively. In particular the problems of detecting a void in a trachite stone wall and in a concrete pillar are considered. Finally, the last example presented here is concerned with the detection of a hole in a metallic specimen. In fact, the ultrasonic inspection technique is widely employed in the steel industry due to favorable propagation conditions that such materials ensure. All the simulations have been done using COMSOL as finite element analysis, solver, and simulation software.

3.1 Introduction

The use of ultrasounds for nondestructive testing (NDT) is based on the fact that when a perturbation is applied to the ends of a body, the disturbance propagates in it in a finite time in the form of sound wave.

Over recent decades, the NDT methods have had a rapid development in all its aspects:

S. Carcangiu (✉) · A. Montisci · M. Usai
Dipartimento di Ingegneria Elettrica ed (DIEE), University of Cagliari, Cagliari, Italy
e-mail: s.carcangiu@diee.unica.it

A. Montisci
e-mail: amontisci@diee.unica.it

M. Usai
e-mail: musai@diee.unica.it

P. Burrascano et al. (eds.), *Ultrasonic Nondestructive Evaluation Systems*,
DOI 10.1007/978-3-319-10566-6_3

- Standards and criteria of use [1, 13]
- Accuracy, resolution, and ability to investigate more and more complex issues [2, 14]
- Recording, postprocessing, and interpretation automation [3, 15].

NDT based on propagation can be differentiated into “microseismic tests” [19] and “ultrasonic tests” [11]. These tests have different ways of dealing with the problems of experimental analysis due to some substantial differences in the material under examination. In fact, while “microseismic tests” are suitable for nonhomogeneous materials, the “ultrasonic tests” are suitable for testing homogeneous materials, which can, therefore, be considered easily penetrable by the waves. The differences between the two cases substantially consist of using different frequencies and intensities of vibrations, different application criteria and interpretations, as well as different types of instrumentation. For materials which cannot be considered homogeneous, such as concrete, mortar, stone in general, wood, masonry, and so on, the development of the corresponding technologies has been significantly slower than that for homogeneous material, such as metals. This is due to both theoretical and technological difficulties and the problems deriving from the low transparency due to their interior intrinsic inhomogeneities.

Currently, the researchers who are dedicated to these specific issues are increasing. Furthermore, dedicated instrumentation, more performing, is appearing in the market [14, 17] and a number of standards and recommendations have been drawn up [1].

The state of the art of NDT provides useful information on the type of transducers and on the number of tests to be conducted in order to obtain satisfactory mappings on the surfaces of the elements under test and to trace shape and size of defects.

The mostly used techniques are the direct transmission technique (DTT) [18] with emitter and receiver arranged in corresponding points on two opposite surfaces, the "pulse-echo" technique with the transducer that is emitter and receiver at the same time and the tomography.

DTT has several disadvantages, namely, the defect cannot be located or identified, the component surfaces must be parallel, and there must be access to both sides of the specimen. The major limit of this technique consists of describing the velocity field in each section of the object using only one value of velocity for each path, i.e., hypothesizing that the mean is homogeneous along each wave path.

The tomography technique overcomes the previously cited limit combining the results of several measurements for transparency for a sharper and reliable investigation of the internal structure of an object, using numerical analysis as a real measurement instrument.

As regards the pulse-echo technique (see Fig. 3.1), when a signal is applied to the external surface of a tested object, if no internal inhomogeneity (e.g., due to a defect) is present, the opposite surface reflects said signal, and this reflected signal can be measured in the same point where the stimulus has been applied. If the material and the geometry of the object are known, the delay of the reflected signal can be predicted, and it will be equal to the ratio between the sound velocity and the length of the path. If an unexpected echo is measured, a defect could be present in the medium.

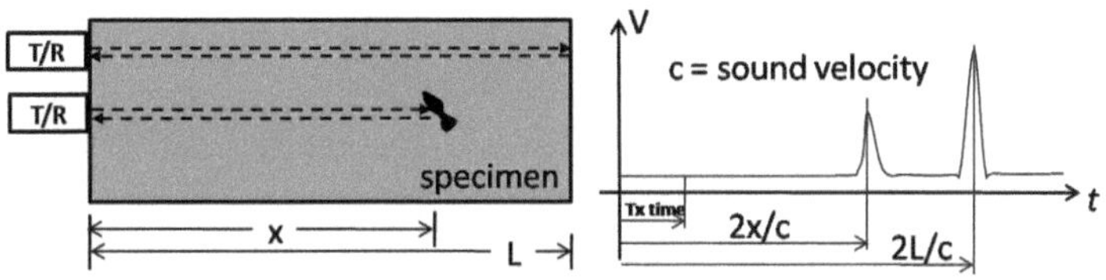

Fig. 3.1 Pulse-echo testing principle description

An important aspect to take into account is the relative dimensions of ultrasonic beam and defect. If the latter is smaller than the former, only a part of the beam is reflected by the defect, so that a reflected wave will be registered with the same delay of the object without defects, but another reflected wave with a smaller delay will be registered. Theoretically, the anticipated reflection carries information about both position and dimension of the defect. Indeed, the delay gives information about the distance of the defect from the emitter, while the energy is related to the part of the beam which has been intercepted by the defect and then to its size. In practice, deducing information about the size of the defect on the basis of the energy of the reflected signal is a difficult task, and this problem can hardly be solved analytically.

In many NDT techniques it is necessary to simulate the tested system before applying the test to the material object. The main reason is that in the test in general a great number of parameters have to be calibrated in order to maximize the information captured by means of the measurements. For example, in many cases the possibility to detect the presence of a defect is based on the possibility to recognize the corresponding measured patterns. One way to do this would be reproducing a representative set of cases. Creating a real set of defects is excessively costly and often very difficult. On the other hand, obtaining said set by means of simulations is much easier. In order to make the method suitable, the precision of the model has to be guaranteed.

In the following sections, some examples of ultrasonic NDT applications are given, with specific recipes that can help to develop suitable models of the tested system.

3.2 Example 1: Detecting a Jelly Bean in a Fish Tank

The first example that we are going to describe concerns the simulation of ultrasonic wave propagation within a fish tank (Fig. 3.2). This is a tutorial example as all the steps followed for the simulation will be described in detail taking into account what has already been discussed in the previous chapter (see Sect. 2.4).

A two dimensional COMSOL-based finite element method (FEM) model has been developed.

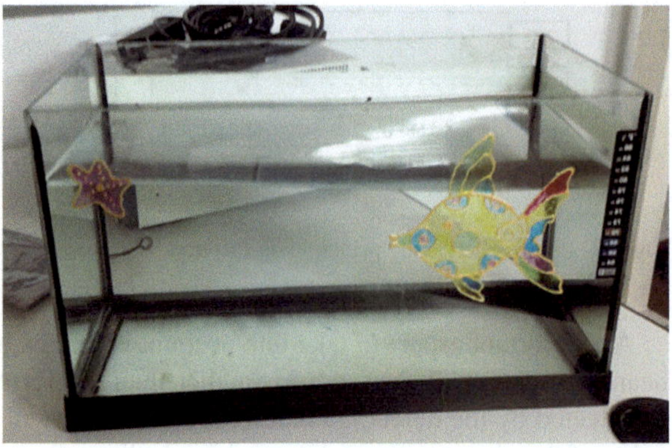

Fig. 3.2 Fish tank filled with water

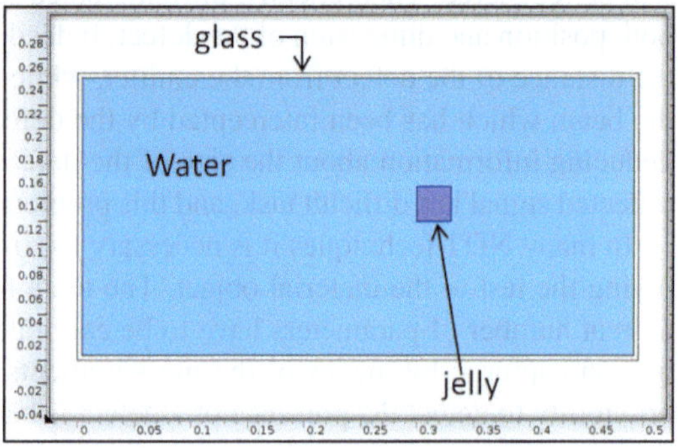

Fig. 3.3 Model of fish tank filled with *water* and a *jelly candy*

- Model Geometry
  The fish tank, filled with water, is 493 mm wide, 298 mm high, and 248 mm thick; it is made of glass 4 mm thick. A jelly candy, with quadratic shape and side of 30 mm has been inserted in the water in order to obtain known anomaly.
  Figure 3.3 shows the model geometry.
- Material Properties
  While the physical properties of water and glass are well known in literature, it is much more difficult to recover the speed of sound and density of jelly candy. In [12], the through-transmission ultrasonic computed tomography is used for measuring the longitudinal wave velocity c_L and pulse-echo transient elastography is used for measuring the shear wave velocity c_S. For materials similar to the jelly, an average density of 1277 kg/m^3 was used, and using the pulse-echo transient elastography, sound velocities between 1450 and 1520 m/s have been calculated. The materials used in the model are then listed in Table 3.1.
- Meshing
  The free triangular elements available in COMSOL are used to mesh the domain. The "user-controlled mesh" option has been set in order to choose the opportune element size l_e of the mesh. In fact, in order to get an accurate solution the mesh should be fine enough to both resolve the geometric features and the wavelength λ.

Table 3.1 Materials properties

Material	Sound velocity c_s (m/s)	Density ρ (kg/m^3)
Glass	3962	2200
Water (20°)	1481.44	999.615
Jelly	1520	1277

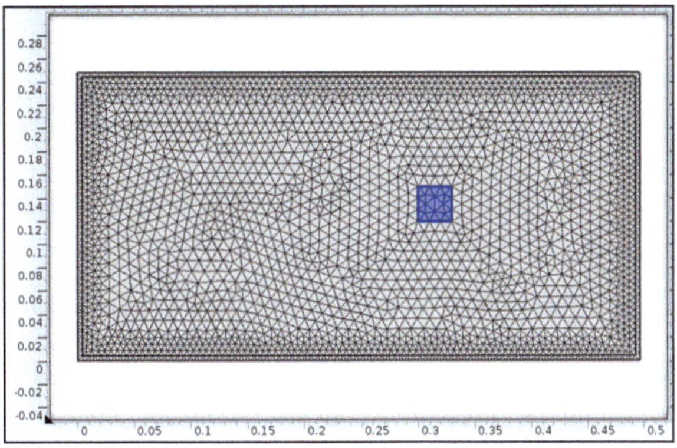

Fig. 3.4 Meshed model

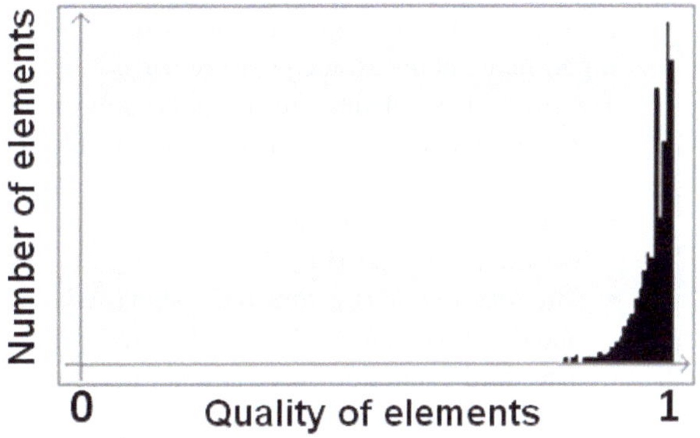

Fig. 3.5 Element quality histogram

The maximal mesh size should be less or equal to λ/N. In this case $N = 10$ has been chosen.

Considering a pulse frequency f_0 equal to 10 kHz the maximum element size l_e is equal to $c_{\text{water}}/f_0 = 0.1481$. In order to improve the quality of the mesh in this simulation the maximum element size has been set equal to $l_e \cdot 0.75$ and the minimum element size has been set equal to $l_e \cdot 0.5$ have been set. In this way 4702 elements are obtained (see Fig. 3.4). The minimum element quality is 0.8218 and the average element area ratio is about 0.1349.

This mesh is analyzed for quality in COMSOL. Quality is how well a particular mesh represents the geometry being modeled. The histogram plot of element quality is shown in Fig. 3.5.

The x-axis is the quality of the element from 0 to 1. The y-axis shows the number of elements at each quality level. The value 0 is assigned to low quality elements,

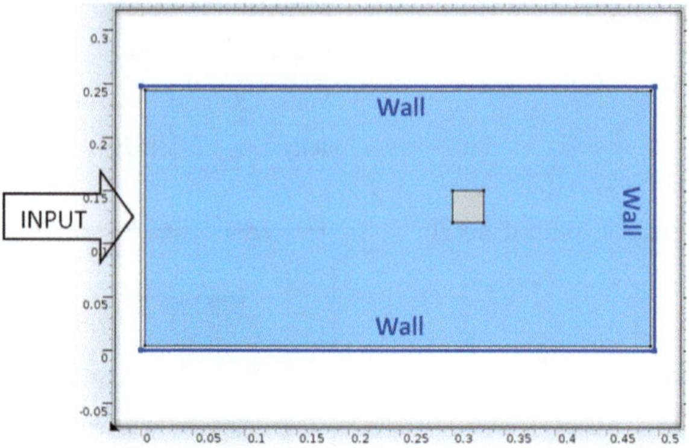

Fig. 3.6 Boundary conditions

and the value 1 is assigned to high quality elements. In this case the average element quality is 0.9725, then the models used are well meshed.

- Application Mode and Boundary Conditions
  Using the "Pressure Acoustics, Transient" interface the classical wave equation is solved with appropriate boundary and initial conditions. There are two boundary conditions used in this model (see Fig. 3.6):
  - Fish tank wall: sound hard boundary $\boldsymbol{n}^{*}(-\nabla p) = 0$
  - Input: Incident ultrasonic pressure wave
    The transducer operates in the pulse-echo mode. The excitation waves are pressure signals that are applied on the left side of the model with displacements along the x-direction. Three different waves have been considered:
    - The first one is a pressure impulse with a maximum amplitude of 3 Pa and duration of 9.3 μs (Fig. 3.7).
    - The second one is a sinusoidal signal with a maximum amplitude of 5 Pa and duration equal to $2 \cdot \frac{1}{f_0} = 0.2$ ms (Fig. 3.8).
    - The third excitation wave is a six-cycletone burst enclosed in a Hanning window as described by Eq. (3.1).

$$y(t) = \frac{1}{2}\sin(2\pi f t) \cdot \left[1 - \cos\left(\frac{2}{15}2\pi f t\right)\right] \tag{3.1}$$

    The parameter f is the excitation frequency equal to 54 kHz, which corresponds to the characteristic frequency (resonance frequency) of the emitting transducer.
    This wave has been chosen to reduce the leakage phenomena.
    The time domain representation of input acceleration used in the third simulation is shown in the Fig. 3.9.

- Study
  A transient analysis has been performed. Taking into account the condition (2.24) in Chap. 2, which dictates the relation between the time step size and the minimal mesh size l_e, a Courant–Friedrichs–Lewy (CFL) number of 0.05 has been chosen and a time step equal to 0.5 μs is considered.

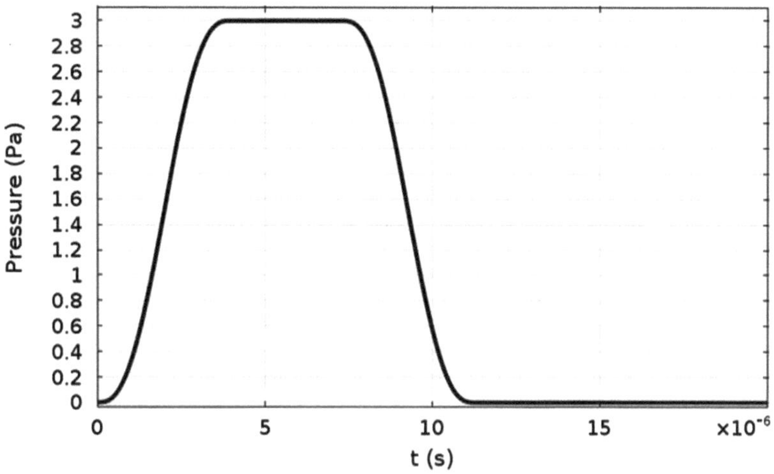

Fig. 3.7 Input: impulsive signal

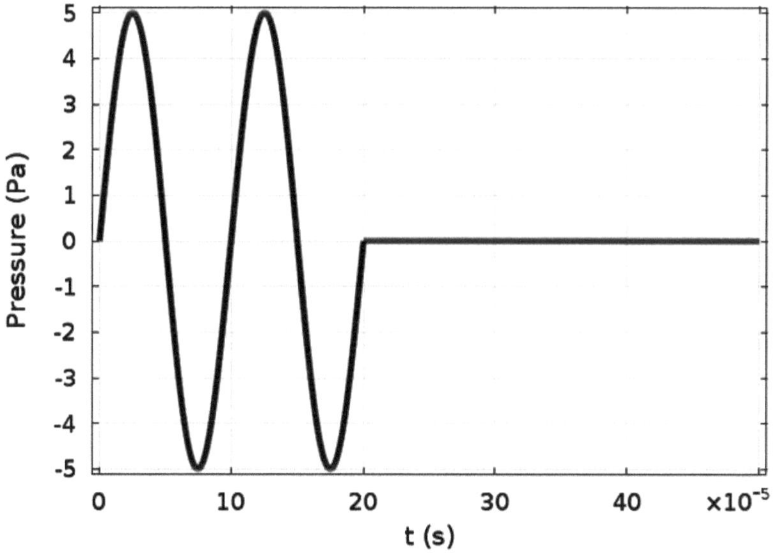

Fig. 3.8 Input: sinusoidal signal

The direct linear solver available in COMSOL is used for obtaining the solution. The duration of time span for the solution was chosen to be 2 ms such that the complete signal and its echo are captured at the point of observation.
"Generalized alpha" scheme has been used as the integration scheme for time marching solution.
All the simulations were carried out in the 64-bit Windows 7 environment.

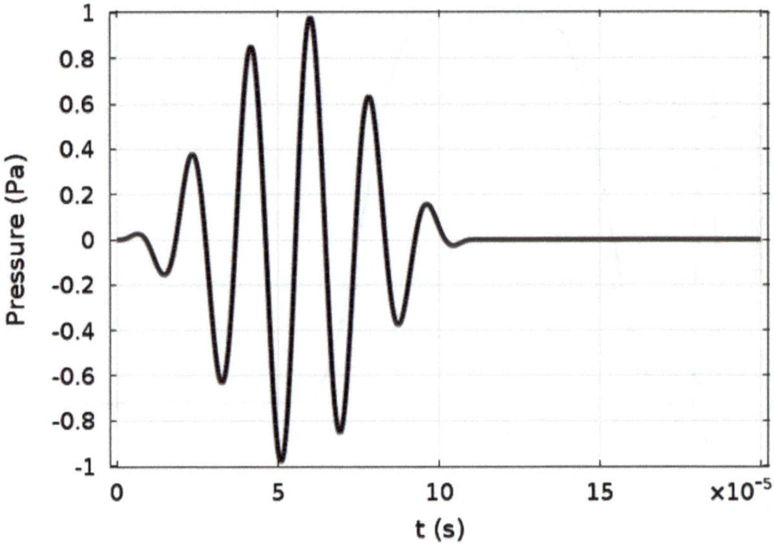

Fig. 3.9 Input: six-cycletone burst enclosed in a Hanning window

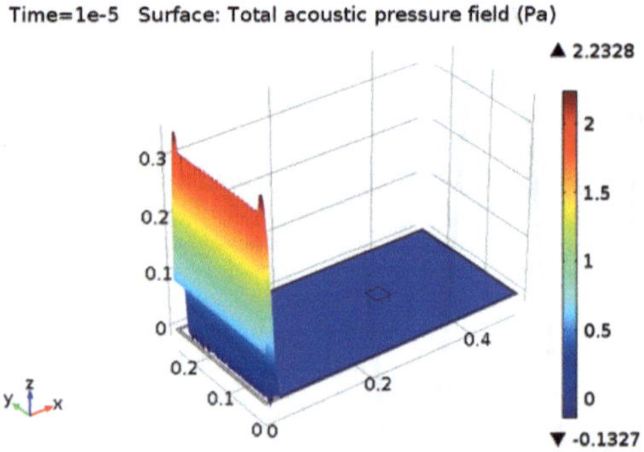

Fig. 3.10 Impulsive signal at the time $t = 10\ \mu s$

- Results
  In the following, the results obtained by the numerical analysis with varying input signal are shown. For each study both the presence and the absence of the defect (jelly candy) are analyzed.
  The total acoustic pressure fields at time 10 μs for the impulsive input (Fig. 3.10), at time 0.2 ms for the sinusoidal input (Fig. 3.11), and at time 120 μs for the third signal (Fig. 3.12) are shown.
  This pressure wave propagates through the water. As can be noted in Figs. 3.13, 3.14, and 3.15, the wave behaves differently depending on the presence or absence of the jelly that acts as defect. If there is the jelly, the discontinuity gives rise to

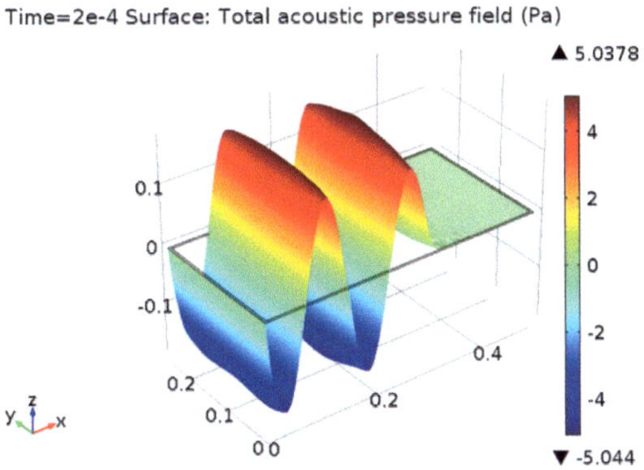

Fig. 3.11 Sinusoidal signal at the time $t = 0.2$ ms

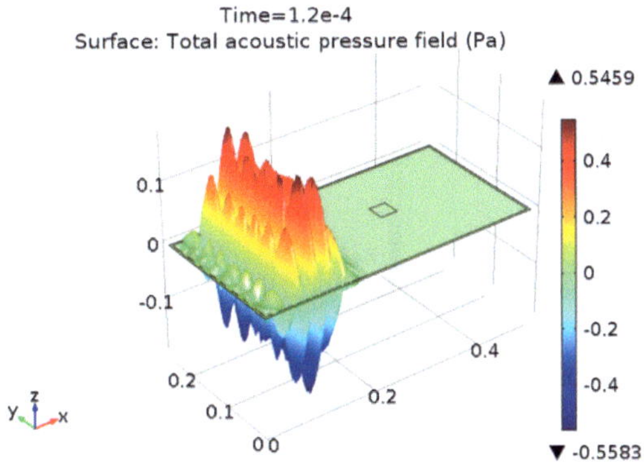

Fig. 3.12 Six-cycletone burst enclosed in a Hanning window at the time $t = 120\ \mu s$

reflections which add up, obtaining a different course from that which would prevail in the absence of discontinuities.

As shown in Fig. 3.6 the defect is positioned at the distance of 30 cm from the input side. Then the first echo of the defect must be detected after 0.4 ms.

Similarly, the opposite side of the fish tank is at 49.3 cm, then its first echo must be detected after 0.67 ms (see Figs. 3.16, 3.17, and 3.18).

The comparison between the two received signals shows in more detail the first echo due to presence of the defect (Figs. 3.19, 3.20, and 3.21).

Finally, all three signals used as inputs allow to obtain good results, but the third signal is the best one because, as already said, it allows to reduce the leakage phenomena, obtaining a better signal.

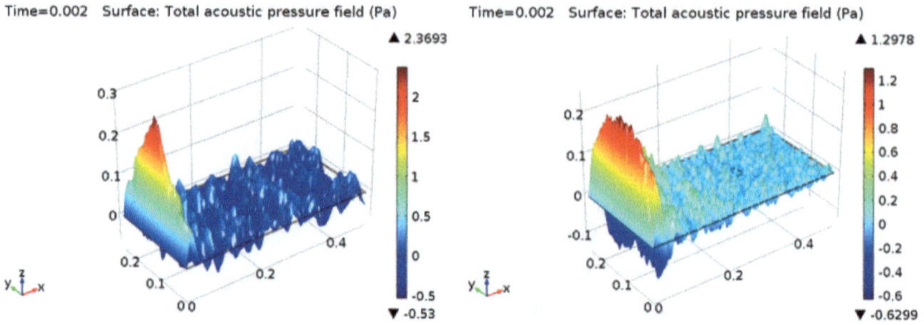

Fig. 3.13 Total acoustic pressure field at time $t = 2$ ms in presence (*left*) and in absence (*right*) of defect (input: impulsive signal)

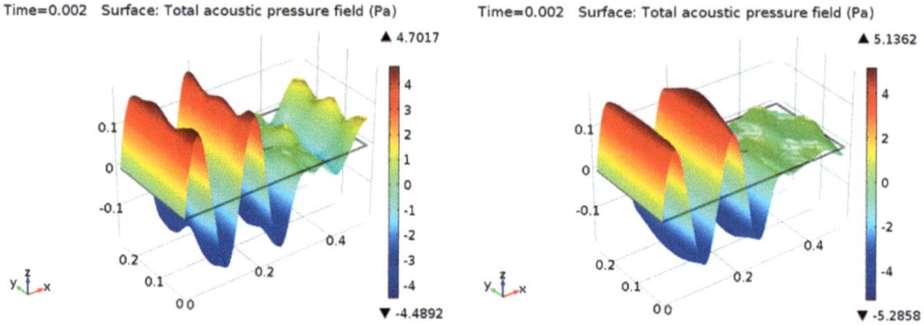

Fig. 3.14 Total acoustic pressure field at time $t = 2$ ms in presence (*left*) and in absence (*right*) of defect (input: sinusoidal signal)

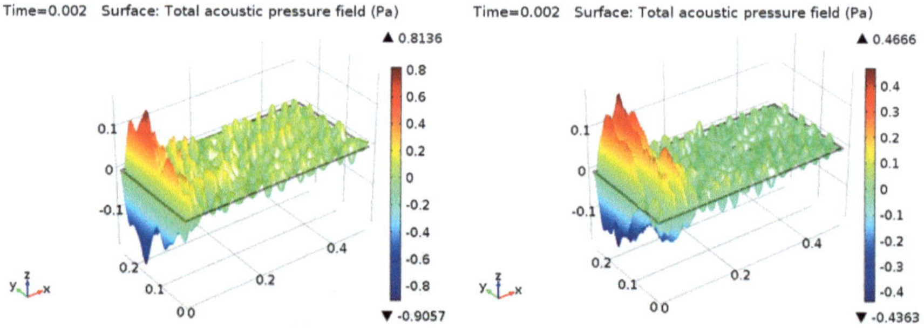

Fig. 3.15 Total acoustic pressure field at time $t = 2$ ms in presence (*left*) and in absence (*right*) of defect (input: six-cycletone burst signal)

3.3 Example 2: Detecting a Void in a Trachite Stone Wall

Let us now consider the case of detecting a void in a trachite stone wall. In this case we will simulate the transducers as well as the trachite stone wall. In fact the experimental setup includes besides the examined block, the emission measurement system. The

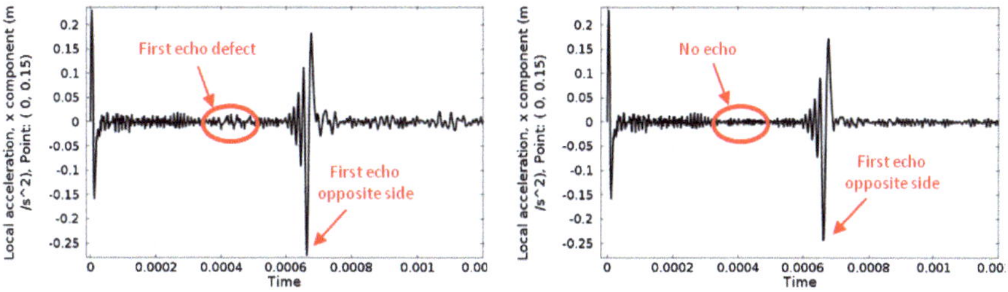

Fig. 3.16 Acceleration wave in presence (*left*) and in absence (*right*) of defect (input: impulsive signal)

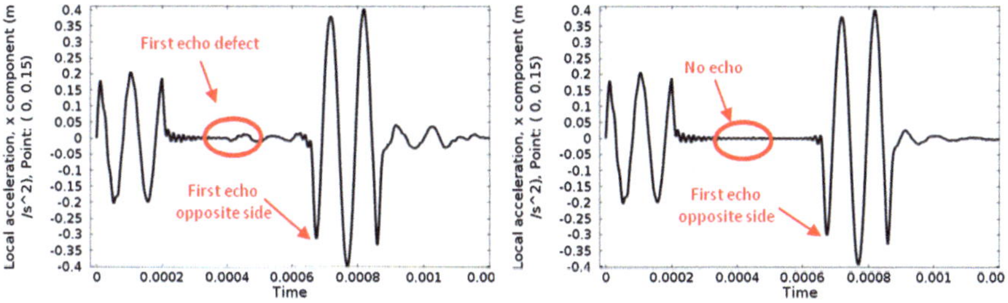

Fig. 3.17 Acceleration wave in presence (*left*) and in absence (*right*) of defect (input: sinusoidal signal)

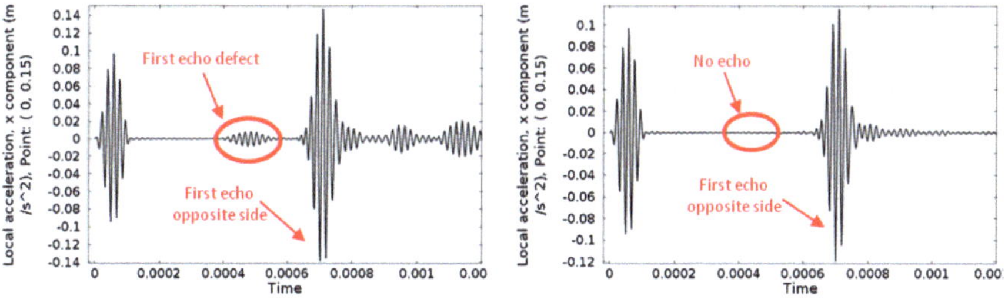

Fig. 3.18 Acceleration wave in presence (*left*) and in absence (*right*) of defect (input: six-cycletone burst signal)

aim is to develop a FEM model of the whole setup, including the transmission and the receiving transducers. In particular, in this example the FEM is applied to the problem of NDT in order to visualize the real acoustic wave propagation into a stone wall, made up of bricks in trachite cemented by mortar. A reliable numerical model of generation and reception of the ultrasonic waves through piezoelectric transducers has been developed.

Experimental measurements are carried out using the equipment PUNDIT (portable ultrasonic nondestructive digital indicating tester), manufactured by CNS Farnell of London [16]. This test instrument allows online data acquisition, waveform analysis, and full remote control of all transmission parameters. Furthermore, the measuring device has the advantage of providing a signal high enough not to necessitate the use of an amplifier. The emitter and the receiver are transducers made of

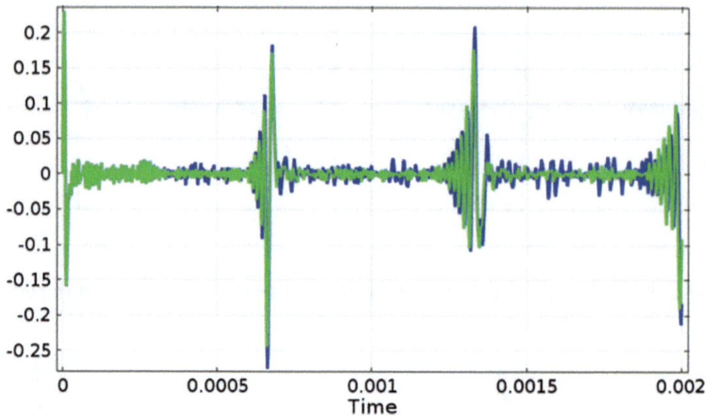

Fig. 3.19 Comparison between the two signals: in presence (*blue*) and in absence (*green*) of defect (input: impulsive signal)

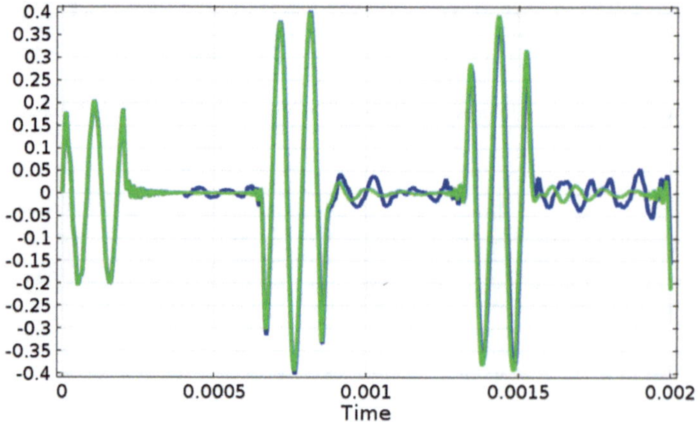

Fig. 3.20 Comparison between the two signals: in presence (*blue*) and in absence (*green*) of defect (input: sinusoidal signal)

lead zirconate titanate (PTZ4) ceramic piezoelectric elements in stainless steel cases with a natural frequency of 54 kHz. The emitter is connected to the signal generator of the PUNDIT equipment. Both transducers are connected to a digital oscilloscope Agilent DPO3000.

The analysis is carried out on a trachite stone masonry with an inside cavity. Figure 3.22 shows the masonry structure sample, made by hand in the laboratory tests of the Department of Civil Engineering, Environmental and Architecture at the University of Cagliari.

The wall is 90 cm wide, 62 cm high, and 38 cm thick, and it is made of trachite blocks sized $20 \times 38 \times 12\ \text{cm}^3$ (see Figs. 3.23 and 3.24) and jointed with cement lime mortar. The block assigned to the central position of the wall was not settled,

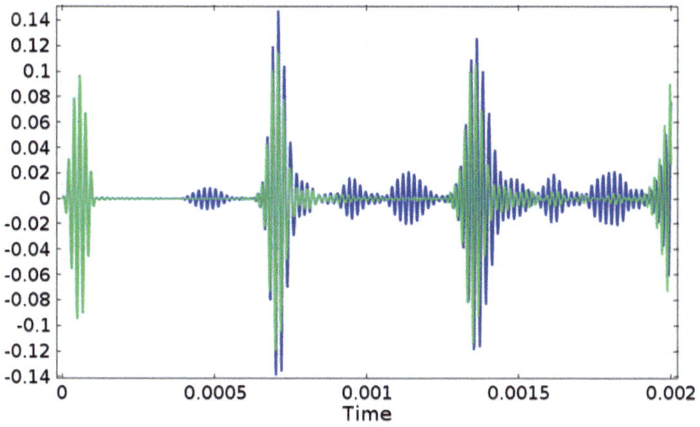

Fig. 3.21 Comparison between the two signals: in presence (*blue*) and in absence (*green*) of defect (input: six-cycletone burst signal)

Fig. 3.22 The sample wall in perspective view

thus realizing a macrocavity with the same size of the missing block (Fig. 3.24), and assumed as a known anomaly. Mortar joints have been assumed to be 1 cm thick, but since the wall is manually built, actual dimensions are not so precise.

The ultrasonic wave is transmitted through the structure and received by a second transducer positioned on its opposite side, according to the DTT. Changes in received signal provide indications of variations in material homogeneity.

3.3.1 *Time Domain Simulation*

In order to simulate the wave propagation through the structure, we develop a bidimensional model using the commercial code COMSOL Multiphysics 4.2 [10].

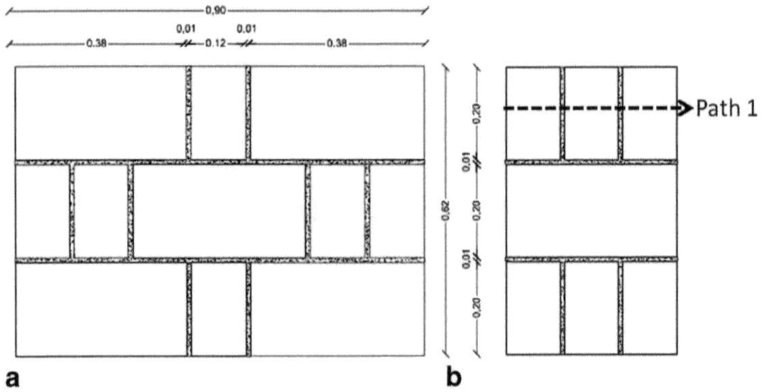

Fig. 3.23 The sample wall. **a** Front view. **b** Lateral view with the indication of a path of transmission

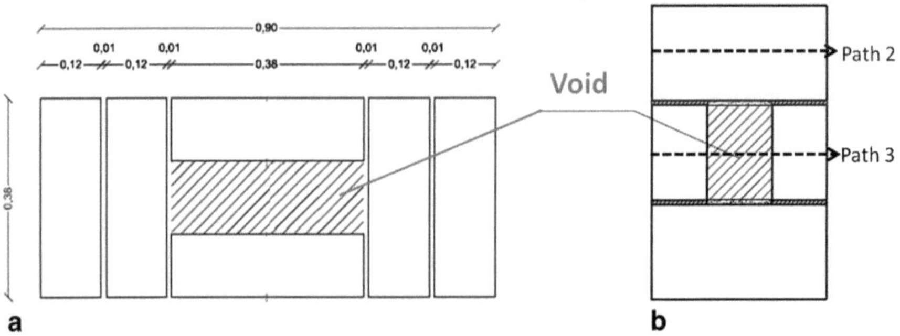

Fig. 3.24 Section that intercepts the cavity. **a** Horizontal plane. **b** Vertical plane with the indication of different paths of transmission

The numerical analyses are carried out simulating the use of piezoelectric transducers for emitting and receiving the signal [5]. In this case study, three emitters and three receivers have been arranged on opposite surfaces of the wall in order to intersect different materials (see paths in Figs. 3.23 and 3.24):

- Path 1: trachite–mortar–trachite–mortar–trachite
- Path 2: trachite
- Path 3: trachite–air–trachite

In order to calibrate the model, a reference bar of aluminum (supplied with PUNDIT equipment) with a known transmission time of 24 μs has been firstly simulated. To reduce the calculation time, an axisymmetric bidimensional model is chosen, exploiting the cylindrical geometry of the transducers, and with a certain approximation, the volume of material crossed by the ultrasonic wave.

The model of the aluminum bar, sized 153 mm × 50 mm with the transmitter and receiver transducers is shown in Fig. 3.25.

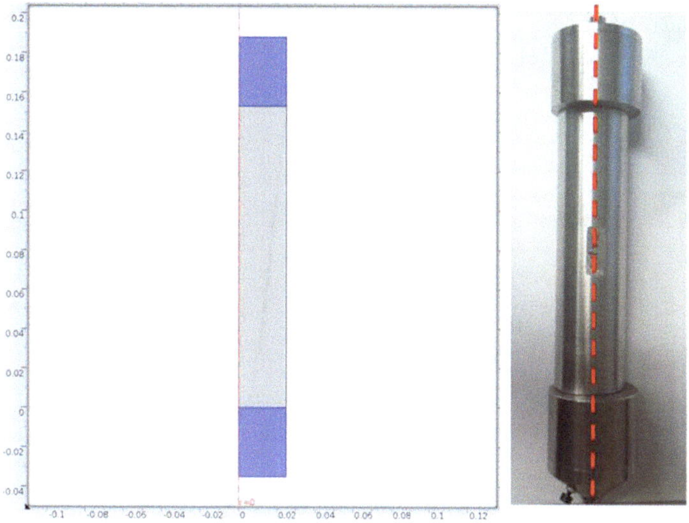

Fig. 3.25 Transducers and reference bar of aluminum: model and PUNDIT equipment

Table 3.2 Materials properties

Material	Sound velocity c_s (m/s)	Density ρ (kg/m^3)	Young modulus E_Y (Pa)	Poisson's ratio
Aluminum	6420	2700	70×10^9	0.33
Trachite	1968	1750	6.1×10^9	0.2
Mortar	1768	1800	5.5×10^9	0.1
Air (20°)	340	1.2	1.5×10^{-40}	0.1
PZT4	–	7500	7.8×10^{10}	0.31

A multiphysics acoustic piezoelectric interaction model is adopted (see Sect. 2.6) and a transient analysis is performed, where the unknowns of the problem are the acoustic pressure p, the displacement vector $\mathbf{u}$, and the voltage V.

The density and speed of sound are assumed to be constant because they vary with time on scales much larger than the characteristic acoustic wave period.

During the simulation the first piezoelectric element (the emitter) is electrically charged in order to transform the voltage V in a vibration which, when applied to the wall, propagates into the material according to the equation of motion of the structural mechanics.

Each material is characterized by its own density, Poisson ratio, Young's modulus, and sound velocity as shown in Table 3.2.

The excitation wave is an acceleration signal, obtained by the piezoelectric transducer, by applying the impulsive voltage signal shown in Fig. 3.26.

This electric pulse with a maximum amplitude of 500 V and duration of 9.3 μs, is applied to the upper part of the emitter transducer, while the bottom part of both the transducers is grounded.

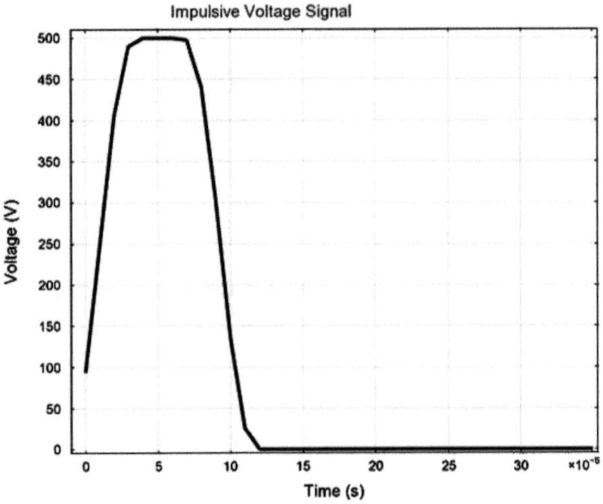

Fig. 3.26 Voltage signal for the excitation of the emitter transducer

For the piezoelectric material (PTZ4), the default global coordinate system is used in order to have a material oriented in the rz plane. The boundary conditions are "axial symmetry" on the z-axis and "sound hard boundary (wall)" on the outside, where the normal component of the pressure vanishes. At the interface between the emitter and the bar of aluminium, the boundary condition is set to "normal acceleration" (see Eq. (2.22)). The other interfaces are continuity boundaries, by default.

The domain is meshed using free triangular elements. Considering the pulse frequency of the transducers, that is $f_0 = 54$ kHz, and a wavelength

$$\lambda = \frac{c_{\text{aluminum}}}{f_0} = \frac{6420 \text{ m/s}}{54{,}000 \text{ Hz}} = 0.1189 \text{ m},$$

the maximum element size of the mesh is

$$l_e = \frac{\lambda}{10} = 0.01189 \text{ m}.$$

In Fig. 3.27 the obtained meshed model is shown.

The model is solved in the range 0–0.1 ms. Using the condition (2.24) and considering a CFL number of 0.05, the time step of the simulation is set equal to

$$\Delta t = \frac{\text{CFL} \cdot l_e}{c} = \frac{0.05 \times 0.01189 \text{ m}}{6420 \text{ m/s}} = 9.26 \times 10^{-8} \text{ s}.$$

In Fig. 3.28 the signal received by the second transducer is shown.

As can be noticed, the travel time is equal to recorded transmission time, i.e, 24 μs.

The simulated pressure fields in the aluminum bar for two different time instants are shown in Fig. 3.29.

Afterward the input voltage signal is applied on three points of the trachite stone wall. In Fig. 3.30 the simulated and meshed models for the three different paths are reported.

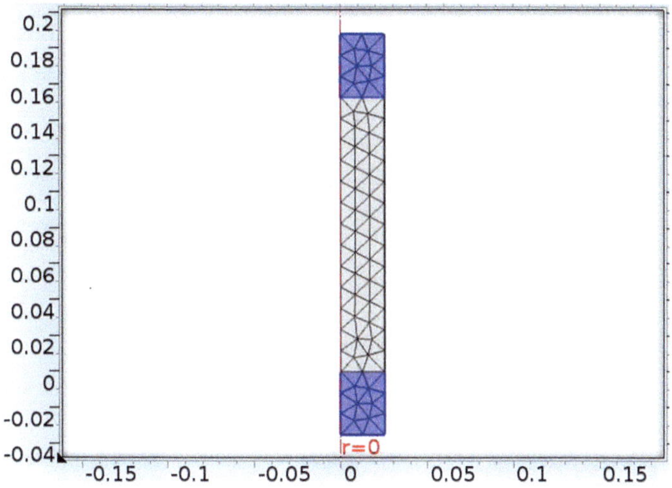

Fig. 3.27 Meshed model

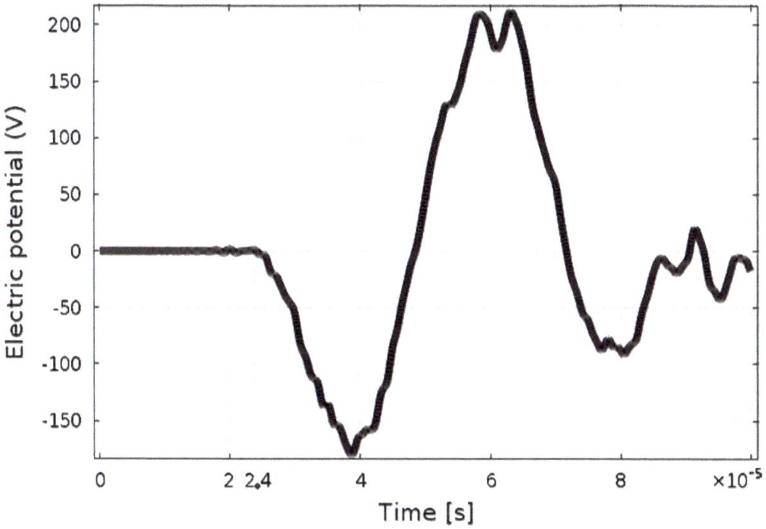

Fig. 3.28 Simulated received signal

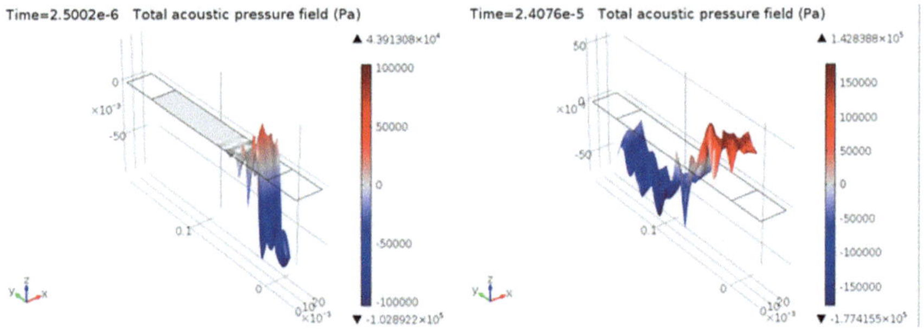

Fig. 3.29 Pressure field at time 2.5μs (*left*) and 24 μs (*right*)

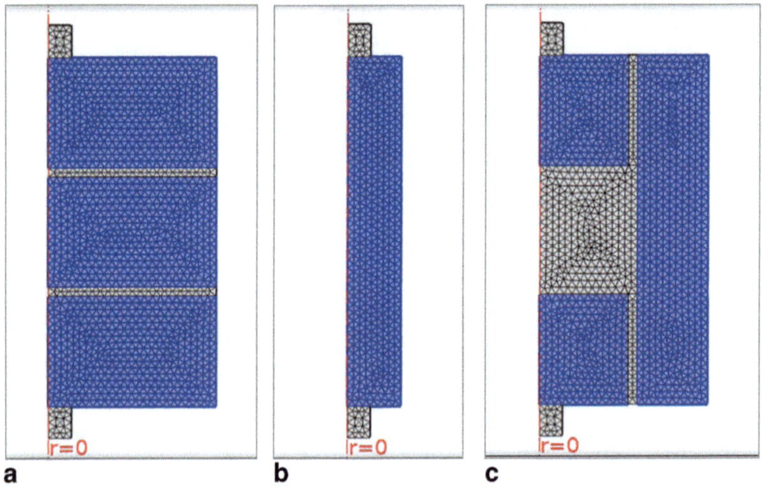

Fig. 3.30 Meshed models for path 1 (**a**), path 2 (**b**), and path 3 (**c**)

The simulated received signals and the pressure field corresponding to three different paths in the wall are reported in Figs. 3.31, 3.32, and 3.33. As can be noted, the waveform corresponding to path 3 is delayed due to the presence of the cavity (Fig. 3.33).

Using the experimental equipment, three delayed signals are received on the opposite surface of the wall. The experimental output signals are all attenuated, but for each of the three paths we detect a different harmonic content and a different attenuation.

Unfortunately, the experimental waveform is influenced by the coupling between structure and transducer. Indeed, the transducer is pressed by the operator directly against the wall, and coupling is achieved by the presence of a thin fluid layer inserted between them. Probably the rough surface of the wall does not allow good contact coupling.

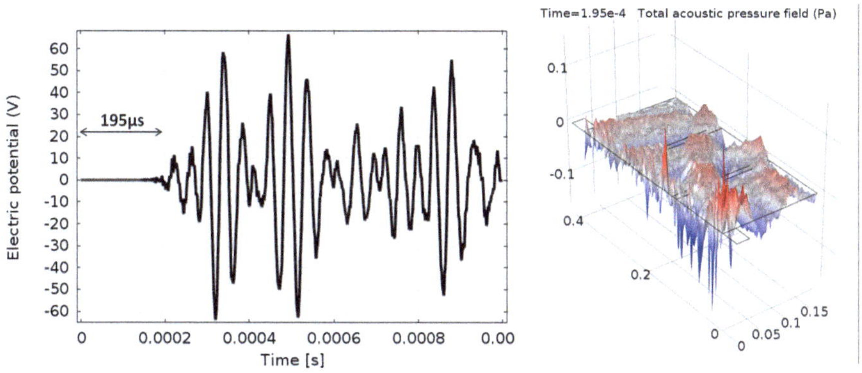

Fig. 3.31 Simulated signal detected (*left*) and pressure field at time 195 μs (*right*) for path 1

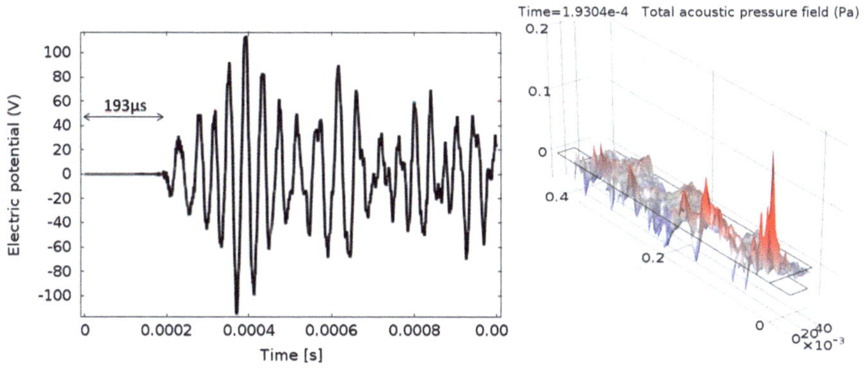

Fig. 3.32 Simulated signal detected (*left*) and pressure field at time 193 μs (*right*) for path 2

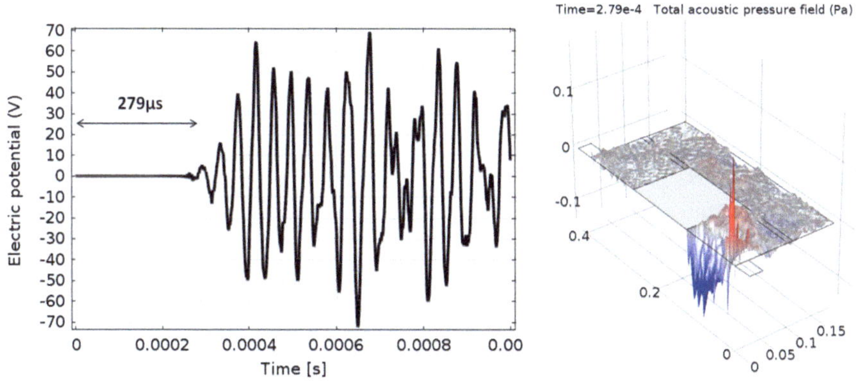

Fig. 3.33 Simulated signal detected (*left*) and pressure field at time 279 μs (*right*) for path 3

Table 3.3 Measured and simulated travel times

Path (μs)	Measured travel times	Simulated travel times (μs)
1	179	195
2	159	193
3	279	279

Thus, it is crucial to eliminate the variability associated with this contact coupling in order to assure the repeatability of the tests.

Nevertheless, there is a substantial agreement among the measured and simulated travel times. In Table 3.3 the travel times simulated and measured with the PUNDIT instrument are reported.

3.3.2 *Frequency Domain*

In order to reduce the processing time and computational errors, the frequency domain rather than the time domain can be used to solve the model [7].

Even in the frequency domain, different types of signal paths, corresponding to different materials region in the wall, have been simulated and analyzed (see Fig. 3.34):

- Trachite path (point 2)
- Trachite–mortar path (points 1 and 3)
- Trachite–air path (point 4)

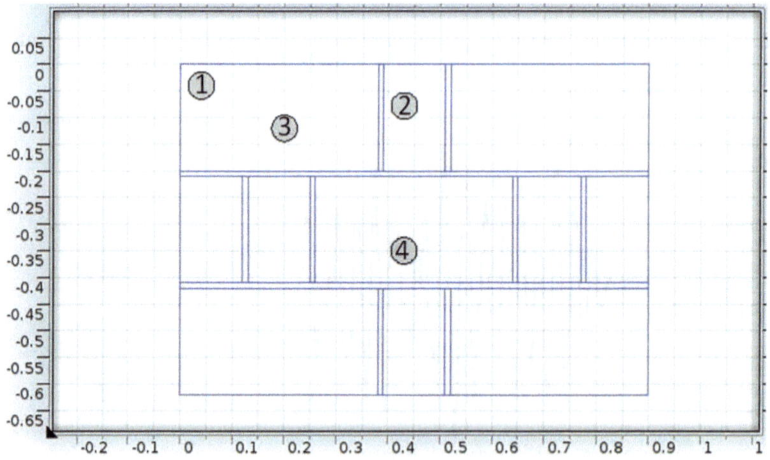

Fig. 3.34 Model of the wall and indication of the signal paths

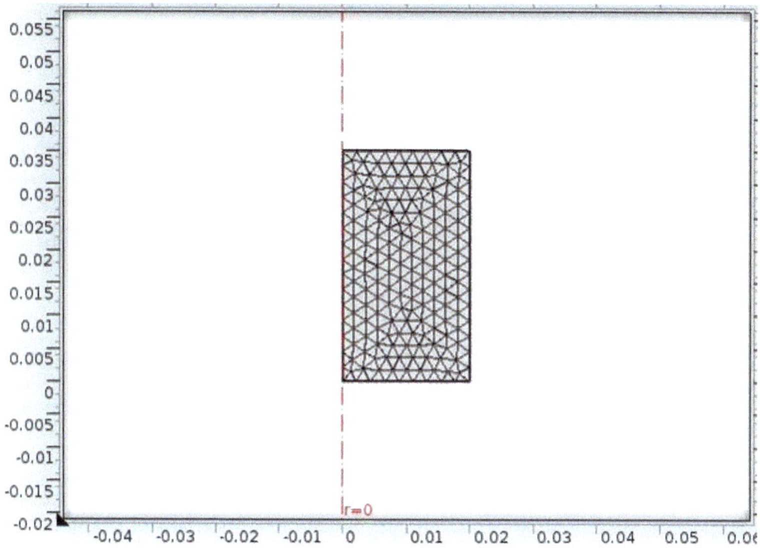

Fig. 3.35 Model and mesh of the transducer

To improve the quality of received signal, the chirp signal (see Sect. 4.5.1) is used as excitation wave. In this way a greater rate of energy through the sound waves is transferred.

The results here described have been obtained by a numerical analyses that use a Pressure Acoustics model and the Piezoelectric Devices Interface belonging to the COMSOL Acoustics Module [9]. The sound field is then described and solved by the pressure p, so we can simulate only one transducer (the emitter).

A two-dimensional axisymmetric model (Fig. 3.35) and a three-dimensional model (Fig. 3.36) are created for the emission transducer and for the wall, respectively. For the entire model a mesh dense enough to accurately describe the propagating perturbation has to be chosen (Figs. 3.35 and 3.36).

Two submodels are then coupled using the Extrusion Model Couplings option of the software.

Considering the minimum wavelength $\lambda_{\min}$ of frequency components of the transmitted signal, a length equal to $l_e = \lambda_{\min}/20$ for the elements of the mesh is assumed.

At the interface between the transducer and the wall, the boundary condition for the acoustics interface states that the pressure is equal to the normal acceleration of the solid domain (see Eq. 2.22). This drives the pressure in the wall domain.

The harmonic content of the signals takes place around the resonant frequency of the transducer (54 KHz) and differs for each path. In the frequency range related to the harmonic components of the received signal, the average gain of the emission transducer is:

$$\frac{\text{force}}{\text{voltage}} = \frac{2\ \text{N}}{10\ \text{V}} = 0.2\ [\text{N/V}].$$

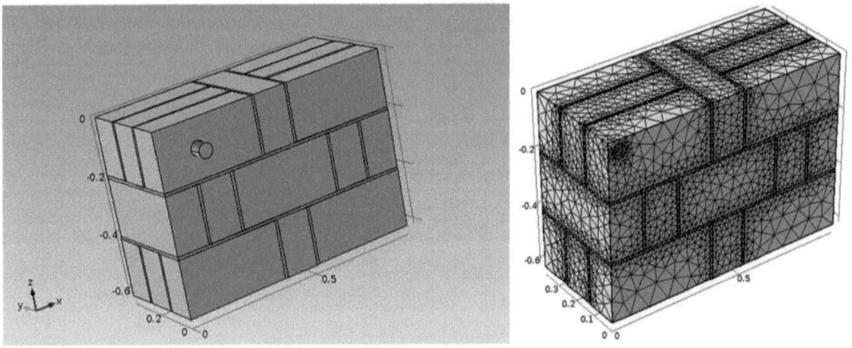

Fig. 3.36 Three-dimensional model and mesh of the wall

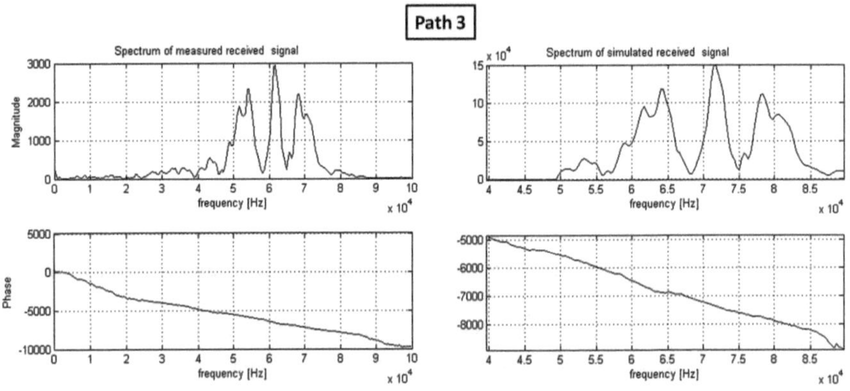

Fig. 3.37 Measured (*left*) and simulated (*right*) spectra for path 3

Figures 3.37 and 3.38 show the harmonic content of the signals received and simulated through the path trachite–mortar and trachite–air respectively. As can be noted, to different paths correspond different signals with unlike amplitudes and harmonic content. Simulated signals show a performance in good agreement with experimental signals.

Spectral analysis shows that the harmonic content is strongly linked to the type of location and the proximity of this to interfaces and edges. Spectral bandwidths are shifted to higher values.

Solved the acoustic problem in the frequency domain, the acoustic pressure $p(t)$ in the time domain is obtained considering the following function of the spectral components:

$$p(t) = \sum_{i=1}^{N} [p_{\text{amp}}(f_i) \cdot \cos(2\pi f t + p_{\text{phase}}(f_i))] \qquad f_i = [f_{\min} \div f_{\max}] \tag{3.2}$$

where N is the number of components of the spectra in the frequency range.

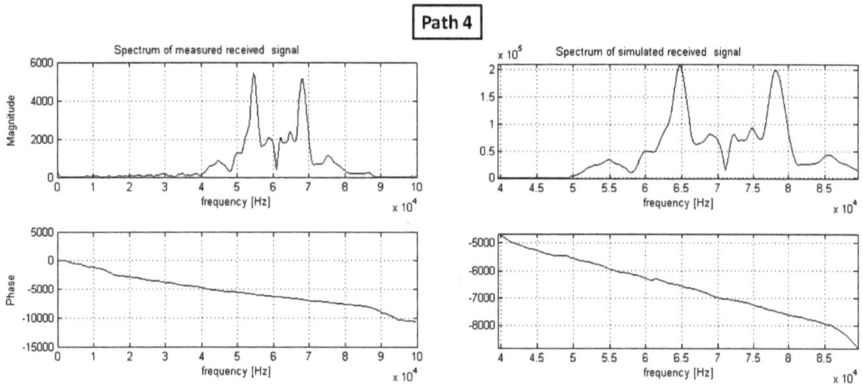

Fig. 3.38 Measured (*left*) and simulated (*right*) spectra for path 4

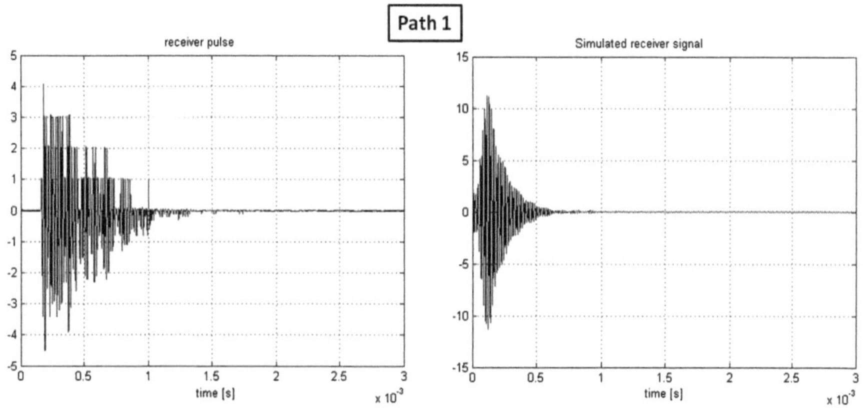

Fig. 3.39 Experimental (*left*) and simulated (*right*) signal crossing path 1 (time domain)

The measured and simulated signals in the time domain for the four paths are show in Figs. 3.39, 3.40, 3.41, and 3.42.

Amplitudes in the time domain are scaled to the average gain of the emission transducer.

Simulated signals have duration of about 1.5 ms. The transition time is very small even when the signal is applied to air regions, in which the signal covers a longer path along the boundary of this region. Signal amplitudes are different depending on the paths. The maximum amplitude occurs in the trachite path (path 2) while the lower amplitude is in correspondence of the trachite–air path (path 4). The amplitude decreases in the trachite–mortar path where there are two interfaces involving reflections and attenuation (path 3). Finally, for paths crossing the same region (paths 1 and 3) there is a greater reduction in amplitude in correspondence of the points nearest to the interfaces and to the edges, indeed they are more subject to the reflections of the transverse or shear waves and in general to all edge effects.

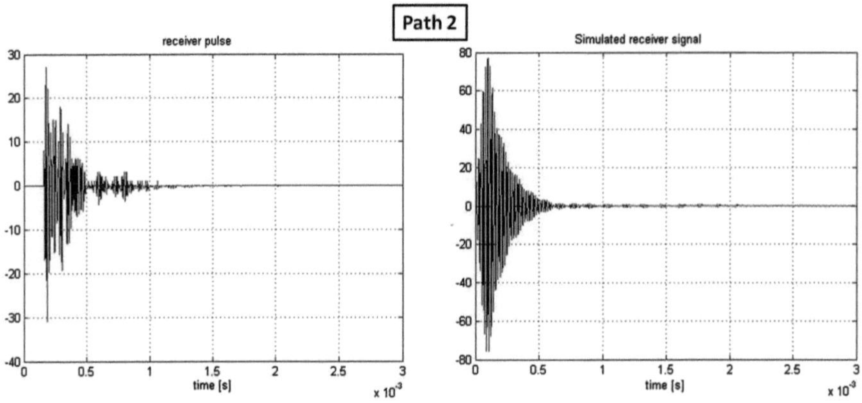

Fig. 3.40 Experimental (*left*) and simulated (*right*) signal crossing path 2 (time domain)

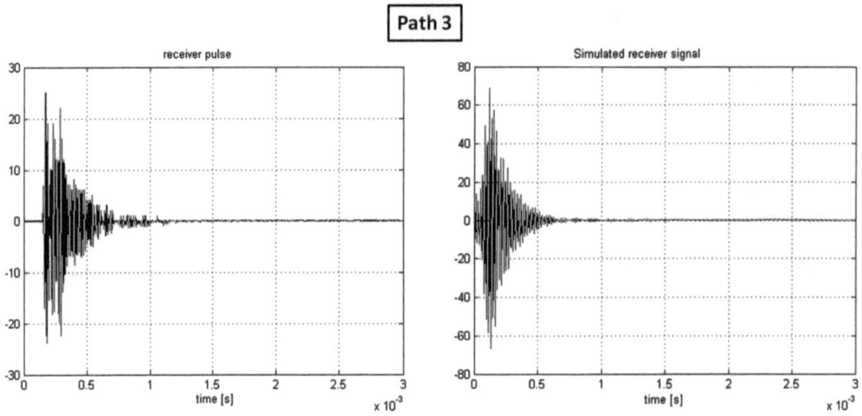

Fig. 3.41 Experimental (*left*) and simulated (*right*) signal crossing path 3 (time domain)

Similarly to what happens for the measured signals, the simulated signals exhibit different behavior for the different types of paths both in frequency and time domain. The simulated signals can be then used to obtain, through postprocessing analysis, maps for detecting the presence of macro-defects. It is possible to calibrate the model by comparing a significant number of simulated signals to the correspondent signals obtained from the experimental sessions. Finally, results show that the implemented model is suitable to effectively simulate the ultrasonic signals transmission to the stone wall.

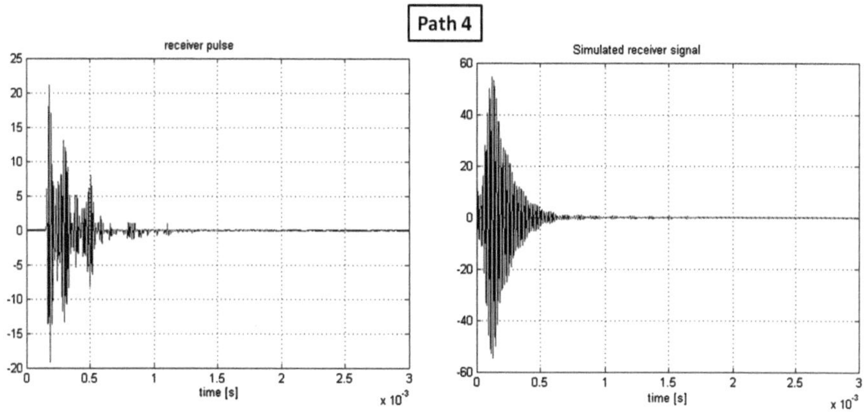

Fig. 3.42 Experimental (*left*) and simulated (*right*) signal crossing path 4 (time domain)

3.4 Example 3: Detecting a Void in a Concrete Pillar

Let us now consider the problem of checking for the presence of internal voids in a pillar. A number of measurements will be performed changing any time the position of both transmitter and receiving transducers, and the diagnosis will be done by combining all such acquisitions. In this way a tomographic problem has to be solved.

The first problem to solve is how to distribute the measure points on the surface of the pillar [8]. Preliminarily to assess how the size of the measurement grid may affect the reconstruction of the shape and the evaluation of the size of the defect, a simple model of concrete pillar with an internal cavity can be simulated with a commercial FEM software. COMSOL Multiphysics [9] has been used for modeling the trial.

Let us now describe a numerical application of the tomography to a concrete pillar. The pillar of this example is 40 cm wide, 162 cm high, and 40 cm thick. Internally a macro-cavity sized $10 \times 10 \times 14\ \mathrm{cm}^3$ is realized and assumed as a known anomaly. In the real case such cavity can be realized by using an empty box of polystyrene. The tomography to an horizontal plane section crossing the wall in order to intercept the void is applied. The investigated section is sized $40 \times 40\ \mathrm{cm}^2$.

We will consider two submodels: one for the emission probe (two-dimensional axisymmetric) and another for the wall (two-dimensional). The two models have to be coupled to perform the acoustic propagation in the wall (modeled by the acoustic module) of the signal generated by the probe (modeled by the piezoelectric device module).

The two-dimensional model and the mesh of the tested section are represented in Fig. 3.43.

The materials used in the model are listed in Table 3.4.

In order to solve the tomographic problem, firstly we have to divide the section into a grid of $N = P \times Q$ cells (pixels). The objective is to obtain the mean velocity

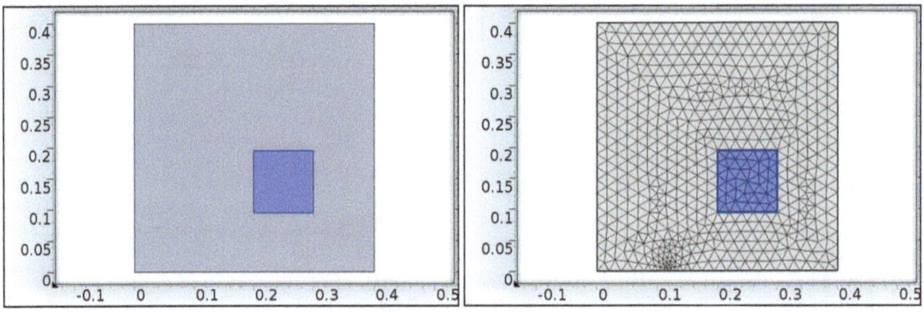

Fig. 3.43 Model and mesh of the cross section of the pillar

Table 3.4 Materials properties

Material	Sound velocity c_s (m/s)	Density ρ (kg/m^3)
Concrete	3600	2300
Air (20°)	340	1.2
PZT4	–	7500

v of each pixel starting from the knowledge of M flight times measured along a series of paths joining couples of transducers located on opposite or adjacent sides of the section. Then the tomographic problem consists of solving a system of M linear equations (equal to the number of measured flight time) in N unknowns (number of cells of the section) [4, 6]. Under the assumption that the transducers are located at the center of the cells, there is a maximum number of ray paths that can be traced, between each couple of transducers, that is, $M_{\text{max}} = P^2 + Q^2 + 4PQ$.

In this example, two different grids have been considered: the first one consists of 25 cells measuring 8 cm × 8 cm, whereas the second one is composed of 64 cells measuring 5 cm × 5 cm. M measurements can be performed with $25 \leq M \leq 150$ in the first case and $64 \leq M \leq 384$ in the second one. In both cases a number of paths equal to 76 is considered. Emitters and receivers have been alternatively positioned along the side of the section. Each measure is run activating the receivers positioned in the sides of the wall parallel and adjacent to the side where the emitter is positioned. Figure 3.44 shows the position of emitters and transmitters for the two cases analyzed.

Considering an excitation frequency equal to 54 kHz, which corresponds to the characteristic frequency (resonance frequency) of the emitting transducer, the input is obtained by the piezoelectric transducer (made of the crystal PZT4), by applying as voltage signal a six-cycletone burst enclosed in a Hanning window (Fig. 3.45) described by the Eq. (3.1).

This electric pulse is applied to the upper part of the emitter transducer, while the bottom part of both the transducers is grounded.

Changes in the received signal provide indications of variations in material homogeneity. In fact, as shown in Fig. 3.46, if we consider two paths of the same length

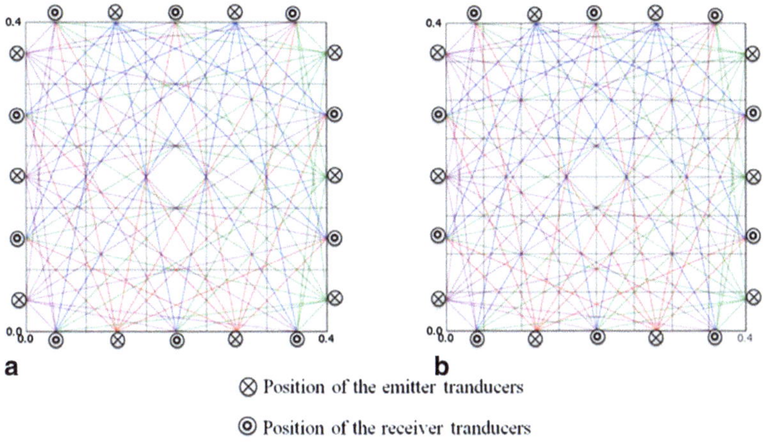

Fig. 3.44 Section crossed by 76 paths. **a** 25 cells. **b** 64 cells

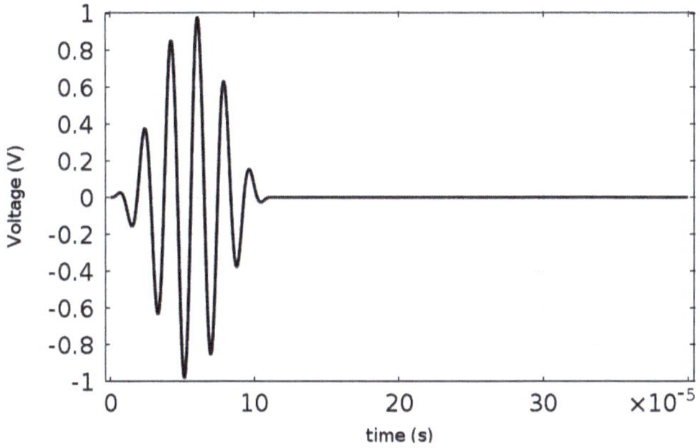

Fig. 3.45 Voltage signal applied by the emitter transducer

that cross the concrete without defect in the first case and with defect in the second one, the received signal is attenuated and delayed if the air cavity is intersected.

All the travel times t, which give us the constant terms of the equations system, are deduced by the FEM simulations, by applying a suitable detection threshold to the signal observed in the receiving point.

The obtained maps of velocities are shown in Fig. 3.47. The maps are represented by a 256-level grayscale diagram, where the highest level (white) corresponds to the maximum velocity (concrete), and the lowest level (black) corresponds to the minimum velocity (air). The maps are represented by using the Kriging method [20]. As can be noted, the internal cavity is clearly identified in both cases, but using thicker grids, it is possible to define the shape of the defect with smaller dimensions, thus achieving a sharper resolution.

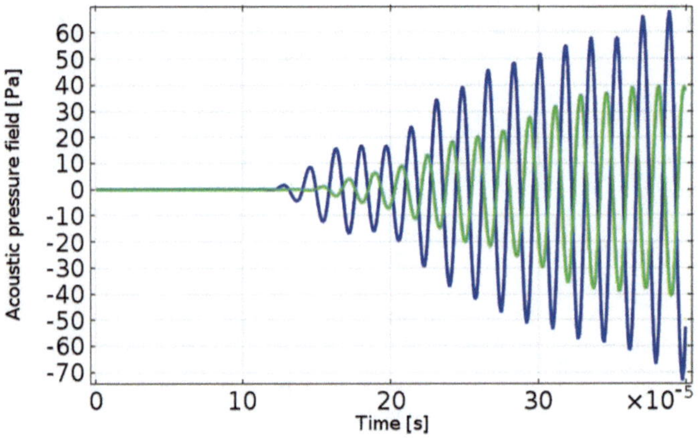

Fig. 3.46 Received pressure wave. *Blue*: path through only concrete; *Green*: path that intersects the cavity

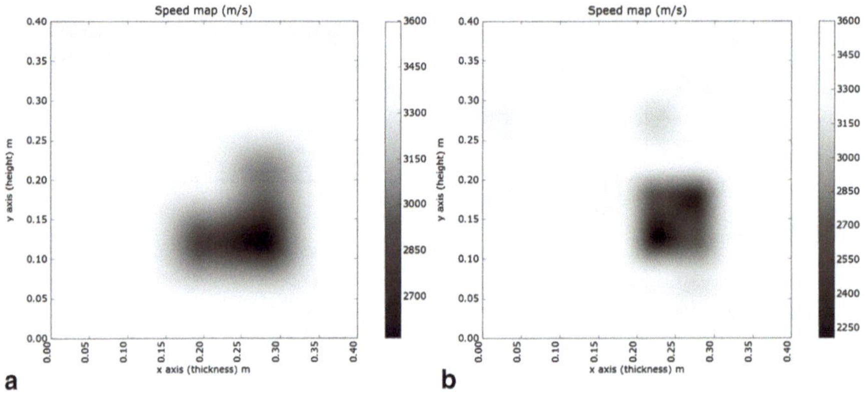

Fig. 3.47 Map of the velocities in the cross section considering 76 paths: **a** 25 cells, **b** 64 cells

3.5 Example 4: Detecting a Hole in a Metallic Specimen

The ultrasonic inspection technique is one of the most common NDT techniques employed in the steel industry due to the favorable propagation conditions that such material guarantees. In particular, ultrasonic NDT is extensively exploited in verifying the quality of forged steel to be used in the energy industry. The pulse-echo method is the method most widely used in industrial and laboratory applications.

The ultrasonic principle is based on the fact that solid materials are good conductors of sound waves. The method of ultrasonic waves uses frequency above 20 kHz. However, the NDT in industrial engineering field prefers waves of variable frequency between 1 and 20 MHz while uses lower frequency waves in civil engineering or in the restoration of monuments. With the frequencies above 1 MHz the resulting wavelength is in millimeter.

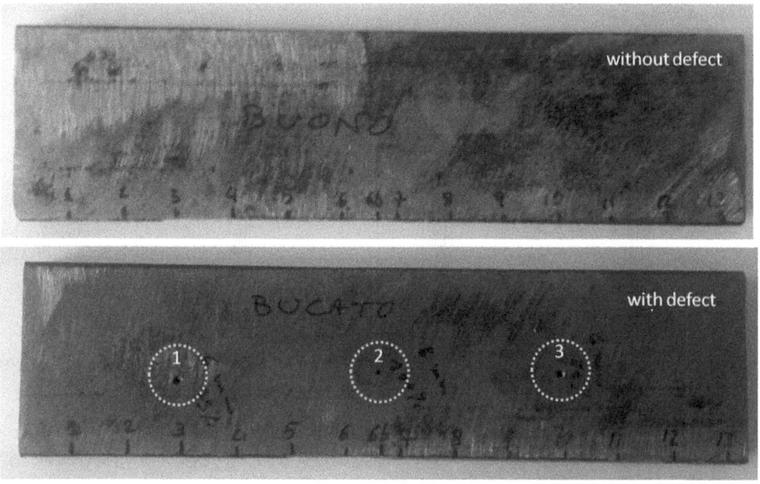

Fig. 3.48 Image of the specimens without and with holes

In this last example the ultrasonic pulse-echo method is applied to the problem of inspecting a metallic specimen in order to verify the presence of internal irregularities or flaws. Two specimens of steel of dimensions $34 \times 135 \times 8\ \text{mm}^3$ with and without defects (Fig. 3.48) have been used. Similar specimens are used in the industry for calibration of the instruments. The defects have been realized in the specimen using a precision drill. In this way three holes of 1 mm diameter have been created in different positions and depth:

- Point 1: $x = 30$ mm, $y = 13$ mm, depth 6 mm
- Point 2: $x = 65$ mm, $y = 15$ mm, depth 7.8 mm
- Point 3: $x = 100$ mm, $y = 15$ mm, depth 4 mm

To characterize the material and evaluate the attenuation and the speed of transmission of signals, we make preliminary experimental tests, using the standard pulse-echo technique (the transducer is emitter and receiver at the same time). The experimental setup is composed by a transducer "Olympus V203-RM" of 10 MHz and a signal generator "500PR Pulser-Receiver." By means of the front panel controls we set pulse height, waveform damping, receiver gain, and pulse repetition rate, as well as operating mode (pulse-echo or DTT).

By examining the applied signal and the relative echo (Fig. 3.49) of the specimen free of defects the speed of sound is determined:

$$c_{\text{specimen}} = \frac{0.008\ \text{m}}{0.0015\ \text{ms}} = 5333\ \text{m/s}\,.$$

As we have mentioned earlier, acoustic waves propagate in different manners through solid materials and cavities or defects, thus enabling fault detection. The small dimensions of the thickness of the specimen and of the defects, makes

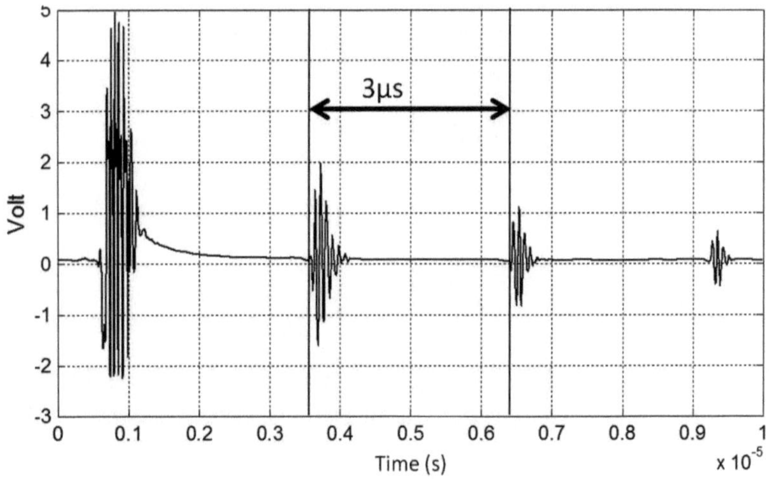

Fig. 3.49 Measured echo signals

the measures troublesome. In this case study we use a high frequency signal ($f = 10$ MHz). If the size d of the defect is much greater of the wavelength λ, the reflection occurs according to the rules of geometrical acoustics. In this case study $d = 1$ mm and $\lambda = 0.5$ mm.

Furthermore the ultrasonic beams diverges with consequent energy dispersion. Due to media dissipative effect, elastic waves are strongly attenuated, thus the emission cone of the signal source has to be as less divergent as possible. The beam spread is a function of wavelength. The diameter, the frequency of the transducer, and the angle of the cone of emission have been chosen in order to reduce the dispersion of energy. In this case study the angle α of the cone of emission is:

$$\sin \alpha = \frac{1.2\lambda}{D} = 0.1 \rightarrow \alpha \approx 6^\circ \tag{3.3}$$

Using the COMSOL multiphysics software, the simulations have been performed in the time domain. Two submodels have been considered: one for the emission transducer (two-dimensional axisymmetric) and another for the metallic specimen (two-dimensional). The two models have then been coupled to perform the acoustic propagation in the specimen (modeled by the acoustic module) of the signal generated by the transducer (modeled by the piezoelectric device module). The diameter of the transducer is assumed equal to 6 mm.

Four possible paths are considered: one without defects and three paths that involve the holes. As the thickness of the specimen is very little (8 mm) and the frequency of the analysis is 10 MHz, the mesh must be very fine in order to resolve both the geometric features and the wavelength λ. The "user-controlled mesh" option has been set and the maximal mesh size l_e is set equal to

$$\frac{\lambda}{N} = \frac{c_{\text{iron}}/f}{N} = \frac{5333\ \text{m/s}^2/10\ \text{MHz}}{4} = 0.13\ \text{mm}.$$

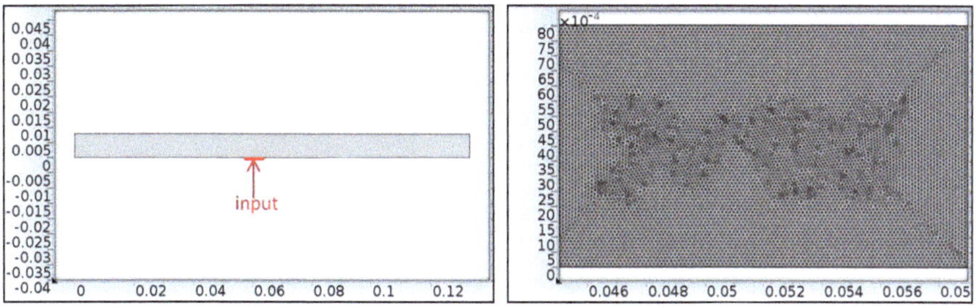

Fig. 3.50 Model and mesh of the metallic specimen without defect

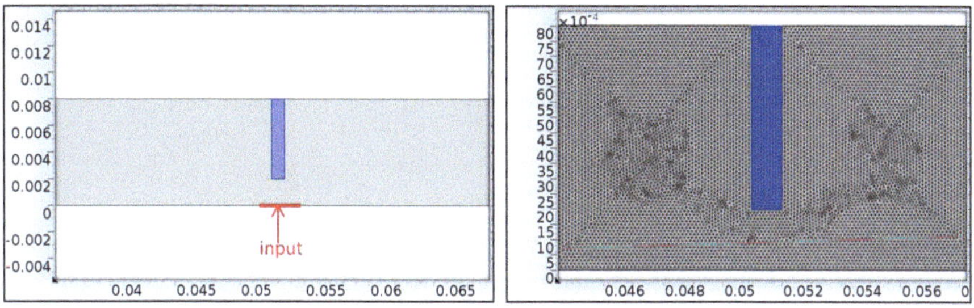

Fig. 3.51 Model and mesh of the metallic specimen with defect of 6 mm

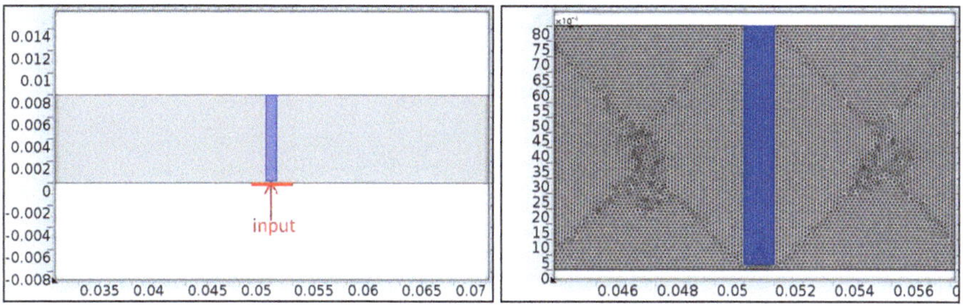

Fig. 3.52 Model and mesh of the metallic specimen with defect of 7.8 mm

Figures 3.50, 3.51, 3.52, and 3.53 show the geometric model and a zoom of the mesh of the four models where the arrow indicates the position where the input signal is applied.

The materials used in the model are listed in Table 3.5.

A sinusoidal signal multiplied by the Hanning function (see Eq. 3.1) is applied as an excitation signal by means of the transducer. The used frequency is 10 MHz and the obtained input is shown in the Fig. 3.54.

The simulations of the four configurations are performed in the time domain with a duration time equal to 10 μs and a time step of 0.01 μs. The obtained pressure trends along the iron specimen at time 1.5 μs are shown in Fig. 3.55. As can be noted

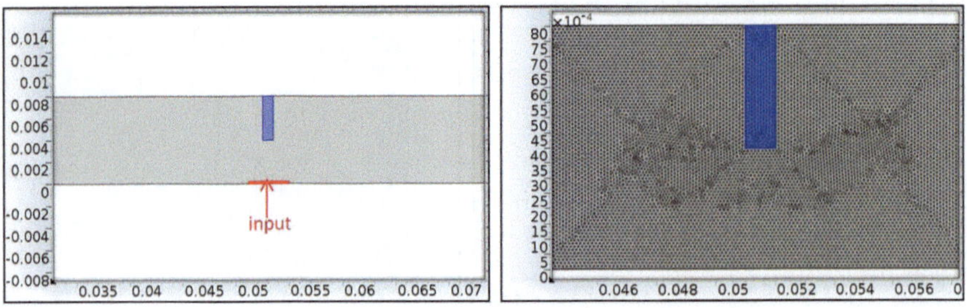

Fig. 3.53 Model and mesh of the metallic specimen with defect of 4 mm

Table 3.5 Materials properties

Material	Sound velocity c_s (m/s)	Density ρ (kg/m^3)
Iron	5333	7870
Air (20°)	340	1.2
PZT5	–	7500

several reflections of the transmitted wave at the interface between air and iron and the pressure wave are caused by the presence of holes.

These reflections are detected by the transducer and in the measured signal there are echoes in different positions allowing to identify the presence and the position of the defects. Figure 3.56 shows the received pressure signal in the pulse-echo mode relative to the path without defect. As can be noted the first echo is detected after 3 μs as obtained in experimental test. Due to the presence of defects the arrival of the first echo is anticipated (Figs. 3.57 and 3.58).

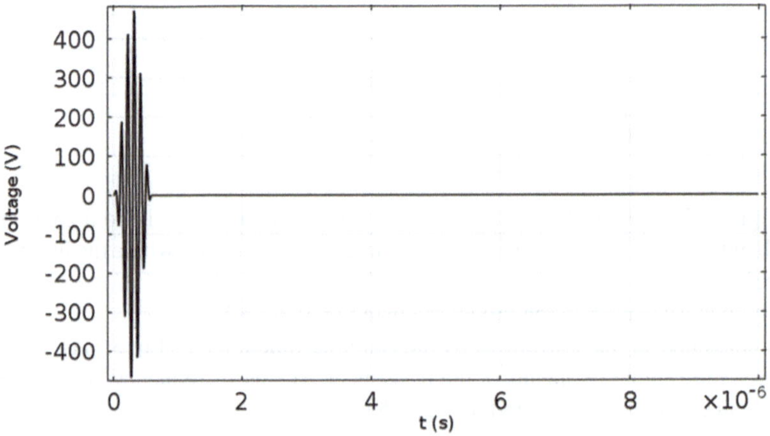

Fig. 3.54 Input signal obtained with a frequency equal to 10 MHz

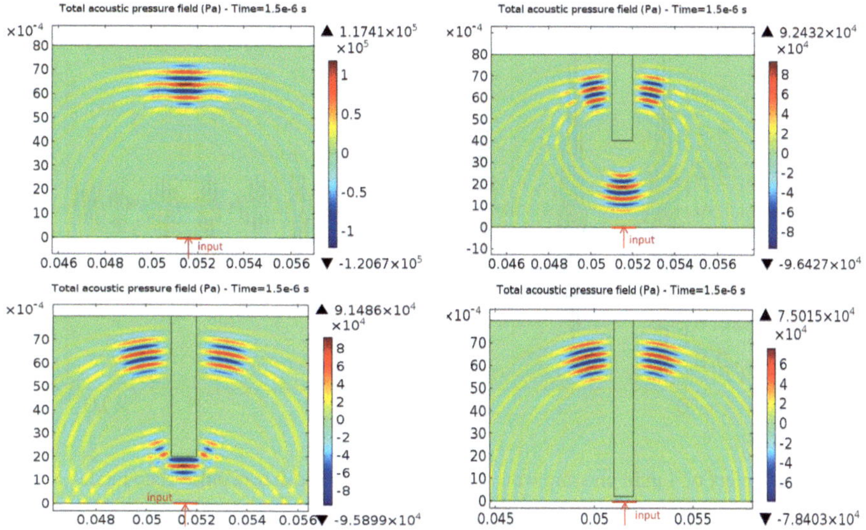

Fig. 3.55 Acoustic pressure at time 1.5 μs through the iron specimen without and with defects

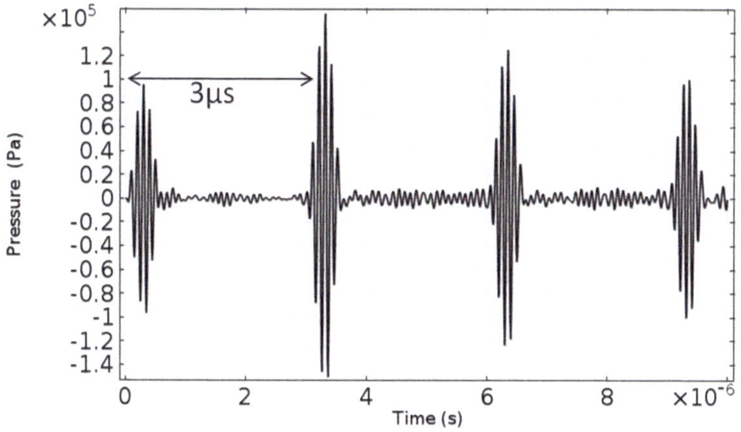

Fig. 3.56 Iron specimen without defect: pressure wave detected by the transducer in pulse-echo mode

When the discontinuity (interface between air and iron) is close to the point of application, echoes appear before the one reflected by the side opposite to the transducer and the amplitude of subsequent echoes are greatly reduced due to the presence of air (Fig. 3.59).

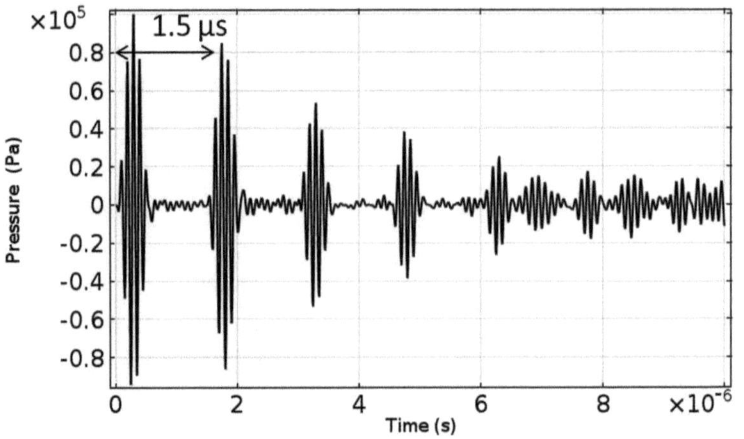

Fig. 3.57 Defect of 4 mm: pressure wave detected by the transducer in pulse-echo mode

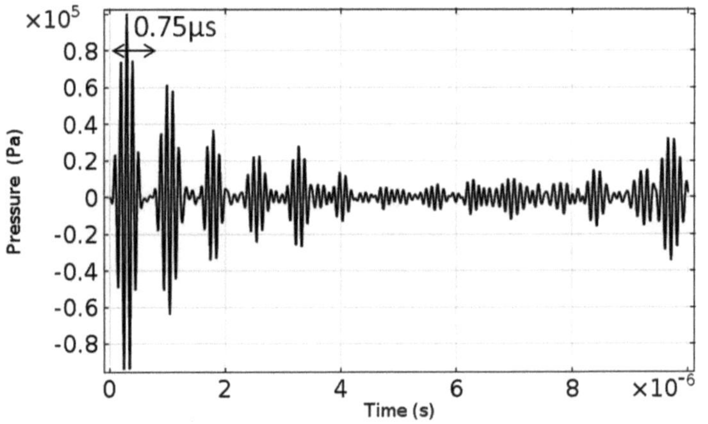

Fig. 3.58 Defect of 6 mm: pressure wave detected by the transducer in pulse-echo mode

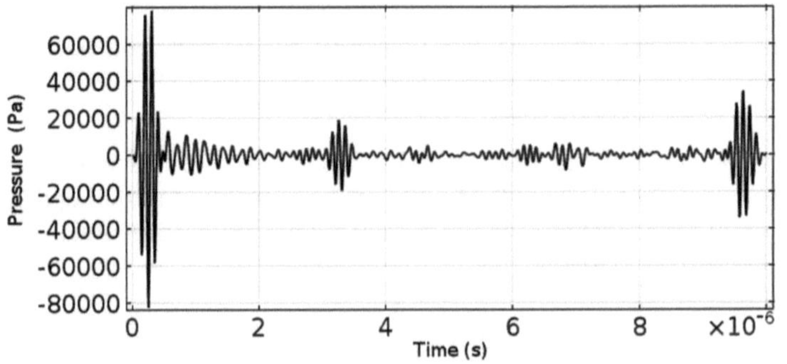

Fig. 3.59 Defect of 7.8 mm: pressure wave detected by the transducer in pulse-echo mode

3.6 Conclusion

The ultrasound-based techniques have acquitted themselves as one of the most popular way for internal inspection of materials, due to the high sensitivity of the waves to the presence of anomalies in the propagation mean. On the other hand, the association between a measured signal and the corresponding defects is in general a difficult task, and it cannot be done analytically. For this reason, before applying such techniques it is necessary to perform a series of trials in order to characterize the behavior of the system in several possible scenarios. To this end, a numerical approach is mandatory, because a physical realization of the cases needed to characterize the behavior of the system would be impractical. In this chapter four different problems are shown where FEM analysis is used to characterize the system under study in order to properly fit the diagnostic set up. The four examples are respectively, the localization of a jelly bean in the water, the detection of a void firstly in a stone wall and then in a concrete pillar, and finally the detection of a hole in a metallic specimen. In each of these cases, FEM analysis is used to study the behavior of ultrasonic waves in absence of the defect to detect and in presence of defects of different size or position. On the basis of the modifications caused by the defects on the acquired signals, the diagnostic system can be designed by selecting the most significant features. In this perspective, the FEM analysis allows one to optimize the parameters of the diagnostic system, such as the properties of the stimulus signal and spatial resolution of the test points. Finally, the same analysis allows one to assess the potentiality of the diagnostic system in terms of the detectability of defects.

References

1. Berger H, Mordfin L (1992) Nondestructive testing standards—present and future, vol 1151. ASTM, Philadelphia
2. Boogaard J, Van Dijk G (2012) Non-destructive testing. Elsevier, Burlington
3. Busse G, Van Hemelrijck D, Solodov I, Anastasopoulos A (2008) Emerging technologies in NDT. Balkema—Proceedings and monographs in engineering, water and earth sciences. Taylor & Francis, New York
4. Camplani M, Cannas B, Carcangiu S, Concu G, Fanni A, Montisci A, Mulas M (2008) Acoustic tomography for non destructive testing of stone masonry. In: Gervasi O et al (eds) Computational science and its applications-ICCSA 2008 (Lecture notes in computer science), vol 5073. Springer, Berlin, pp 596–605
5. Cannas B, Carcangiu S, Concu G, Fanni A, Usai M (2011) Numerical simulations of ultrasonic non destructive techniques of masonry buildings. Proceedings of COMSOL conference 2011, Stuttgart
6. Cannas B, Carcangiu S, Fanni A, Forcinetti R, Montisci A (2012) Hops: a new tomographic reconstruction algorithm for non destructive acoustic testing of concrete structures. In: Chang S, Bahar S, Zhao J (eds) Advances in civil engineering and building materials, vol 831. Taylor & Francis, New York, pp 259–263
7. Carcangiu S, Concu G, Usai M (2012) Calibration of ultrasonic testing for faults detection in stone masonry. Proceedings of COMSOL conference 2012, Milan

8. Carcangiu S, Montisci A, Raugi M, Tucci M (2013) Optimization of the set of path-rays in linear tomography. 18th international conference on Digital Signal Processing (DSP), 2013, pp 1–6
9. COMSOL (2011) Acoustics Module User's Guide for COMSOL 4.2. COMSOL
10. COMSOL (2011) Multiphysics Reference Guide for COMSOL 4.2. COMSOL
11. Ensminger D, Bond LJ (2011) Ultrasonics: fundamentals, technologies, and applications (Dekker Mechanical Engineering), 3rd edn. Taylor & Francis, New York
12. Glozman T, Azhari H (2010) A method for characterization of tissue elastic properties combining ultrasonic computed tomography with elastography. J Ultrasound Med (Official Journal of the American Institute of Ultrasound in Medicine) 29(3):387–398
13. Hellier C (2001) Handbook of nondestructive evaluation. Mcgraw-hill, New York
14. McCann DM, Forde MC (2001) Review of NDT methods in the assessment of concrete and masonry structures. NDT E Int 34(2):71–84
15. Paipetis AS, Matikas TE, Aggelis DG, Van Hemelrijck D (2012) Emerging technologies in non-destructive testing V. Taylor & Francis, London
16. Pundit Lab (2013) Pundit Lab/Pundit Lab+ ultrasonic instrument—operating instructions. Proceq SA. http://www.abmbv.nl/files/proceq_pundit_lab_operating_instructions_en.pdf. Accessed June 2014
17. Thurston RN, Pierce AD, Papadakis EP (1998) Reference for modern instrumentation, techniques, and technology: ultrasonic instruments and devices I, vol 23. Academic, New York
18. UNI ISO (2004) Natural stone test methods—determination of sound speed propagation. Published standard EN 14579:2004, UNI ISO
19. Vallen H (2002) AE testing fundamentals, equipment, applications. J Nondestruct Test 7(9): 1–30
20. Watson D (1992) Contouring: a guide to the analysis and display of spatial data. Computer methods in the geosciences, Pergamon, Oxford

Part II
Improving the Ultrasonic Systems Performance: Signal Processing Tools

In this part, we discuss how signal processing techniques can improve the performance of an ultrasonic nondestructive test (NDT) system. The section consists of two chapters: Chap. 4 introduces the topic of signal processing in the context of a single-channel ultrasonic system; Chap. 5 extends the main aspects introduced in Chap. 4 to the multi input multi output (MIMO) systems context.

The starting point is the pulse-echo model described in the introductory chapter. We start clarifying what we mean by operating *at the signal processing* level and why we are mostly concerned with the excitation signals used in the ultrasound NDT and in the deconvolution process.

The definition of a quantitative evaluation of the US system performance calls for the identification of a mathematical model of the entire ultrasonic inspection process. In order to allow a manageable mathematical treatment, only the fundamental aspects are identified and retained in the definition of the model, and thus, the so-called *minimal channel model* is introduced: the pulse-echo system is revisited using the proposed *minimal channel model*.

In order to overcome the conflicting choice regarding the excitation energy and the time resolution, the *compression* approach is defined starting from an analysis of the limits of the pulse-echo model in terms of signal-to-noise ratio (SNR); the *compression* approach allows to escape the SNR/resolution trade-off. Pulse compression systems are introduced and motivated, and *chirp* excitations are used to exemplify this *compression* approach, although the possibility of using binary sequences as an alternative basis for pulse-compression is remarked.

In order to verify the improvements obtainable by means of signal processing techniques, and to compare the different processing procedures, mathematical tools are necessary to quantitatively evaluate the performance of the overall system we model. The importance of introducing application-related merit factors is discussed: merit factors based on a probabilistic system view are proposed.

Choosing the appropriate transducers is a fundamental aspect in ultrasonic inspection. The effect of transducers in the presence of the complex signals required by the pulse compression processinng techniques is discussed. The aspect of how the characteristics of the transducers adopted impact a system's performance is analyzed, together with the related aspect of how the signal can be designed in order to adapt to the features of a specific probe.

In Chap. 5, the discussion is opened to the MIMO case, to show how a proper choice of excitations can reduce the mutual interference by using a *code division* approach.

Chapter 4
Excitation and Deconvolution in Ultrasound Nondestructive Testing Systems

Salvatore Caporale, Sergio Callegari, David Arthur Hutchins, Stefano Laureti, Pietro Burrascano and Marco Ricci

Abstract The pulse-echo (PuE) model based on short pulses is introduced as a starting point. Its limits in terms of signal-to-noise ratio (SNR) are reviewed, together with the tradeoff between the excitation energy and the time resolution. It is shown how signal processing can improve the overall performance in terms of system sensitivity and resolution as long as the exciting waveform is chosen taking into account the system characteristics and is in agreement with the adopted processing techniques. The degrees of freedom on which one can operate are the time evolution law of the excitation signal and the processing technique adopted on the receiving end. In their exploitation, the designer should consider the physical characteristics of the hardware devices being employed in order to optimize the overall performance. The importance of introducing merit factors is discussed. By offering a quantitative measure of system performance, they are indispensable tools to drive formal optimization

S. Caporale (✉)
Advanced Research Center on Electronic Systems for Information and Communication Technologies "E. De Castro" (ARCES), University of Bologna, Bologna, Italy
e-mail: salvatore.caporale@unibo.it

S. Callegari
Department of Electrical, Electronic and Information Engineering "G. Marconi" (DEI) and Advanced Research Center on Electronic Systems for Information and Communication Technologies "E. De Castro" (ARCES), University of Bologna, Bologna, Italy
e-mail: sergio.callegari@unibo.it

D. A. Hutchins · S. Laureti
School of Engineering, University of Warwick, Coventry, UK
e-mail: d.a.hutchins@warwick.ac.uk

S. Laureti
Dipartimento di Ingegneria, Polo Scientifico Didattico di Terni, University of Perugia, Terni, Italy
e-mail: s.laureti@warwick.ac.uk

P. Burrascano · M. Ricci
Department of Engineering, Polo Scientifico Didattico di Terni, University of Perugia, Terni, Italy
e-mail: pietro.burrascano@unipg.it

M. Ricci
e-mail: marco.ricci@unipg.it

P. Burrascano et al. (eds.), *Ultrasonic Nondestructive Evaluation Systems*,
DOI 10.1007/978-3-319-10566-6_4

strategies and for comparing different approaches. Application-related merit factors are introduced for trying to overcome some current limitations and to prepare the ground for the extension of ultrasonic techniques to multiple input, multiple output (MIMO) systems, which are considered in the following chapter.

4.1 Introduction

This chapter deals with the problem of overcoming the inherent limits of the standard ultrasonic measurement schemes discussed in the introduction, namely, the pulse-echo (PuE) and through-transmission (TrT) approaches based on short pulses. The focus will be on aspects such as signal-to-noise ratio (SNR) and resolution, giving specific consideration to the tradeoffs and the conflicting choices that exist in view of their improvement. The discussion is opened by clarifying what is intended by operation *at the signal processing level* and why the excitation signals used in the ultrasonic nondestructive testing (NDT) and deconvolution process are fundamental for performance optimization.

Since any ultrasonic measurement can ultimately be considered as an estimation of the impulse response of a linear time invariant (LTI) system, a minimal channel model is preliminary defined to support the following discussion. Here, only the essential parts of the underlying phenomena are taken into account and all those aspects that may be deemed as secondary with respect to signal design issues (like, for instance, propagation details and scattering) are removed. In this way, a minimal model suitable for the analysis and the optimization of the overall system is attained, to be used to revise the classical PuE and TrT schemes.

The pulse-compression (PuC) approach is then introduced and motivated as a method to break the SNR/resolution tradeoff. Various PuC schemes are presented and their performances are described with the help of the channel model. Through the whole of the chapter, *chirp* excitations are used to exemplify the PuC approach and taken as a benchmark, remarking that they are a fundamental type of stimulation signal used in practical applications. The inherent pros of chirps are highlighted in an application-related perspective, focusing on their constant envelope and continuous-phase properties. The possibility of using binary sequences as an alternative basis for PuC schemes is then discussed, and different approaches for generating suitable sequences are considered. Advantages and disadvantages in the use of pseudo noise (PN) sequences with respect to chirps are intuitively presented.

The importance of introducing *application-related merit factors* is discussed. The essential difficulties in proposing merit factors are evidenced. Specifically, it is shown that: (i) it is quite hard to summarize any application-related behavior by a limited number of scalars; (ii) many different kinds of set-ups and applications exist; and (iii) existing standards are more related to how measurements are taken than to a system view. For instance, PuE and TrT procedures are usually designed for different tasks requiring the optimization of different parameters, such as sidelobe levels or range resolution [3, 43]. A review of the merit factors that are readily available and relatively well acceded is therefore provided. Then, novel merit factors based on a probabilistic system view are proposed, taking into account that the discussion focus is threefold, aiming at a good tradeoff between: (i) SNR, (ii) time-resolution, and (iii) sidelobes. The merit factor framework is based on the minimal system model which lets its formulation be relatively lightweight and suitable for future extensions to more complex setups (e.g., multiple input, multiple output (MIMO) systems as illustrated in Chap. 5). The reasons why the approach allows for a probabilistic interpretation related to the chances of missing the detection of some ultrasonic paths are presented.

After the introduction of the PuC framework and the analysis of its pros and cons with respect to the minimal channel model, the physical phenomena that were initially removed are be progressively re-added. The effect of probe features, summarized in their transfer function, is therefore considered together with other frequency-related effects, including those deriving from the sample under test (SUT) material. A discussion is proposed on how the signals can be designed in order to second the probe features. Two aspects are brought under consideration. First of all, pushing power in the frequency bands where the transducers are inefficient at translating it to and from the acoustic domain can result in SNR loss. Consequently, the convenience of having excitations that fit the probe's amplitude response shape is illustrated. Second, the interaction of the probe transfer function with the PuC method is described. These aspects are used to support the idea that signal processing methods and excitations should be adapted to the system features. Transducer bandwidth is identified as a fundamental aspect. The merit factors previously introduced are used to compare different PuC schemes while varying the bandwidth of the transducers in order to highlight the peculiar features of each code and to introduce the opportunity of developing spectrum-shaping-based codes.

4.2 A Minimal Channel Model

In order to better formalize the concepts introduced so far, to use them in analysis and design, a model of the *channel* is needed. The choice of this word is not coincidental and aims at underlining some similarities of the NDT setup to a communication system. In the latter, an unknown message is sent through a medium and the receiver needs to estimate it, possibly taking advantage of some knowledge of the medium properties. In ultrasound NDT, a specially crafted message is sent through a medium and the receiver needs to estimate the medium properties, exploiting the fact that it knows the message.

By channel we mean the whole *transmission path* between the transmitting transducer (TX) and receiving transducer (RX), where the latter can be physically distinct or not. Consequently, the channel model must take into account many physical effects, such as: the propagation of acoustic waves inside the SUT; the behavior of the interfaces between the SUT and the transducers; the wave reflections at the interfaces inside the SUT (possibly indicating defects) and at its boundaries; and more. All these aspects have already been dealt with in Part I of this book. The reader is invited to

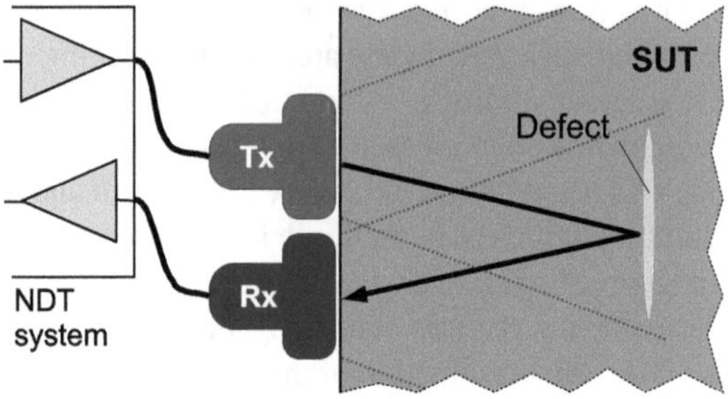

Fig. 4.1 Basic channel elements

refer to it and particularly to Chaps. 1 and 2 for a review and also to [43]. Here, a much more abstract approach is taken. The development of the channel model is, so to say, *subtractive*. Details are *removed*, until a *minimal* description is distilled, still capable of capturing the interesting effects. This approach is indispensable in order to obtain mathematical formulations manageable enough for design tasks.

Recall that the designer work is commonly aimed at the optimization of merit figures. In this task, knowledge of how the available parameters influence the merit factors is indispensable, which is why a clear system model is required. Furthermore, the more straightforward the ties and constraints implied by the model, the more likely the possibility to understand trends and adopt formal tools and well-codified methods. All this is highly desirable not just in view of design *quality*, but also of *reproducibility* and *assessability*. Consequently, the model construction is quite a delicate task, where simplification should be pursued as much as it can, but not so much to become misleading. In this book, the distillation of a minimal model is also important in view of the complications that will be layered on top of it in the next Chap. 5 where strategies enabling the simultaneous use of multiple TXs are introduced.

To start with, one can refer to the setup in Fig. 4.1. This is relatively similar to Figs. 1 and 2 in the introduction, but for the more explicit indication that a propagating medium exists.

Along the signal path, the following elements can be recognized:

1. a power amplifier inside the NDT control unit, feeding the TX
2. the TX itself
3. an (acoustic) interface between the TX and the SUT
4. a propagation path through the SUT
5. an interface inside the SUT where a reflection occurs
6. a second propagation path through the SUT
7. in interface between the SUT and the RX
8. a low noise amplifier (LNA) fed by the RX inside the NDT apparatus, possibly followed by an analog-to-digital converter (ADC) for the digital acquisition of the response.

Items 3–7 constitute the ultrasound path proper.

In order to build a *minimal* model, some preliminary assumptions must be made. An initial premise is that all the elements can be dealt with independently and that their models can be simply cascaded. A second premise is that all elements can be described in linear terms. This is a rather strong assumption, since many items are clearly nonlinear. For some of them, linear approximations can be used as long as the excitations are not too strong. For instance, the TX can be considered linear if its elastic parts are not stressed and there are no significant thermal effects associated to self-heating. For other elements, an additive error component can be substituted for the nonlinearity (e.g., quantization noise in the ADC). Then, as a third premise, the behavior can be considered time-invariant. A fourth assumption is that phenomena happening at a spatial scale smaller than the acoustic wavelength can either be ignored or summarized into a noise superposition. With this, things like *diffusion* and *scattering* due to the SUT grain structure can be reduced to more tractable terms. Also, propagation phenomena can be characterized just in terms of attenuation and delays. As a final assumption, all the dynamical effects that are nondominant can either be ignored or attributed to neighboring elements. Note that there is a pattern in this modeling strategy. Things are made to fit in simple forms. Whatever escapes these forms gets treated as noise.

Under these premises, the power amplifier feeding the TX can be considered ideal. The TX and RX reduce to a transfer function. What happens at the TX to SUT and SUT to RX interfaces can be modeled as a mere signal attenuation. The signal journey through the SUT becomes a combination of three effects: attenuation; delay proportional to the traveled distance r via the $\Delta t = r/c_{\text{medium}}$ relationship (where c_{medium} is a single propagation speed); and noise. In fact, the transducers typically compel the excitation waveform to fit in a bandwidth small enough to let all the frequency components travel with very similar speed and attenuation in the SUT. Note that some care is necessary with this latter hypothesis since some material–transducer combinations may make it weak. For instance, in some cases such as guided waves ultrasonic inspection, the dispersion of the group velocity is actually used to infer information about the SUT. Finally, the LNA becomes a gain factor and a noise insertion point, since its dynamic effects can be considered nondominant with respect to those in the transducers. Finally, the ADC can be modeled as a (quantization) noise insertion point.

In this discussion, the defect deserves some special treatment. An ideal defect, represented by a single, perfect, infinitely thin interface, would operate as a pure reflector. However, real world defects tend to be anything but ideal objects. They have non-null thickness, they refract in addition to reflecting, and, most important, they are often composed of a multitude of internal interfaces layered through their depth. Consequently, a defect often causes a plurality of reflections slightly displaced in space and time. Thus, while bouncing over a defect, a short ultrasound burst gets spread in time. This situation can be modeled by associating an impulse response $h_d(t)$ to the defect. With this, if $x_i(t)$ is an arbitrary waveform incident onto the defect, one can obtain the reflected waveform as $x_r(t) = \int_0^t x_i(\tau)h_d(t-\tau)\mathrm{d}\tau$. The issue with this approach is that $h_d(t)$ cannot be known in advance, since gathering information about the defects is the very purpose of the NDT technique itself. Thus,

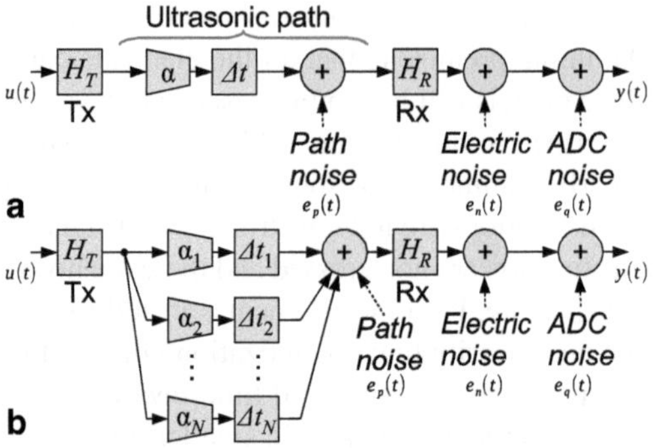

Fig. 4.2 Block diagram representing the channel model. **a** Model corresponding to a single reflection. **b** Model extended to multiple reflections

it is convenient to assume that defects are ideal, even if they are not, restricting the related set of parameters to a minimum. Consequently, defects are modeled merely through an attenuation and, again, some noise.

Cascading the submodels, one may immediately note that there are multiple points where attenuation occurs, as well as multiple points where noise gets superimposed. Due to linearity, all the attenuation points can collapse into a single one. Furthermore, the many noise sources can be collapsed with good approximation into three noise insertion points. One refers to the noise introduced along the ultrasound path and is mainly due to the SUT grain and other dispersion effects. One is related to the electrical or electro-acoustic parts of the systems and is mainly due to the RX and the LNA. The final one is the quantization noise. The cascaded model is shown in Fig. 4.2a where the three components are indicated as $e_p(t)$, $e_n(t)$ and $e_q(t)$, respectively.

The diagram helps introduce the following notation: $H_T(s)$ is the TX transfer function and $h_T(t)$ is the corresponding impulse response. $H_R(s)$ is the RX transfer function (and $h_R(t)$ the corresponding impulse response), α is the ultrasound path attenuation and Δt its delay, $u(t)$ is the excitation (here indicated in this way to avoid confusion with the Laplace variable s), and $y(t)$ the measured response. The ultrasonic path dynamical model can be summarized in a transfer function $H_u(s)$ and in the corresponding impulse response

$$h_u(t) = \alpha\delta(t - \Delta t) \tag{4.1}$$

where $\delta(\,\cdot\,)$ is the Dirac delta. Altogether one has

$$y(t) = [([u * h_T * h_u] + e_p) * h_R](t) + e_n(t) + n_q(t) \tag{4.2}$$

where the asterisk indicates convolution.

In the last expression, the noisy terms may deserve some further consideration. In fact, $e_p(t)$ and $e_n(t)$ have rather different natures, revealed by the fact that $e_p(t)$ scales in power with $u(t)$, while $e_n(t)$ typically does not. Actually $e_p(t)$ should be better considered as an *unwanted* signal instead of a proper noise and in general, without having knowledge of $h_d(t)$, it cannot be attenuated or removed by means of standard techniques such as *Wiener filter* and *matched filter* [48, 52]. An example of this phenomenon is encountered in diagnostic ultrasounds where $e_p(t)$ represents the *speckle* noise due to the coarse grain of the tissues that acts as a huge number of microdefects. Analogously, in the inspection of forgings, the steel grain causes *speckle* noise that increases as the average grain size increases [4, 11]. Without employing ad hoc filtering strategies, it is quite difficult to reduce these noisy components since they are generated by the same physical processes that produce defects signals, i.e., reflection and refraction [25, 38, 55]. On the other hand, $e_n(t)$ can typically be assumed to be an additive white Gaussian noise (AWGN), therefore independent from $u(t)$, and this makes it much easier to deal with. In practice, in many setups (and particularly in those where the acoustic paths have a large attenuation) $e_n(t)$ can dominate $e_p(t)$. For this reason, it is desirable to attribute all the noisy effects to $e_n(t)$ even if this may not be totally correct and the discussions reported in the following sections rely on this assumption. It is worth to mention that, in the most general case where $e_n(t)$ and $e_p(t)$ are taken into account simultaneously, one should first try to remove almost completely $e_n(t)$ and then apply more sophisticated denoising methods (wavelet filtering, independent component analysis, etc. [8, 27, 40]) to optimally separate the *desired* signal components from the *unwanted* one $e_p(t)$.

Finally, $e_q(t)$ is also worth mentioning. This is a wide band noise that takes a uniform value distribution rather than a Gaussian one. In many cases, the ADC can be good enough to make this term nondominant. However, there are systems where quantization noise is well visible. If this is the case, $e_q(t)$ can be treated similarly to $e_n(t)$.

Unfortunately, the model given so far is insufficient since there will certainly be multiple ultrasonic propagation paths between the TX and RX couple. In fact, inside the SUT there can be multiple defects, and each of them can operate as a reflector. Second, the very boundaries of the SUT are *interfaces* where a discontinuity in the *acoustic impedance* exists, so that reflections are introduced. Fortunately, the model allows a straightforward extension, as shown in Fig. 4.2b. This leads to:

$$y(t) = \left[h_R * \left(\sum_{i=1}^{N} [u * h_T * h_{u_i}] + e_p \right) \right] (t) + e_n(t) + e_q(t) \qquad (4.3)$$

where N propagation paths are considered and their impulse responses are subscripted as $h_{u_i}(t)$ for $i = 1, \ldots, N$. Similarly, the path attenuations and delays can be subscripted as α_i and Δt_i to give $h_{u_i}(t) = \alpha_i \delta(t - \Delta t_i)$. With this, the overall response, limited to the ultrasonic paths, is $h_u(t) = \sum_{i=1}^{N} h_{u_i}(t) = \sum_{i=1}^{N} \alpha_i \delta(t - \Delta t_i)$, i.e., a train of pulses with various amplitudes and timings. As for the single path model, also in this case, it is desirable to ignore $e_p(t)$ assuming that $e_n(t)$ is dominant.

How many paths should be considered? Given that the SUT boundary itself creates reflections and that it can be a rather extended interface, one may presume that the overall number of paths can be quite large. However, dealing with a large N can complicate the channel estimation. Furthermore, many paths can have very strong attenuation (particularly if they are long), making their effects hardly visible. Finally, many paths can be completely uninteresting. Most likely, one may not care about the

SUT boundary. Similarly, one might be uninterested in defects that are very far from the probes, provided that multiple probe placements can be tried. In fact, defects that are hardly "visible" from one side of the SUT may be much more conveniently spotted from another side. To keep N low, it can be convenient to introduce an *attenuation threshold* α_t and to assume that all paths characterized by $\alpha_i < \alpha_t$ can be ignored (i.e., merely contribute to system noise).

Once the basic model is defined, a few more considerations are worth giving. First of all, in estimating the channel features, one will certainly see the whole of it and not just the ultrasonic portion. Thus, it is convenient to introduce an overall

$$h_{\text{ch}}(t) = [h_T * h_u * h_R](t) \tag{4.4}$$

impulse response for it (with the corresponding $H_{\text{ch}}(s)$ transfer function). Yet, ultimately, it is the $h_{u_i}(t)$ that one is interested in. For reasons that will become evident shortly, when trying to estimate an individual path, all other paths become "interference". This self-interference is a completely different phenomenon from the noise components $e_n(t)$ and $e_q(t)$. For once, its power scales proportionally to that of $u(t)$. Second, it can be relatively large and it cannot be neglected as other terms.

Another important point is that in the model all signals appear as *continuous time*. This is reasonable, since the system under consideration is a physical one. However, the signal processing chain is likely to contain an ADC which implies a time discretization. Thus, quantities depending on time may get sampled at instants $t = nT_{\text{ADC}}$ with $n \in Z$, where T_{ADC} is the ADC sampling period (while $f_{\text{ADC}} = T_{\text{ADC}}^{-1}$ is its rate). In the following, when dealing with sampled signals, for the generic signal $x(t)$ the notation $x(n)$ may be used in place of $x(n\,T_{\text{ADC}})$ as a shorthand.

4.3 Filtering Noise

In an ideal, noiseless world, estimating the channel should be quite easy. As long as $H_T(s)$ and $H_R(s)$ can be ignored, in the case of impulsive excitation, the channel output simply delivers the superposition of multiple-delayed (and attenuated) copies of its input since in this case $h_{\text{ch}}(t) = \sum_{i=1}^{N} \alpha_i \delta(t - \Delta t_i)$. When $H_T(s)$ and $H_R(s)$ are taken into account, $h_{\text{ch}}(t)$ is not anymore the superposition of ideal impulses, as these get spread in time by the two transfer functions. However, the spreading is generally modest enough to permit the recognition of pulse-like shapes and the readout of their timings. This fully justifies the traditional PuE NDT approach based on short bursts that make the best practicable approximation of an ideal excitation pulse. In this scenario, finding the path delays Δt_i should be a mere matter of seeking peaks in the response, or better in the response envelope to remove modulation due to the typical band-pass behavior of the transducers.

Note that using pulses with a non-null length is indispensable not only to have signals that *can physically be generated*, but also to ensure that *most of them can pass through TX*. In fact, an ideal pulse has an infinitely large bandwidth, while the transducers can only carry a finite portion of it. If an ideal pulse was used, most of its energy would be dissipated in heat in the TX, which would be a waste and could also compromise the operation of the TX itself (e.g., self-heating might damage the TX, or large components below its lower corner frequency could make it saturate). Using pulses that match the bandwidth of the TX, these issues are automatically relaxed. Due to time-frequency duality, this requires fixing a minimal pulse length. Restricting excitations to the transducer bandwidth comes with a little price. It is not anymore possible to fully estimate the part of the channel dynamics due to $H_T(s)$ and $H_R(s)$ out of the signal band since frequencies corresponding to their roll-off regions are not exited anymore. This is not an issue, though, since one is not that interested in $H_T(s)$ and $H_R(s)$, but in $H_u(s)$ and the associated Δt_i. On the other hand, using pulses that can pass almost unfiltered through the TX and RX assures that what is received at the output is the superposition of almost exact copies of the input. In other words, the task of assessing the Δt_i can now be seen as the *seeking of copies of an input template* in the output in place of just peaks. As it will shortly be shown, this change of perspective will be quite important when noise is taken into account.

Example 4.1
Imagine having a TX and RX with a center band frequency $f_0 = 1\text{MHz}$ and a bandwidth $B = 0.5\text{MHz}$, i.e., corner frequencies $f_l = f_0 - B/2$ and $f_h = f_0 + B/2$. To deal with them, one could use excitation pulses built as chunks of sine waves, with frequency f_0 and duration $T \approx 1/B$, namely

$$u(t) = \begin{cases} \text{Re}\left(\mathrm{e}^{2i\pi f_0 t}\right) & \text{for } 0 < t \leq T \\ 0 & \text{otherwise} \end{cases} \tag{4.5}$$

Now, if B is sufficiently large, the duration T becomes very small and the band of $u(t)$ almost corresponds to that of the transducers. With this, an ideal noiseless channel should return multiple time-displaced and attenuated copies of this very excitation. To show this, Fig. 4.3 illustrates how sinusoidal bursts with the same frequency f_0 but different duration T are transformed by passing through the channel. Specifically, plot a represents the transfer function of the overall TX–RX system while plot b the corresponding impulse response. Plots c and d show how a chunk consisting of a single period is transformed. Eventually, plots c and d show how a chunk consisting of three sine periods is transformed. In this case, the input signal fits better the channel transfer function and, consequently, is less affected by the channel.

As a little technicality, note that if $H_T(s)$ and $H_R(s)$ have a nonflat magnitude response or a phase response whose behavior in band is different from linear, the response may contain *slightly distorted* copies, and not just delayed copies, of the excitation burst. In this case, the template to be sought should in principle be given by $[u * h_T * h_r](t)$.

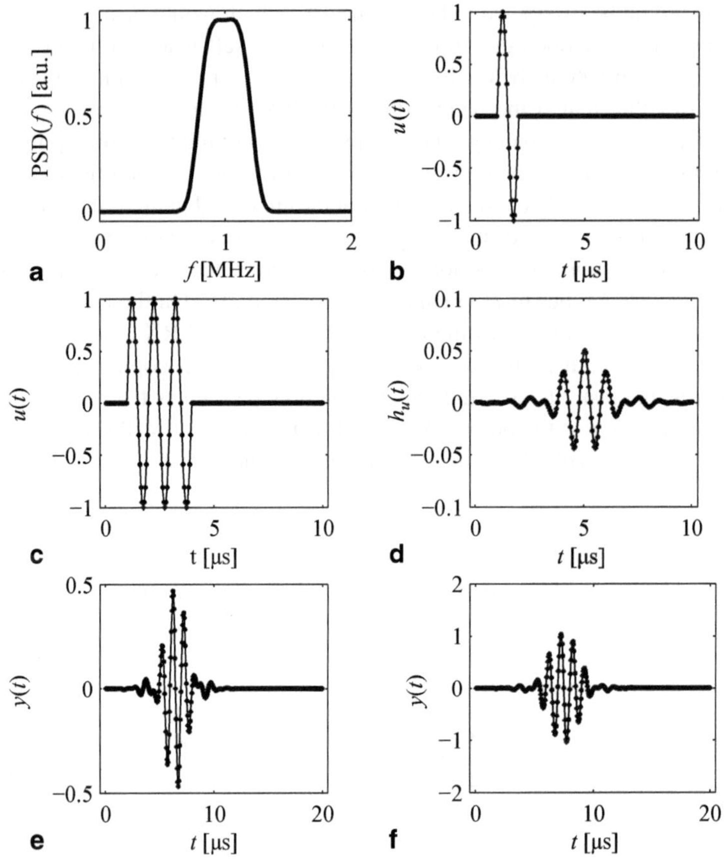

Fig. 4.3 Illustrations relative to Example 4.1

It is now possible to consider how noise can affect the framework introduced so far, addressing in deeper detail a topic mentioned in the introduction. Since the received copies of the input burst can be strongly attenuated, there is an evident risk that they fall below $e_n(t)$ becoming virtually invisible. Because their power is proportional to the power of the excitation injected into the system, an apparent solution is to simply increase the amplitude of $u(t)$. However, as mentioned in introduction, this may hit on the transducers technological limits. Another trivial solution requires repeating several times the measurement, and then taking the average of the collected data as in:

$$\overline{y}(t) = \frac{1}{N_{\text{avg}}} \sum_{j=1}^{N_{\text{avg}}} y_j(t) \tag{4.6}$$

where N_{avg} is the number of measurements and $y_j(t)$ is the signal collected at the jth measurement. Under the Gaussian hypothesis, it is well known that the standard deviation σ_{e_n} of $e_n(t)$ decreases as $1/\sqrt{N_{avg}}$, while the averaged value of $\overline{y}(t)$ tends to the actual value. The result is a relative enhancement of the signal power, and thus of the SNR, since the noise power, tied to σ_{e_n} decreases, even if each individual measurement is executed with the same, bad, SNR. Although the averaging strategy does not introduce any approximation or distortion in the measure of $y(t)$, its efficiency is quite low, especially where long impulse responses, typical of reverberating structures, have to be measured. For instance a gain of 20 dB in the SNR implies an extension of the measurement time by a factor 100 at least, that could be irreconcilable with on-field applications.

Fortunately, there are other possible strategies to increase the final SNR of the measurement without increasing the input power or repeating the same measurement a lot of times. To gain deep insight on these methods, a fundamental preliminary question is whether power is the real key of the issue or not. Indeed, it is not, so let a fundamental point be underlined: what counts is the excitation *energy*, not its power. One of the reasons why it is so is that the noise floor depends on the bandwidth of the signals being considered. In fact, the power of $e_n(t)$ can be approximately obtained by integrating the power spectral density (PSD) of noise introduced at the RX and the LNA through the frequency range required to process the system response. As long as this PSD can be assumed to be flat, the noise power is approximately proportional to the bandwidth required to process the copies of the excitation template, i.e., the excitation bandwidth. Now, suppose that an excitation pulse is doubled in length, while halving its power (thus keeping its energy constant). As long as the burst is brief, doubling duration means approximately halving the bandwidth and thus also halving the bandwidth required by the signal processing chain after the LNA. Consequently, the noise floor can also be halved, leaving the final SNR unchanged.

The considerations above suggest that as long as the excitation burst amplitude cannot be increased, increasing its length T can be an alternative measure to assure that its copies in the received output are visible in noise. Unfortunately, merely increasing the burst length may have adverse consequences on the system range-resolution unless other appropriate actions are taken. In fact, there may be multiple paths with similar associated delays. If the ith and jth paths have $|\Delta t_i - \Delta t_j| < T$, the corresponding copies of the input burst run together in the output. Unless appropriate strategies are used, this overlap may make it impossible to distinguish the two copies of the input template in the output. These strategies shall be the topic of the next Sect. 4.4. For now, let noise be the discussion focus. One more reason why it is the excitation burst energy that counts rather than power is that signal processing techniques can be used to isolate the useful signal out of noise. Due to signal processing, one can actually detect copies of the input burst even when their instantaneous power falls below that of $e_n(t)$. In fact, appropriate processing allows to exploit a priori knowledge on the useful signal to make it stand out. Many signal processing techniques exist for dealing with signals embedded in noise. However, all share a common trait. Rather than looking at *punctual* signal features (i.e., values taken at single instants in time), multiple features corresponding to larger time spans are simultaneously used. Intuitively, the wider the available time spans for consideration, the higher the ability to identify useful components in noise. This makes time duration counts as much as amplitude.

To widen the time spans available for the isolation of useful components, two different approaches can be taken. As just said, one of them consists in using longer excitation bursts. The other one consists in taking advantage of the time-invariant nature of the system, repeating the experiment multiple times and *averaging* (or otherwise jointly processing) the results, a technique that has already been hinted at. In the first case, there is an explicit increase in the excitation energy. In the second, there is a virtual one, since the multiple experiments are in many ways equivalent to a single experiment where the excitation power is superimposed onto itself multiple times. In this section, focus is on the first approach, while hints at the second one are given in Sect. 4.4.2. Note that the two approaches can even be combined. Also note that the first approach is generally cheaper with respect to the overall measurement time. In fact, repeating an experiment has an immediate impact on it, while enlarging T has not, since the measurement time is dominated by the need to wait for the largest ΔT_i.

It is now possible to consider actual signal processing techniques. The most straightforward one is linear filtering. In practice, one may place a filter $H_f(s)$ after the LNA (or equivalently $H_f(z)$ after the ADC) to obtain a *better* $y_f(t)$ out of $y(t)$. What filter should be adopted? For sure, it should eliminate all those frequency components that are not expected in the useful output. Since the noise is almost white and the useful signal component is band limited, this can already remove a large amount of noise. Indeed, this is already sufficient to let the noise power scale with the signal bandwidth, as suggested in a previous example. Yet, this operation alone would not be *optimal*, as it would only use a very limited amount of the available a priori knowledge on the input signal (the two corner frequencies). Conversely, the more a priori knowledge can be used, the more successful the filter can be. For instance, relying on knowledge of the PSD of the useful signal, the Wiener filter could already be much more efficient than a mere band restriction [52].

In the problem at hand, the amount of a priori information is huge. One has a whole template for the signal to be sought. For this particular problem, the filter known to be optimal is the so-called *matched filter* [48]. In cases where the noise is AWGN, like the current one, this reduces to an $H_f(s)$ whose associated impulse response $h_f(t)$ is given by the time reversal of the template. Namely,

$$h_f(t) = \mathrm{Rev}[u * h_T * h_R](t) = [u * h_T * h_R](-t) \tag{4.7}$$

where the operator Rev indicates reversal of the reference axis. Note that some care is required, since the definition in the previous equation returns an impulse response that is non-null for $t < 0$, namely, a filter that is noncausal. Hence, in practice, one may need to use

$$h_f(t) = \mathrm{Rev}[u * h_T * h_R](t + \tau_r) = [u * h_T * h_R](-t - \tau_r) \tag{4.8}$$

with a suitable τ_r. Then, τ_r needs to be compensated in the estimation of the Δt_i associated to the ultrasonic paths. This is a technicality that shall be taken for granted in the following, avoiding the explicit indication of τ_r. Also note that in many cases $h_f(t)$ can simply be set at $u(-t)$ ignoring the effects of $H_T(s)$ and $H_R(s)$.

An explanation of the effectiveness of the matched filter is that this filtering is inherently equivalent to computing a correlation. Operation is like making the template waveform progressively slide along the recorded response, evaluating the resemblance between the two at each point. Maxima in the filter output $y_f(t)$ correspond to best matches, and when they are sufficiently large, they indicate the finding of the template in the signal. Obviously, this explanation does not constitute a proof of optimality for the approach, and the interested reader is invited to check those available in the signal processing and estimation literature (see for instance [48, 49]). From the filter construction and the explanation above, it should be clear that the matched filter does not return the signal component that is sought, but a strongly distorted version of it: $y_f(t)$ does not resemble $u(t)$, but is a different waveform, that peaks when the signal component is recognized in the noise. Thus, one merely needs to look for local $y_f(t)$ maxima that stands well out of the noise floor, the latter being given by $e_f(t) = [h_f * e_n](t)$, i.e., the filtered $e_n(t)$. From the timings of such maxima, one can estimate the delays Δt_i of the propagation paths. Furthermore, from their amplitude γ_i, the path attenuations α_i can also be estimated. In fact, the detected maxima are as large as

$$\gamma_i = \alpha_i \cdot [h_f * u * h_T * h_R](0) \tag{4.9}$$

so that α_i can be estimated as

$$\tilde{\alpha}_i = \frac{\gamma_i}{[h_f * u * h_T * h_R](0)} . \tag{4.10}$$

Expression (4.9) is also useful to evaluate the quantities on which SNR depends after the matched filter. Ignoring for simplicity $H_T(s)$ and $H_R(s)$, the terms γ_i in (4.9) have amplitudes proportional to $[h_f * u](0)$, that is $R_{uu}(0)$, the 0-lag autocorrelation of $u(t)$, namely, its energy. Hence, at the times of interest, the instantaneous power of $y_f(t)$ is proportional to the squared energy of the excitation burst. On the other hand, the power of $e_f(t)$ is proportional to the energy of the filter impulse response, that is the energy of $u(t)$. Thus, the SNR is proportional to the *energy* of the excitation burst. This is one reason why the matched filter is so effective. In one of the previous examples, the doubling of the excitation burst duration could lead to an SNR advantage via the halving of the signal bandwidth, that allowed the noise floor to be halved too. With the matched filter, the doubling of the excitation burst length does not need to be coupled to a bandwidth reduction to cash the same advantage. The ability to increase the SNR without reducing the excitation bandwidth is indispensable to preserve a good resolution, even when using long excitation bursts, as it will be better illustrated in Sect. 4.4.

Example 4.2
It has been said that the matched filter is the optimal choice to maximize the SNR for detecting the time at which a template is reproduced on a noisy track.

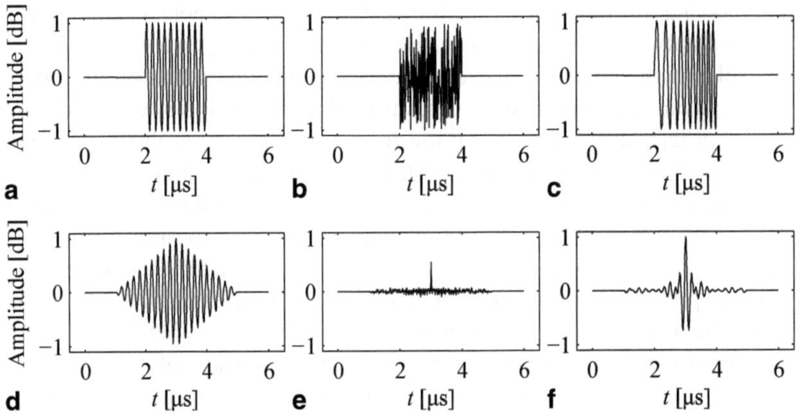

Fig. 4.4 Illustration of Example 4.2

The SNR peak value scales with the energy of the signal and is inversely proportional to the bandwidth of the collected noise. Figure 4.4 shows what happens at various signals when the matched filter is applied. Plots a, b, and c show a sinusoidal burst, some random noise, and a chirp or sweep signal, respectively. Plots d, e, and f show the corresponding filtered signals.

Before concluding this section, a final remark is due. The matched filter is not the sole signal processing technique that can be adopted. Other approaches exist, among which *sparse deconvolution* and *matching pursuit* are certainly worth mentioning [28, 50]. While quite complex, these technique are, broadly speaking, based on the very simple concept of adding further a priori knowledge to the noise removal process. This basically consists in the assumption that there can only be a very limited number of *templates* embedded in the received signal.

4.4 Pulse Compression Systems

The previous section established two important points. The first is the need for a filter $H_f(s)$ after the RX. The second is that, by appropriate filtering, the SNR can scale with the excitation burst *energy*, namely, the excitation burst power times the burst duration. As it has already been underlined in multiple occasions, power cannot be augmented beyond certain limits. Thus, augmenting the burst duration is desirable, particularly in setups where there is a large attenuation (e.g., with air-coupling). Unfortunately, when the excitation burst is a sinusoidal chunk, increasing its length causes a resolution degradation. This has already been shown considering the raw received signal and the signal filtered through the matched filter (Example 4.2).

The issue is that, with an apparent paradox, one would like to have a *long* excitation burst at the transmitting (in order to maximize the energy injected into the SUT) and a very *short* one at the receiving end, where the problem is to distinguish two paths with similar associated delays. How can these two desiderata be conciliated? The idea is to start with a long pulse and to shorten it when its length becomes an issue. In fact, such an operation can be conceived and practically implemented.

The first deployment of this concept was based on the idea of giving to the receiving filter $H_f(s)$ the ability to delay the initial parts of the received pulses more than the last parts, so achieving a *compression in time* of the pulses. The delay that a filter achieves on a narrowband signal is the so-called *group delay*. In a time-invariant filter, this quantity can depend only on frequency. This means that a possible way to differentiate between the initial and final parts of a signal can be achieved by making such parts different in their spectral content. For instance, one can use an excitation signal whose frequency varies in time following some monotonic profile (e.g., lower frequencies first, higher frequencies later). This signal is *locally* narrowband. Then one can apply an $H_f(s)$ with a group delay dependency on frequency tuned to the same profile (e.g., larger delay for low frequencies, smaller delay for high frequencies). This idea has an immediate implication: the excitation, considered *as a whole*, must be wideband.

The easiest way to practice the approach is to base the excitation on a sinewave whose frequency increases linearly in time, namely, a *linear chirp* (LChirp)

$$U_{\text{chirp}}(t) = \sin\left(2\pi f_0 t + \pi k t^2\right) \tag{4.11}$$

where f_0 is the initial frequency value and k is the rate of change so that the instantaneous frequency is $f_{\text{inst}}(t) = f_0 + kt$. This leaves the designer with the need for a filter whose group delay decreases linearly with frequency. Since the group delay is the opposite of the derivative of the phase response with respect to frequency, this sets the requirement for a quadratic profile in the (unwrapped) phase response. Quite interestingly, an *all pass* filter with such phase behavior has an impulse response that is an LChirp itself, where instantaneous frequency decreases linearly over time. This observation can be formally proven via the inverse Fourier transform. On a more intuitive basis, one can consider that the ideal impulse has a white spectrum, i.e., contains at time $t = 0$, all the possible frequency components. Thus, a filter transforming an ideal impulse into a chirp with instantaneous frequency $f_{\text{inst}}(t)$ is actually delaying the component with frequency f to time $t = f_{\text{inst}}^{-1}(f)$. As an example, Fig. 4.5 shows a chirp, an inverse chirp, and the group delay of a filter whose impulse response is the inverse chirp.

An interesting point is that this intuitive view also holds for wideband signals that are not LChirps as long as an instantaneous frequency can be identified. Their inverse necessarily defines a filter whose group delay function provides some PuC. This fact is quite appealing, since it means that the matched filter is also a *pulse compressing filter*.

From the considerations in the previous paragraph, it may seem that, by the matched filter, a signal with length T whose instantaneous frequency monotonically

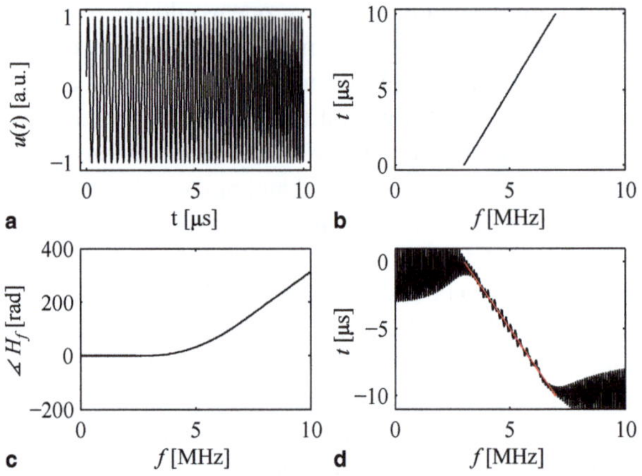

Fig. 4.5 Illustration of the pulse compression principle. **a** An LChirp excitation. **b** The evolution of instantaneous frequency over time in the excitation. **c** The (unwrapped) phase response of a filter whose input response is the reverse of the excitation in **a**. **d** The group delay versus frequency corresponding to the phase response in **c**. Here, the opposite of the curve in **b** is superimposed in *red*, to show how this filter can time-compress the excitation signal by delaying those frequency components that come earlier and delaying less the frequency components that come later

increases or decreases should always get *ideal compression*, i.e., should be reduced to a null duration. In fact, the component appearing at time $\hat{t}$ in the signal should appear at time $\hat{t} + (T - \hat{t}) = T$ after the filter for all $\hat{t}$. Unfortunately, this is never the case, since the considerations provided so far are intuitive rather than formal. Section 4.4.1 provides a modern discussion of PuC explaining the relevant phenomena and showing how it can be applicable also to signal where the idea of instantaneous frequency does not apply. For now, it is important to underline a notable point. The success of the matched filter in providing PuC depends on the specific excitation being used. Thus, an appropriate design of the excitations is quite important to ultrasound NDT systems.

4.4.1 Fundamental Concepts on Pulse Compression

As just mentioned, PuC is a procedure for estimating the impulse response $h(t)$ of LTI systems that relies on the contextual exploitation of two key points:

- the theory of matched filter to optimize the SNR at the receiver stage;
- the use of broadband signals to assure also a good resolution at the receiver stage;

This implies that a signal suitable for PuC is *long* with respect to the requested time resolution of the ultrasonic measurement and is also broadband to cover almost uniformly the TX–RX channel bandwidth. How much a signal is *good* for PuC applications can be quantified by a parameter, called time bandwidth product (TBP), which measures the product of duration and bandwidth of the signal. The larger the

TBP, the higher is the gain that a PuC procedure can provide. For instance, standard short pulses such as boxcar, Gaussian pulse, sinusoidal burst, etc., exhibit TBP values close to 1 since their bandwidth is inversely proportional to their duration and indeed they are not fitted for the use in a PuC procedure, as anticipated in the previous sections. On the contrary, more complex waveforms like signals achieved by frequency modulation (FM) or PN sequences, can be defined to have TBP $\gg 1$. In the following sections, several waveforms suitable for PuC will be described, also giving information about their TBP, but, beforehand, it is useful to review the PuC measurement procedure by assuming that the TBP $\gg 1$ condition is satisfied.

Formally, the PuC technique can be illustrated as follows: if a pair of signals $u(t)$, $\Psi(t)$ exist such that their convolution $[u * \Psi](t)$ is a good approximation of the Dirac delta $\delta(t)$, namely, $[u * \Psi](t) = \hat{\delta}(t) \approx C\,\delta(t)$, where C is a suitable constant (usually ending up with a value close to TBP), then the pair $\{u(t), \Psi(t)\}$ can be employed in a PuC scheme and an estimate of $h(t)$ can be retrieved by executing the following two steps:

STEP I excite the LTI system with the signal $u(t)$;
STEP II convolve the output signal $y(t)$ with $\Psi(t)$.

In fact,

$$
\begin{aligned}
y(t) &= [h * u](t) \\
[u * \Psi](t) &\approx C\,\delta(t) \\
\hat{h}(t) &= [y * \Psi](t) = [h * u * \Psi](t) \approx C\,[h * \delta](t) = C\,h(t)
\end{aligned}
\tag{4.12}
$$

where $\hat{h}(t)$ is the impulse response estimation obtained by the proposed technique. Clearly, the better the approximation of $\delta(t)$ via the $\{u(t), \Psi(t)\}$ couple, the higher the quality of the estimation of $h(t)$.

As said, in presence of AWGN, i.e., when $y(t) = [h * u](t) + e(t)$ (where $e(t)$ is the noise term), the $\Psi(t)$ that maximizes the SNR is the time-reversed replica of the input signal $\Psi(t) = \mathrm{Rev}[u](t)$, namely, the matched filter [48]. With this choice, that is the most widely adopted in actual implementations, the approximation $\hat{\delta}(t)$ of δ(t) coincides with the autocorrelation function $R_{uu}(\tau)$ of the input signal $u(t)$. From the Wiener–Kinchin theorem [34], it is known that the Fourier transform of the autocorrelation function $R_{uu}(\tau)$ equals the energy spectral density (ESD) of $u(t)$, so that the δ-like condition depends on the spectral content of the signal. The approximation is quite good if $u(t)$ is a wideband signal with a bandwidth that spans as much as possible the entire frequency range of response of the LTI system.

More generally, $\Psi(t)$ can differ from $\mathrm{Rev}[u](t)$ in order to optimize merit factors other than SNR, as will be discussed further. Furthermore, note that a more formal description of the PuC theory lies beyond the scope of the book and the reader is invited to refer to the literature (see for example [15, 23, 44]). Here, we limit to give some information about the most common waveforms used in PuC and moreover we highlight that the δ-like feature of the signal autocorrelation can be assured both by "single-shot" or periodic signals, leading to two main PuC schemes: acyclic pulse

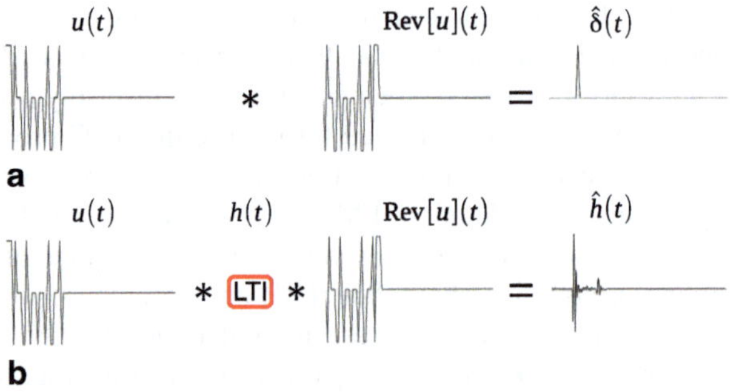

Fig. 4.6 Examples of the APC approach. **a** The ideal convolution condition between the signal $u(t)$ and the matched filter Rev$[u](t)$ is depicted. **b** The application of the APC scheme to the characterization of the impulse response of an LTI system is shown

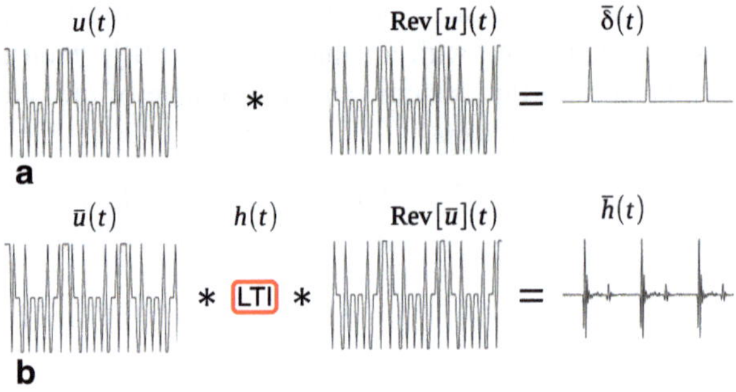

Fig. 4.7 Examples of the CPC approach. **a** The ideal convolution condition for the periodic signal $u(t)$ and the corresponding matched filter Rev$[u](t)$ is depicted. **b** The application of the CPC scheme to the characterization of the impulse response of an LTI system is shown

compression (APC) and cyclic pulse compression (CPC), as depicted in Figs. 4.6 and 4.7. The particular features of these two approaches are described below, together with a computational representation of the two procedures in terms of the discrete Fourier transform (DFT), to provide a general framework. Then, in Sect. 4.5, the properties of the various codes will be reviewed in the light of the differences between the two measurement methods.

4.4.2 *Pulse Compression Measurement Procedures: Single-shot vs Periodic Excitations*

The δ-like properties of the autocorrelation of $u(t)$ depend on its ESD, as previously said. But, when dealing with discrete-time signals, in order to optimize the PuC procedure—namely to have a better approximation $\hat{\delta}(t)$—it also becomes relevant how the autocorrelation can be computed, especially when binary PN sequences are used as excitations. Two main procedures exist, henceforth called APC and CPC, based respectively on acyclic or cyclic convolution between the system output signal and the matched filter. To gain some insight into this process, let us suppose that for the APC and CPC cases, there exist pairs of signals that guarantee that the following equalities hold

$$\begin{aligned} [u * \mathrm{Rev}[u]](t) &= \delta(t) \qquad \text{acyclic} \\ [\overline{u} * \mathrm{Rev}[\overline{u}]](t) &= \overline{\delta}(t) \qquad \text{cyclic} \end{aligned} \tag{4.13}$$

where the bar indicates periodic repetition. Hence, in the acyclic case, $u(t)$ is a coded waveform defined in $t \in [0, T]$, while, in the cyclic case $\overline{u}(t)$ is a periodic-coded waveform with repetition period T. Similarly, $\overline{\delta}(t)$ consist of a periodic train of δ functions spaced by T as illustrated in Fig. 4.7.

In the APC case, if the output signal is convolved with the compression filter, one directly obtains an estimation of the impulse response $h(t)$, while, for the CPC case, the convolution between $h(t)$ and $\overline{\delta}(t)$, is achieved, namely a periodic repetition of the estimated impulse response. At this point, the careful reader may have already noted the main difference between the two schemes. Indeed, if $h(t)$ is shorter than T, the convolution between $h(t)$ and $\overline{\delta}(t)$ produces a periodic train where the copies of $h(t)$ are nonoverlapping (in these considerations the nonideality of the estimation process, that may enlarge the tail of $h(t)$ is neglected, as it can usually be done). Conversely, if $h(t)$ is longer than the input signal period, the convolution between $h(t)$ and $\overline{\delta}(t)$ shows a piling-up of multiple, displaced copies of $h(t)$ from which $h(t)$ is not retrievable. In other words, there is a *time-aliasing* phenomenon. Therefore, the CPC approach should only be used if a priori information about the maximum length T_h of $h(t)$ is available. In this case, by chosing $T > T_h$ to avoid time-aliasing, $h(t)$ can be estimated by merely selecting a single period of the cyclic convolution. Conversely, in APC, the input signal duration T of $u(t)$ is not constrained by the duration of T_h.

From the discussion above, it may seem that the CPC scheme presents a significant limitation with respect to the APC one. Actually, there are many circumstances, as in air-coupled ultrasonic inspection, where the signal length is usually known to be significantly larger than the length of the impulse response to be estimated (or at least larger than the length of that part of the impulse response that produces effects above the noise level). In this case, CPC has no issues. For this reason, henceforth we assume $T > T_h$. Incidentally note that in air-coupled systems, things are purposely set up in this way (i.e., with long excitation signals) to guarantee that the initial excitation energy is large, because a large attenuation exists at the interface between the probes and the SUT. In fact, with air coupling the interface is doubled (probe-to-air and air-to-SUT) and characterized by a large acoustic impedance mismatch that causes a large part of the acoustic energy to be reflected rather than transferred.

In addition to what has been said so far, also from an experimental point of view, there are differences between the APC and CPC approaches. In the APC case, the coded signal $u(t)$ excites the system that produces an output signal of duration $T_y = T + T_h$. If the whole $h(t)$ needs to be estimated, the output signal must be collected for a time $T_{\mathrm{rec}} \geq T_y$. Then, PuC can be implemented by applying the matched filter to the output signal. Conversely, in the CPC case, a periodic signal $\overline{u}(t)$ of period $T > T_h$ is switched on at $t = 0$, and excites the system that produces an output signal $y(t) = [h * \bar{u}](t)$. After a short transient, certainly shorter than T, the output signal becomes periodic with the same period T. By performing the cyclic convolution between a single period of the steady-state part $y_{\mathrm{ss}}(t)$ of $y(t)$ and a single period of the input signal $\overline{u}(t)$, an estimation of $h(t)$ is retrieved. In practice, it is sufficient to transmit two periods of the excitation signal and to record the output corresponding to the second excitation period only.

Mathematically, once that the output signals have been acquired, the procedure is as follows. In APC, the matched filter can be applied by performing standard or DFT-based convolution algorithms. Clearly, computational approaches that can take advantage of fast Fourier transform (FFT) algorithms and techniques such as *overlap-add* and *overlap-save* [17, 22] are generally preferable. In CPC mode, the matched filter is applied by computing the cyclic convolution between the output signal and the matched filter. By exploiting the *convolution theorem for discrete signals*, the cyclic convolution can be efficiently implemented in the frequency domain as follows:

$$[y_{\text{ss}} * \overline{u}](n) = \text{IDFT}[\text{DFT}[y_{ss}] \times \text{DFT}[\overline{u}]](n) \tag{4.14}$$

where the DFT and IDFT operators return the direct and inverse discrete Fourier transform, respectively and can be implemented via the FFT algorithm [6, 22]. For completeness, it should be noted that the convolution theorem for discrete signals can also be used in the single-shot case, provided that the discrete input and output signal are padded with zeros up to double the number of samples of the input signal, having assumed $T > T_h$.

At this point of discussion, it is, therefore, fair to ask whether there are any advantages of CPC over APC. Indeed, there are. The first one concerns the attainable SNR. By considering the optimized APC procedure in which the output is recorded for the minimum time needed to recover the whole of the impulse response, it is straightforward to see that energy is delivered to the system for a fraction $p_{EX} = T/(T + T_h)$ of the measurement time, and that the final SNR is given by $\text{SNR}_{EX} = p_{EX} \cdot \text{SNR}_{\text{max}}$, where SNR_{max} is the limit value achievable for $T \gg T_h$. On the other hand, in the CPC case, the excitation is always active during the measurement, thus saturating the SNR for any given excitation power once the condition $T > T_h$ is satisfied. Moreover, it is common to improve SNR by averaging. To increase the SNR in the APC case for a given p_{EX} value, averages can be taken by sending a train of single-shot excitations, but the minimum repetition time of these must be $T_r = T + T_h$. For CPC systems, averages are easily performed by considering several consecutive periods of the steady-state output signal, making it more time-efficient. From a computational point of view, the input and output signals to be processed are equal in length and usually shorter (if the assumption $T > T_h$ is valid) for the CPC case than the APC one, and therefore, the convolution with the matched filter is less time-consuming. Furthermore, the cyclic convolution admits an efficient implementation through the FFT that is also at the heart of fast convolution algorithms such as overlap-add used in APC mode.

In summary, CPC methods can only be used if an a priori information about the duration of the impulse response to be measured is available. Therefore, its application may be limited when compared to standard APC approaches. Nonetheless, when applicable, it assures the optimal SNR and computational savings for selected measurements as it will be exemplified shortly.

4.5 Pulse Compression Waveform Design

To illustrate how PuC schemes work, it was assumed in the previous section that signals, either periodic or not, exist with an ideal δ-like autocorrelation. Unfortunately, no finite-duration waveform has this property. Hence, strategies have been devised that use signals merely approximating this condition. During the development of PuC applications, starting from the first ones that date back to the 1960s and 1970s, two main families of waveforms have been suggested: FM waveforms in the form of *chirps* and binary PN codes. In the next two sections, excitations belonging to both classes are discussed, together with possible variants. In particular, the performances of LChirps and *nonlinear chirp*s (NLChirps) are evaluated in terms of maximum achievable SNR and resolution and are then compared to two types of binary sequences, namely, maximum length sequences (MLSs) and Golay complementary sequences (GCSs).

Clearly, before getting into performance evaluation, it is necessary to briefly review the main features of these excitations, also highlighting their pros and cons. For the sake of clarity, henceforth the normalized autocorrelation (both acyclic and cyclic) function of the various signals will be denoted either by $\hat{\delta}(t)$ or by $\hat{\delta}(n)$, in the continuous and discrete time, respectively, to indicate that they represent an approximation of the ideal continuous Dirac delta and discrete time Dirac delta (unit impulse) functions, respectively.

4.5.1 *Frequency Modulated Waveforms: Linear and Nonlinear Chirp*

The FM waveform *par excellence* used in PuC applications is the chirp signal, that is a harmonic signal whose phase rather than being linear can assume any continuous and differentiable monotonic profile. Mathematically, a generic chirp waveform is defined as

$$u_{\mathrm{CHIRP}}(t) = A \sin\left(\Phi(t)\right) \tag{4.15}$$

where $\Phi(t)$ is the phase as a function of t [13].

Since the phase is nonlinear, it is not possible to define for the chirp a unique well-defined oscillation frequency but instead one can consider an instantaneous frequency $f_{\mathrm{inst}}(t)$ that is related to $\Phi(t)$ by

$$f_{\mathrm{inst}}(t) = \frac{1}{2\pi}\frac{\mathrm{d}\Phi(t)}{\mathrm{d}t}\,. \tag{4.16}$$

By inverting Eq. (4.16), a chirp signal is obtained via Eq. (4.15), once a proper trajectory $f_{\mathrm{inst}}(t)$ in a given time interval $t \in [0, T_s]$ is defined. The instantaneous frequency is the key property of a chirp signal because it is strictly related to the ESD of the resulting waveform, on which the autocorrelation function depends.

Indeed, if the duration T of the chirp signal is large enough, i.e., for TBP $\gg 1$, the ESD is confined within the frequency interval spanned by $f_{\text{inst}}(t)$ and—at any given frequency—the ESD is inversely proportional to the rate of change of $f_{\text{inst}}(t)$ [51, 30]. For instance, if $f_{\text{inst}}(t) = f_{\text{start}} + \frac{B}{T}t$ with $B = f_{\text{stop}} - f_{\text{start}}$ and $t \in [0, T]$ the usual LChirp is obtained, where $\Phi(t) = 2\pi(f_{\text{start}}\, t + \frac{B}{2T}t^2) + \phi_0$ is quadratic in t and the ESD is almost flat and confined within the interval $f \in [f_{\text{start}}, f_{\text{stop}}]$ (see Sect. 4.4, Eq. 4.11). It is therefore quite easy to define a chirp signal whose ESD covers all the TX–RX channel bandwidth. Yet, the flexibility of chirp signals does not end with the LChirp. In general, the frequency can be swept in either a linear or nonlinear way, leading to an NLChirp in the latter case. With a nonlinear sweep, it is possible to design the waveform so that its ESD matches a prescribed smooth profile. This opportunity can be an advantage to adapt the excitation to the channel when the whole of its magnitude response is known, and not just the corner frequencies [37]. Specifically, in [30] an approximated relation was identified between $f_{\text{inst}}(t)$ and the the amplitude of the ESD of an NLChirp. If $t_{\text{inst}}(f)$ is the inverse relation of $f_{\text{inst}}(t)$, one has

$$-\frac{\mathrm{d}t_{\text{inst}}(f)}{\mathrm{d}f} \approx C \cdot \mathrm{ESD}(f)\,. \tag{4.17}$$

where C is a suitable scaling constant. Incidentally note that in [30], the quantity $t_{\text{inst}}(f)$ is indicated as a *group time delay*, for its relationship with the group delay of the compression filter, already described at the beginning of Sect. 4.4 and illustrated in Fig. 4.5. Also note that the larger is the TBP, the better is the approximation. With the above, it is easy to derive a method for the design of an NLChirp signal having the requested spectrum. This involves two steps.

STEP I. Obtain the required $f_{\text{inst}}(t)$ profile, so that

$$t_{\text{inst}}(f) = \int_0^f C \cdot \mathrm{ESD}(\nu)\mathrm{d}\nu \tag{4.18}$$

and

$$f_{\text{inst}}(t) \quad \text{s.t.} \quad f_{\text{inst}}(t_{\text{inst}}(f)) = f\,. \tag{4.19}$$

STEP II. Obtain the NLChirp expression as

$$s_{\text{CHIRP}}(t) = \sin\left(2\pi \int_0^t f_{\text{inst}}(\tau)\mathrm{d}\tau + \phi_0\right) \tag{4.20}$$

where ϕ_0 is an arbitrary initial phase.

Generally, an analytic solution is not available for the inversion in the first step. Nonetheless, a numerical approximation can easily be obtained.

Equations (4.18)–(4.20) are extremely powerful since they allow FM signals to be designed with an arbitrary power spectrum, opening the way to the possibility of distributing the excitation energy on the characteristics frequencies of the channel. For instance in acoustics, where it can be convenient to distribute the energy unevenly, following a profile based on octaves, NLChirps with exponential instantaneous frequency profiles have been used for several years [14], also being useful in the characterization of nonlinear systems [45].

To gain insight about the procedure summarized by Eqs. (4.18)–(4.20), Fig. 4.8 illustrates how three sample signals, one LChirp and two NLChirps, sweeping within 3–7 MHz and exhibiting different ESDs, can be generated starting from given ideal power spectra. The left-most column of plots gives the target spectra, the center one the excitation signals, and the right-most column shows the achieved spectra. While at first sight the waveforms do not appear very different, their actual spectra indicate that the three signals have rather different properties. This can be further observed by looking at the trajectories followed by $f_{\text{inst}}(t)$ in the time-frequency plane, in Fig. 4.9. This figure corresponds row-by-row with the signals in Fig. 4.8. The left-most column of plots shows the excitation signals. The center column contains plots of the instantaneous frequency as a function of time (with swapped axes), while the right-most column shows the group delay associated to the corresponding matched filters. In particular, the LChirp is associated with a straight line in this type of plot, whereas the trajectory of the NLChirp can be any continuous monotonic curve, defined by the type of frequency sweep being adopted.

As an alternative to NLChirps, some spectral shaping can also be achieved working with LChirps by introducing a proper amplitude modulation on Eq. (4.15):

$$u_{\text{weighted\_CHIRP}}(t) = w(t) \sin\left[2\pi\left(f_{\text{start}}\,t + \frac{B}{2\,T}t^2\right) + \phi_0\right]. \tag{4.21}$$

Due to the biunivocal correspondence between time and frequency in chirps, modulation of the signal envelope implies modulation of the LChirp ESD. This procedure is known in the literature as *windowing* since the modulation function $w(t)$ is typically chosen among the standard windows developed for frequency analysis: Gaussian, Blackman, and Tukey [1, 20]. Indeed, *windowing* is quite common in LChirp PuC applications since a nonwindowed LChirp exhibits abrupt changes in the ESD, in proximity to the limit frequencies f_{start} and f_{stop}, which lead to slowly-attenuating tails of the function $\hat{\delta}(t)$. Conversely, the application of the window function smooths the ESD, thereby producing a faster decay of the tails of $\hat{\delta}(t)$ at the cost of slightly reducing the signal energy, due to a transient loss of the constant envelope feature. Also the effective bandwidth of the signal can get slightly reduced, causing a worsening in the way $\hat{\delta}(t)$ can approximate $\delta(t)$ for t close to 0. In fact, as it will be better explained in the following, the desired $\delta(t)$ is typically rendered as a central peak (the main lobe, that should be as thin as possible) and some side oscillations (side bands). Any bandwidth reduction is reflected in a broadening of the main lobe. To quantify this reduction, the following definition can be used for the effective bandwidth B_{eff} [47]:

$$B_{\text{eff}} = \sqrt{12}\sqrt{\frac{\int_{f_{\text{start}}}^{f_{\text{stop}}} (f - f_c)^2 S(f) \mathrm{d}f}{\int_{f_{\text{start}}}^{f_{\text{stop}}} S(f) \mathrm{d}f}} \tag{4.22}$$

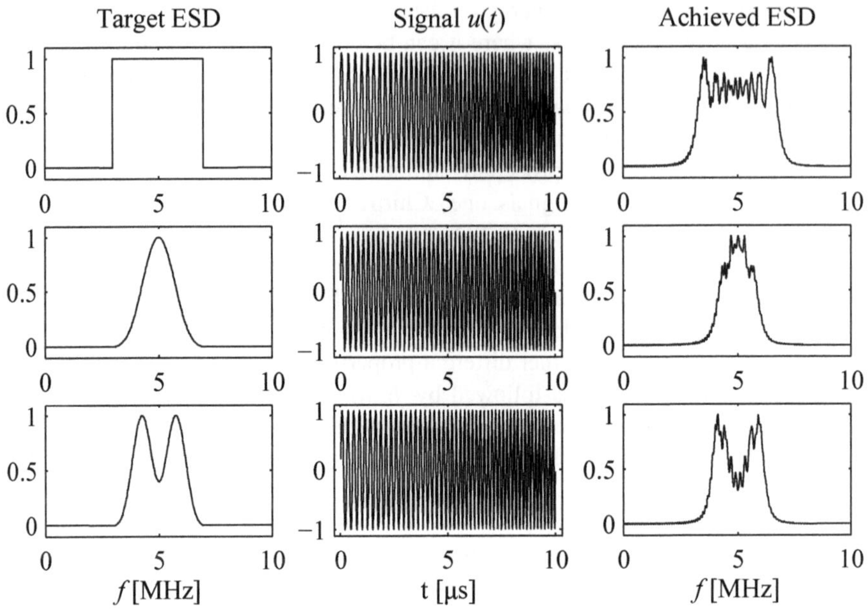

Fig. 4.8 Examples of linear and nonlinear chirp signals obtained from a target ESD following Eqs. (4.18)–(4.20). The *leftmost column* contains plots with three different target spectra. The *center column* shows the corresponding chirp signals and the *right most column* contains their actual spectra

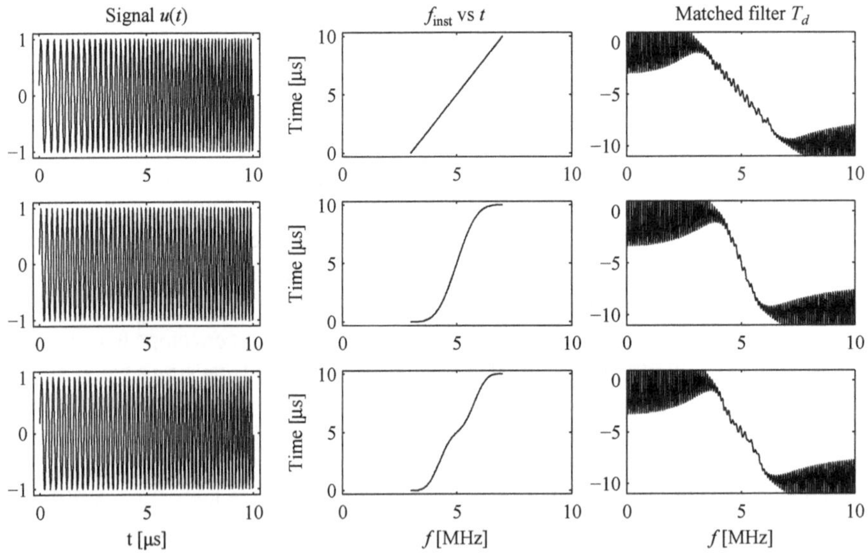

Fig. 4.9 Trajectories in the time–frequency plane of the LChirp (*top row*), an NLChirp defined to have a bell-shaped PSD (*center row*), and an NLChirp having a bimodal ESD (*bottom row*)

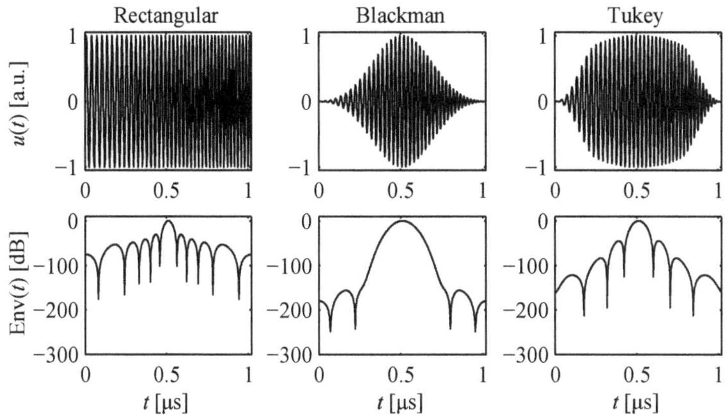

Fig. 4.10 Some sample windows, applied to an LChirp signal. The plots in the *top row* show the windowed signals, while those in the *bottom row* show the *envelope* of the autocorrelation as a function of time

where $S(f)$ is the ESD of the signal under evaluation, f_c is the central frequency and the factor $\sqrt{12}$ is a normalization constant used to achieve $B_{\text{eff}} = B = f_{\text{stop}} - f_{\text{start}}$ in the case of an ideal LChirp with a long duration. As said, there are many different types of window that can be used (see Fig. 4.10 for some examples), nevertheless in the experiment examples reported in Part IV and through this chapter the most frequently adopted window is the Tukey one, that is assumed to assure an optimal tradeoff between side-band reduction, achievable resolution, and SNR [35].

At this point one can wonder if it is better to use a windowed LChirp or instead apply spectral windowing through an NLChirp? Of course, a general rule does not exist since the optimization depends on the specific application. Here, we merely highlight that by using a *windowed* NLChirp, the constant envelope feature is preserved, avoiding the SNR loss and at the same time reducing the tails of $\hat{\delta}(t)$ with respect to an unwindowed LChirp, but if the reduction of the tails is the principal requirement, a *windowed* LChirp guarantees better perfomances.

As an example, Figs. 4.11 and 4.12 compare the effect of time-windowing an LChirp and of spectral-shaping an NLChirp by appropriately choosing the modulation law (which is substantially equivalent to a spectral shaping). The excitations used for the plots sweep in frequency from 150 to 450 kHz, which is a frequency range suitable for air-coupled inspections. Time windowing provides the best reduction in the tails in $\hat{\delta}(t)$, but also exhibits the largest central lobe. On the other hand, spectral shaping ensures a resolution that is almost the same as for the unwindowed LChirp, and provides a significant reduction of the tails in proximity of the main lobe. Far from the main lobe region, the side-band levels almost coincide with those of the *unwindowed* LChirp.

For the sake of completeness, two further applications of the NLChirp are worth mentioning:

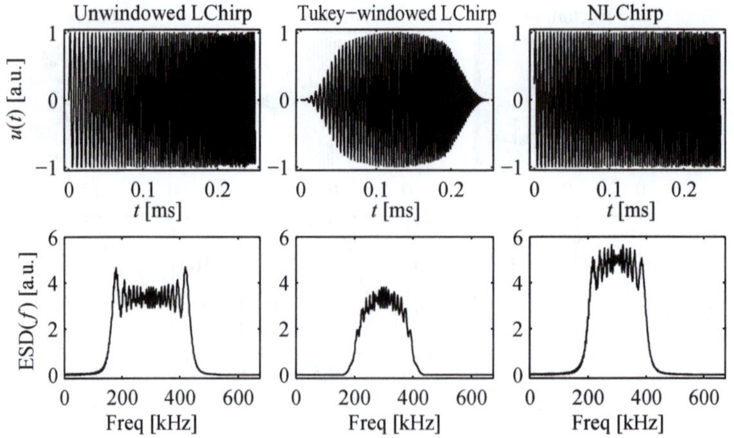

Fig. 4.11 A comparison of time-windowing applied to an LChirp excitation and of spectrum shaping by means of NLChirps. In the *top row*, three chirp signals are shown. *Left to right*: a reference unwindowed (namely, rectangular-windowed LChirp); the same signal after a Tukey time-windowing; and an excitation where spectrum shaping is achieved by changing the LChirp into an NLChirp. *Bottom row*: the corresponding spectra

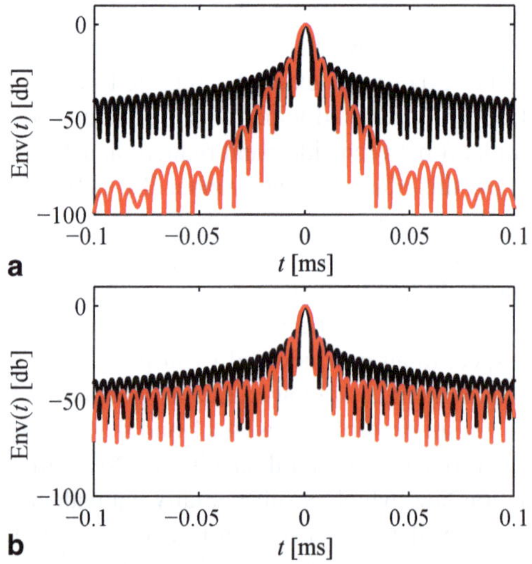

Fig. 4.12 Effect of windowing and spectral shaping techniques on the autocorrelation functions of chirp signals. In both panes, the *black line* corresponds to the *unwindowed* LChirp reported in Fig. 4.11 while the lighter (or red) line shows the autocorrelation when windowing or nonlinear modulations are applied. In the *top plot*, the autocorrelation envelope of the time-windowed LChirp is shown and in the *bottom plot*, the autocorrelation envelope of the NLChirp is shown

- using knowledge of the transfer function of the TX–RX channel to equalize the ESD in order to have the largest effective bandwidth at the cost of wasting some energy;
- using knowledge of the transfer function of the TX–RX channel to match exactly the channel transfer function so optimizing the energy bandwidth product (EBP) of the output signal and assuring an enhancement of the SNR.

Note, however, that in these cases, the measurement resolution might not be optimized. For further details, see [30, 37, 47].

Example 4.3
Generate an LChirp with different windows and evaluate the effective TBP: what is the influence of on the SNR gain with respect to a Gaussian-wavepacket excitation with TBP = 1?

Solution: assume for simplicity that the effective duration T_{eff} and the effective bandwidth correspond to the −6 dB level of the envelope and of the ESD, respectively. With this, the TPB corresponding to different excitations can be recorded from simulation data. Some graphs illustrating the different options are provided in Fig. 4.13.

In summary, chirp signals are extremely flexible and they can thus be tailored for diverse applications. For this reason, they are the most widely used signals for PuC applications, not just in ultrasonic nondestructive evaluation (NDE) but also in radar, optics, eddy-current, and thermography NDT [2, 10, 24, 31, 32, 46].

4.5.2 Pseudo-Noise Sequence-Based Pulse Compression

An approach to PuC alternative to the use of FM waveforms is based on the adoption of binary sequences as excitations. Binary sequences, and in particular PN codes, play an important role in communications, where they are used in *spread spectrum* techniques and error-correction schemes, having been extensively studied since the 1960s. A multitude of different PN sequences exist, each designed to have special mathematical characteristics; here, we focus on only two types of them, namely GCSs and MLSs (or m-sequences), which are known for providing the δ-like autocorrelation property at the heart of the PuC procedure, see Eq. 4.12. As shown in the following pages, PN codes are less flexible with respect to chirp signals, especially in spectrum shaping, and in general exhibit limitations or extra-requirements for the experimental implementation. Nonetheless, they can achieve very high performance, in fact, in some cases better than chirps. The application complexity mainly derives from the fact that the signal takes only two values, which imposes strong constraints in the code design. On the other hand, the binary nature allows a more immediate processing of the sequences by digital circuits. Indeed, for both GCSs and MLSs, efficient hardware schemes exist to generate them and perform correlation.

GCSs consist of couples of sequences, say $u_A(n)$ and $u_B(n)$ with the peculiar property that their out-of-phase aperiodic and periodic autocorrelation coefficients

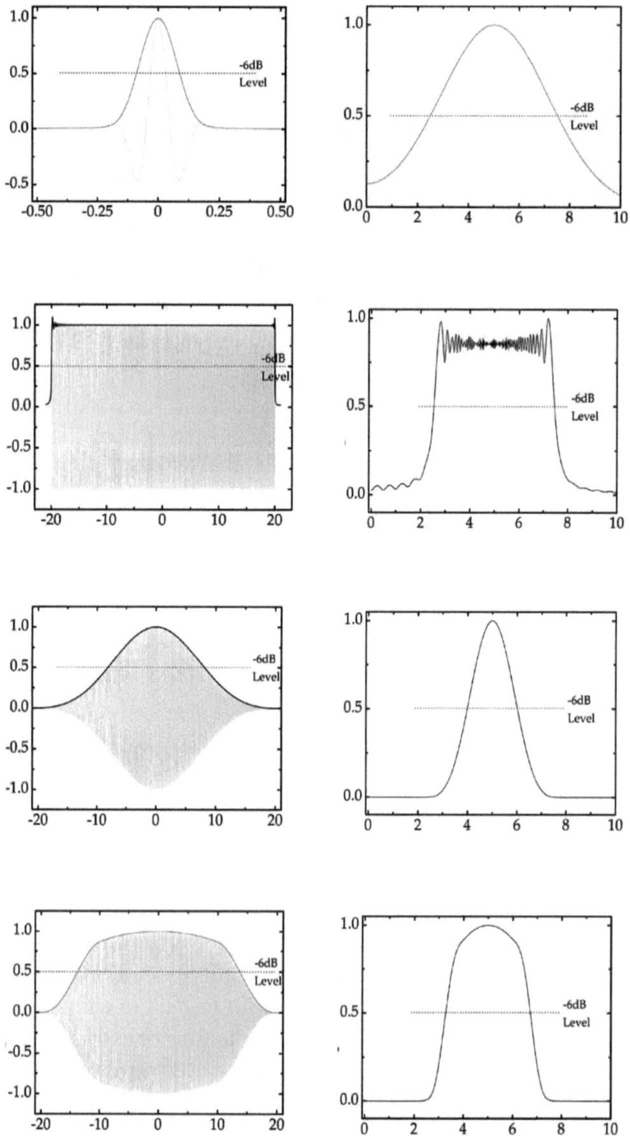

Fig. 4.13 Graphical solution to the problem in Example 4.3 calculated for signals having f_c =5 MHz and B =5 MHz. In the *first row*, Gaussian RF wavepacket having $TBP = 1$ (*left*, both the signals and their envelope are reported) and its corresponding spectrum (*right*) are shown. In the other rows, chirps with various windowing functions and duration $T = 40\ \mu s$ are shown. In particular, rows 2, 3, and 4 depict an LChirp and the corresponding spectra for a rectangular, a Blackman, and a Tukey-Elliptical window, respectively

sum to zero:

$$\begin{aligned} \hat{\delta}_{u_A}(n) &= -\hat{\delta}_{u_B}(n)\ \forall n \neq 0 \\ \hat{\delta}_{u_A}(n) + \hat{\delta}_{u_B}(n) &= 2L\,\delta(n); \end{aligned} \tag{4.23}$$

where L is the sequence length. GCSs were first introduced by Marcel J. E. Golay in 1949 for infrared spectroscopy, consisting in a family of sequences with $L = 2^M$ generated recursively from two seeds (see Example 4.4). Some years later the same author showed that complementary sequences of lengths 10 and 26 also exist [18] while now it is known that the most general GCSs pairs have length $L = 2^M 10^K 26^P$ (where M, K and P are integer coefficients). The complementary property of the autocorrelation can be usefully exploited in PuC by taking advantage of the linear hypothesis at the heart of the method. Operation is as follows:

1. the two complementary sequences are both used as input of the LTI system under evaluation (in succession);
2. the matched filter is applied separately to each output to obtain the two estimates $h_{u_A}(n) = [\hat{\delta}_{u_A} * h](n)$ and $h_{u_B}(n) = [\hat{\delta}_{u_B} * h](n)$;
3. the two filtered outputs $\hat{h}_{u_A}(n)$ and $\hat{h}_{u_B}(n)$ are summed to obtain the ideal impulse response apart for a multiplicative factor $2\ L \times h(n) = (\hat{h}_{u_A}(n) + \hat{h}_{u_B}(n))$.

GCS are renowned for being the only code that theoretically allows a perfect reconstruction of the impulse response at the end of the PuC procedure. In practice, this is never the case, as a "sequence" needs to undergo some changes to be applied to a real world system (some of which, such as hold operations are detailed in the following). Yet, their performance can be quite good and this is why they are popular in ultrasonic measurements [16, 26, 51] even if it is necessary to use two distinct sequences as input in succession. Moreover, as stated above, the complementary feature holds both for APC and CPC cases, so one can decide to choose any of the two approaches depending on the specific application.

As a further important feature of GCSs, it is worth mentioning that a fast algorithm, the so-called *fast Golay correlation*, exists to perform correlation–convolution between a generic signal and a GCS pair with length L. This algorithm exhibits a computational complexity similar to the FFT, while being less demanding in terms of its elementary operations. In fact, due to the binary feature of GCSs, it mainly requires sums and subtractions instead of multiplications [5]. The existance of this algorithm opened the way to the hardware implementation of GCSs PuC procedures for ultrasonic NDT applications [21, 36].

Example 4.4 *(Generation of Golay Complementary Sequences)*
The easiest way to define a GCSs pair is to consider a family of GCSs whose length is a power of 2, namely $L = 2^M$. Such family can be generated recursively starting from the two seed sequences $GCS_A^0(n) = (+1)$, $GCS_B^0(n) = (+1)$

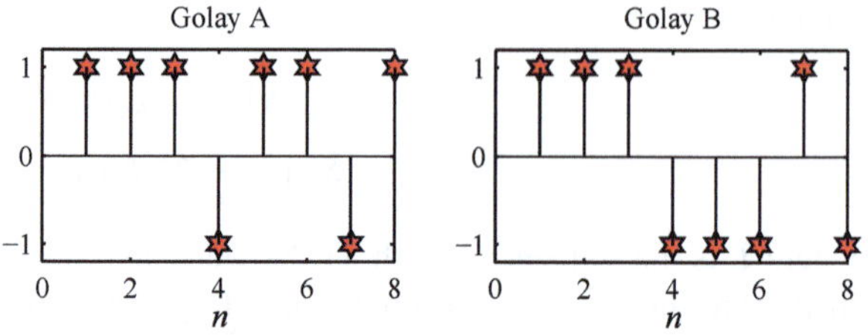

Fig. 4.14 Example of GCSs of order $M = 3$

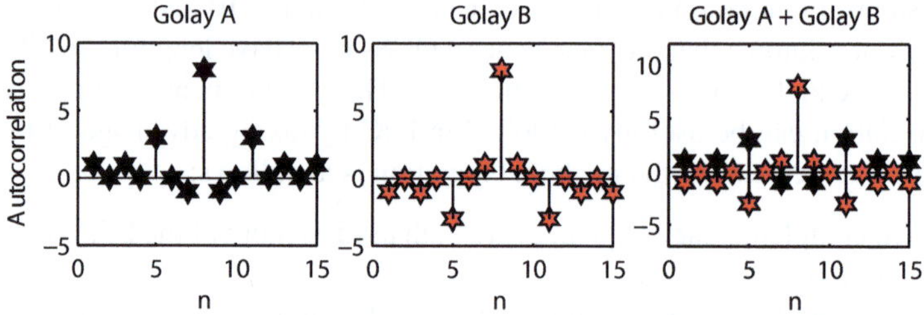

Fig. 4.15 Complementary property of GCSs autocorrelation in the acyclic case for $M = 3$

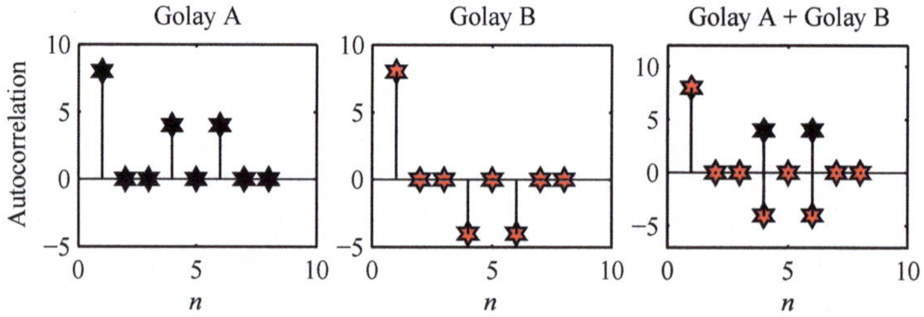

Fig. 4.16 Complementary property of GCSs autocorrelation in the cyclic case for $M = 3$

and by following the rule

$$
\begin{aligned}
GCS_A^1 &= (+\,GCS_A^0, +GCS_B^0) = (+\,1, +1) \\
GCS_B^1 &= (+\,GCS_A^0, -GCS_B^0) = (+\,1, -1) \\
GCS_A^2 &= (+\,GCS_A^1, +GCS_B^1) = (+\,1, +1, +1, -1) \\
GCS_B^2 &= (+\,GCS_A^1, -GCS_B^1) = (+\,1, +1, -1, +1) \\
&\vdots \\
GCS_A^M &= (+\,GCS_A^{M-1}, +GCS_B^{M-1}) \\
GCS_B^M &= (+\,GCS_A^{M-1}, -GCS_B^{M-1})
\end{aligned}
\tag{4.24}
$$

An example of GCSs is given in Fig. 4.14 for $M = 3$, while in Figs. 4.15 and 4.16, the complementary properties of GCSs autocorrelation are graphically represented for the acyclic and the cyclic case, respectively.

Therefore, GCSs represent a valid alternative to chirp excitation in PuC schemes when the use of two distinct excitation waveforms does not hamper the procedure. One can now wonder if there are other PN codes that can overcome this limitation. Fortunately, with some precaution that will be explained below, the answer is positive and a significant case is represented by MLSs, whose main differences with respect to GCSs are here highlighted. MLSs are periodic sequences generated by a linear feedback shift register (LFSR) (see Example 4.5.2). As illustrated by Golomb [19], they exhibit the maximum period length for any given register depth. In particular for a register with N cells, the period of the MLS is $L = 2^{N-1}$. The configuration of the LFSR machine is derived by the theory of *Galois finite fields* and assures several useful mathematical properties to these codes. For ultrasonic PuC applications, the most useful one is related to the normalized cyclic autocorrelation function, which ends up as a train of unit pulses superimposed with a dc bias whose amplitude decreases with an inverse proportionality to L. Within a period:

$$[u_{\mathrm{MLS}} * \mathrm{Rev}[u_{\mathrm{MLS}}]](n) = \delta(n) - \frac{1}{L}. \tag{4.25}$$

If $h(n)$ can be assumed to be zero mean (where $h(n)$, in the ultrasonic NDT application is $h_{\mathrm{ch}}(n)$, also taking into account the transfer function of the transducers), and the considerations about the duration of $h(n)$ are satisfied, then it is straightforward to see that by adopting the cyclic operation a perfect estimate of $h(n)$ can be retrieved. Of course, with respect to this statement, the usual care is necessary due to the already mentioned nonidealities that a real system can exhibit in dealing with sequences. Still, performance can be quite good, and the previous hypotheses apply in almost all cases, so that MLS can be fully exploited in CPC applications. Indeed, after the work of Golomb [44], MLS-based impulse response schemes were successfully introduced into acoustic measurements. Furthermore, even in the most general case, $h(n)$ can be perfectly retrieved by adopting the cyclic protocol as long as specific processing [53] is adopted or by slightly modifying the amplitude of the MLS in order to attain a "perfect sequence" [29]. MLSs can thus be performance-equivalent to GCSs if CPC is possible, and in fact, they may be preferred as only one sequence is necessary. Moreover, as for GCSs, there exists a fast algorithm to perform cyclic correlation with MLSs. This is derived from the Hadamard Transform and is commonly indicated as the *Fast-m-Transform* [12]. Note that, if APC has to be used, MLS methods are not suitable, since consistent noise appears in the aperiodic autocorrelation function and therefore GCSs are highly preferable. A further advantage of MLSs with respect to GCSs, is provided by the so called decimation property and consists in the possibility to undersample the output signal in CPC by some factor $2k$ attaining exactly the original signal after $2k$ periods. This enables the use of ADCs with sampling rates equal or lower than the signal bandwidth itself [41]. Also note that MLSs can be used in multi TX systems by exploiting their pseudo-orthogonality [42, 54] (as shown in the next chapter). It should also be noted that Legendre sequences [44] have very similar properties to MLSs. However, they cannot be generated recursively, and a fast algorithm for their correlation does not exist. Hence, the MLS approach has historically been preferred. Moreover, it has been recently shown that both GCSs and MLSs make PuC schemes extremely robust against the quantization noise introduced by ADCs. Indeed, they both allow the measurement of impulse responses with dynamics well below the quantization step [9, 39].

Example 4.5 *(Generation of Maximum Length Sequences)*
MLSs can be generated by means of a LFSR with M delay taps. Indeed each MLS, $\{u_{MLS}(n) : n \in [1, L], u_{MLS}(n) \in \{-1, 1\}\}$ is associated to a recursive equation such as:

$$u_{MLS}(n + M) = \prod_{i=0}^{M-1} u_{MLS}[n + i]^{\gamma_i} \tag{4.26}$$

where the $\gamma_i \in \{0, 1\}$ are the coefficients of a primitive polynomial on the Galois field GF(2^M). The value M is called *order of the sequence* and for any order M, a different primitive polynomial can be adopted. Each polynomial leads to a distinct sequence. The properties of MLSs have been formulated by Golomb in [19]. The most appealing ones for PuC applications are: (i) the cyclic autocorrelation function of an MLS is a very close approximation to a delta function. Within each period, $\Phi^{\text{cyc}}_{u_{MLS},u_{MLS}}(n) = 2^M\delta(n) - \theta(n)$, where $\theta(n)$ is a small error; (ii) their power density spectrum is flat, with the exception of a near-zero DC term; (iii) a variant of the FFT, called Fast m-Transform and here indicated as FMT, exists to help in the computation of the correlation between an MLS and a signal $y(n)$: $\Phi_{u,y}(n) = \text{FMT}_u[y](n)$ [12, 42].

The name *maximum length sequences* refers to the fact that, for a generic recursive equation such as Eq. (4.26), the generated sequence is always periodic with a period $L \leq 2^M - 1$. MLSs saturate this condition. An example of MLS is given in Fig. 4.17 for $M = 3$ (left), while the other plots represent the MLS autocorrelation for the acyclic and the cyclic case, respectively. It can be seen that in the acyclic case, MLSs are affected by a large noise component, which is inherent and due to mathematical reasons.

For both GCSs and MLSs, it is important to stress that the aforementioned autocorrelation properties hold for numerical sequences but, when changing sequences of numbers into physical signals (that are continuous time), some aspects should be

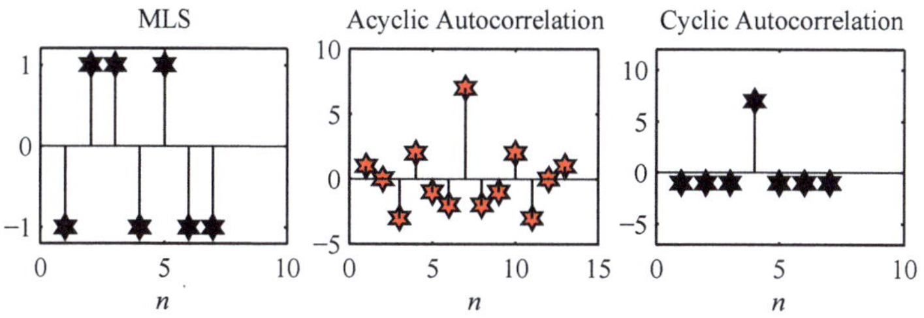

Fig. 4.17 Example of MLS u_{MLS} of order $M = 3$ corresponding to the primitive polynomial $p(x) = x^3 + x + 1$

considered. In particular, each $+1$ or -1 of the codes, in the domain of physical systems, is rendered by a 0-hold operation into a signal step with duration T_P, the latter being the inverse of the sequence generation rate f_{gen}. In other words, a discrete-to-continuous time transformation is applied to the excitation. On the other side of the signal processing chain, operations such as correlation are conveniently practiced in the digital domain. Hence, acquisition of the RX output is practiced by an ADC. By sampling the signal, the ADC brings it back to a discrete-time representation. However, the rate ADC rate f_{ADC} is not necessarily the same as f_{gen}. To correctly apply the PuC approach, f_{ADC} must be a multiple of f_{gen}. Namely, $f_{\text{ADC}} = m f_{\text{gen}}$ with $m \in \mathbb{N}$. Furthermore, if $m \neq 1$ some preprocessing is needed in order to exploit the fast correlation algorithms.

In practice, it is quite common to have $f_{\text{ADC}} > f_{\text{gen}}$. In fact, f_{gen} determines the actual spectrum of the excitation, but f_{ADC} determines how the details of the output signals are captured. Strict requirements may exist in order to avoid aliasing effects, errors in the signal reconstruction or the need for complex analog filters before acquisition. On the other hand, the sequence-based excitations have power spectra close to that of a boxcar pulse of width T_P, so that it is better to choose f_{gen} in order to concentrate the energy in the actual pass-band of the system. Some sample spectra are illustrated in Fig. 4.18. The different requirements placed on f_{gen} and f_{ADC} often make it convenient to select different values for them.

In the following, it will be shown that a good criterion is to take $2f_c < f_{\text{gen}} < 4f_c$. At the same time, as a general rule f_{ADC} should be at least 5–10 times greater than the upper frequency of interest. Of course, it is possible to choose $f_{\text{gen}} = f_{\text{ADC}}$ to consider the input signal as the ideal numeric sequences, but at the cost of exciting a bandwidth larger than the available one. This could waste a lot of energy, due to the band-pass behavior of a typical ultrasonic experimental system.

In contrast to this, chirp-based PuC schemes, even when implemented for generating the excitation in a digital realm, allow f_{gen} and f_{ADC} to be chosen without any reciprocal constrain, provided that the matched filter operates at f_{ADC}.

From an experimental point of view, in most cases, the transducers have a bandwidth that is just a fraction of the central frequency f_c. Moreover, the transfer function of the probes (often quite irregular) is generally better approximated by a bell-shape than by a rectangular function. The spectra of binary sequences thus do not fit very

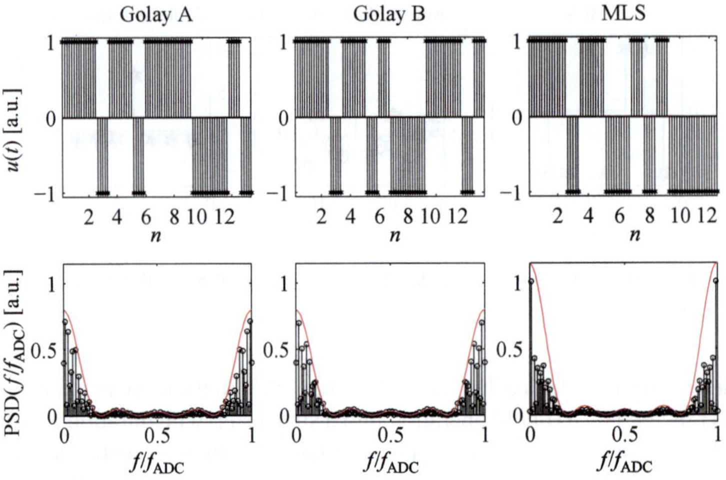

Fig. 4.18 Sample spectra of GCS and MLS waveforms of order $M = 4$, acquired with $f_{ADC}/f_{gen} = 5$. The spectra of the sequences resemble quite closely that of a boxcar pulse. *Top row*: some sample signals. *Bottom row*: the corresponding spectra (with the abscissa reporting frequency normalized over f_{ADC}

well with the frequency range of interest, causing a decrease of the effective amount of energy delivered to the system. For this reason, in order to increase the energy transfer ratio, it has been proposed in literature to modulate the GCSs or MLSs sequences by replacing the $+1$ and -1 values with short sequences of binary values. This enables a limited spectral shaping [33]. The easiest solution is to replace each $+1$ with the subsequence $(+1, -1)$ and each -1 with the opposite subsequence $(-1, +1)$, so achieving a band-pass behavior. This technique is largely adopted in acoustics where it is commonly referred as inverse repeated sequences (IRS) [45]. An example of sequences obtained by this approach is provided in Fig. 4.19.

Figure 4.20 shows some power spectra obtained by the IRS technique to be compared with those in Fig. 4.18. From the comparison, it is evident that IRS can result in signal spectra matching much more effectively the transfer function of a typical ultrasound transducer.

In the following pages, the various signals introduced so far will be compared for use in practical NDT. Particular attention will be given to high-attenuation setups such as air-coupled ultrasound. Performance will be determined for various levels of noise, first numerically—by considering measurement systems with different operation bandwidths—and then experimentally, so that their suitability in NDT can be assessed.

4.6 Merit Factors

As it should be clear from the previous sections, the success in practicing the PuC approach depends on the excitation signal being used. Furthermore, even for those types of excitation that are most successful, after the compression it is never possible

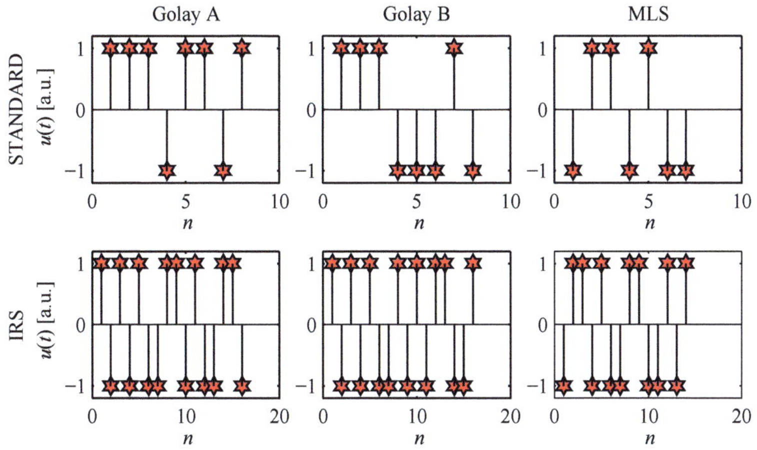

Fig. 4.19 Examples of PN sequences useful for PuC applications derived from GCSs and MLSs. *Top row*: from *left to right*, two standard GCSs forming a complementary pair and one standard MLS. *Bottom row*: the same codes, after the application of the IRS technique. The number of samples of the sequences obtained by IRSs is twice as large as that of the original codes

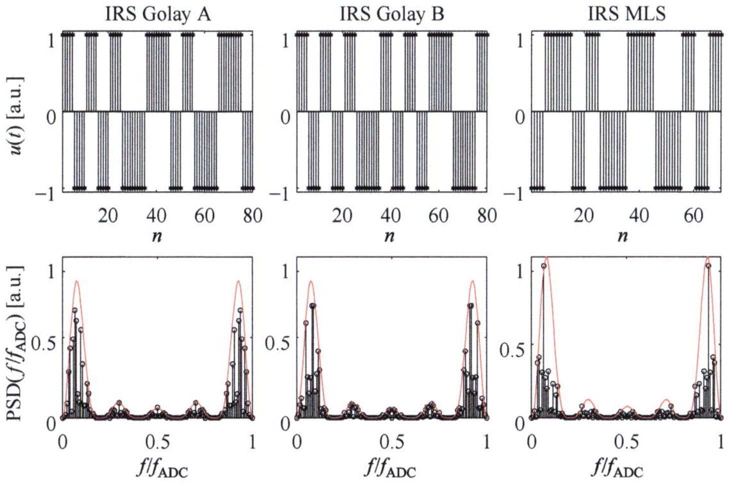

Fig. 4.20 Examples of IRS waveforms and relative spectra. The waveforms are derived from GCSs and MLSs of order $M = 4$, after acquisition with $f_{ADC}/f_{gen} = 5$. In the spectra of IRS-based sequences, the DC term is null and the energy is centered around the center frequency f_c in the pass-band of the TX–RX channel

to concentrate the whole of the excitation energy at a single instant in time. Some energy will always *leak* along a time interval as wide as $2\,T$, centered around the instant where most of the energy is placed. This is true also of excitations based on PN codes that may give perfect compression only as long as they are observed in the discrete time at their own sample rate. As soon as one moves to continuous time

Fig. 4.21 The SUT used for Example 4.6. The figure illustrates the localization of the defect and the probe placement. The *red line* superimposed on the image is meant to give a rough illustration of the signal amplitudes involved in the excitation and in the reflections at the SUT boundaries and at the defect

or samples them at an ADC rate that is a real multiple of the code rate, one can get $\hat{\delta}(t) \neq 0$ for $t \neq 0$ or $\hat{\delta}(n) \neq 0$ for $n \neq 0$.

Typically, this leakage manifests with an autocorrelation function $\hat{\delta}(t)$ that has a strong oscillatory behavior. If one looks at the instantaneous power at the output of the matched filter, this results in a central peaking lobe (commonly referred to as the *main lobe*) that rapidly decays to zero, surrounded by much lower lobes (commonly indicated as *side lobes*) that altogether create two *tails* around the main lobe.

This structure can create problems when there are multiple propagation paths to be detected. As an example, and for the sake of simplicity, suppose that there are only two paths. When the matched filter is applied to detect the corresponding templates in the received signal, it is clearly impossible to recognize the existence of two distinguished templates unless $|\Delta t_1 - \Delta t_2|$ is larger than the width of main side lobe. Hence, it should be clear that the main side lobe determines the available resolution. Furthermore, even if the two paths have delays that are sufficiently different to resolve them, the *tails* of $\hat{\delta}(t)$ can still represent a problem. In fact, the peaks at the output of the matched filter always need to be identified while they are embedded in noise. As long as $|\Delta t_1 - \Delta t_2| > T$, this noise is mainly given by $e_n(t)$, as already discussed in Sect. 4.3. However, when $|\Delta t_1 - \Delta t_2| < T$ each peak also gets embedded in a noisy component given by the PuC tails of the other peak. This situation is illustrated in Example 4.6.

Example 4.6 *(Detection of a Small Defect in Proximity of a SUT Boundary)* To illustrate the importance to have a fast decay of the sidelobes, we report the example of the detection of a small defect in a cylindrical piece of steel with height of ≈ 30 cm and radius ≈ 6 cm. The defect has a radius of ≈ 1 mm and it is artificially placed near one of the plain surfaces of the SUT, as sketched in Fig. 4.21.

When inspecting the SUT from the farther surface with respect to the defect, the signal from the back-wall echo is expected to be significantly larger than that from the defect (approximately by a factor equal to the ratio of the respective surfaces). If the sidelobes of $\hat{\delta}$ are not low enough, the echo from the defect can be completely embedded in the noise. Figure 4.22 illustrates this aspect by comparing how different windows applied to LChirp signals affect the sidelobes.

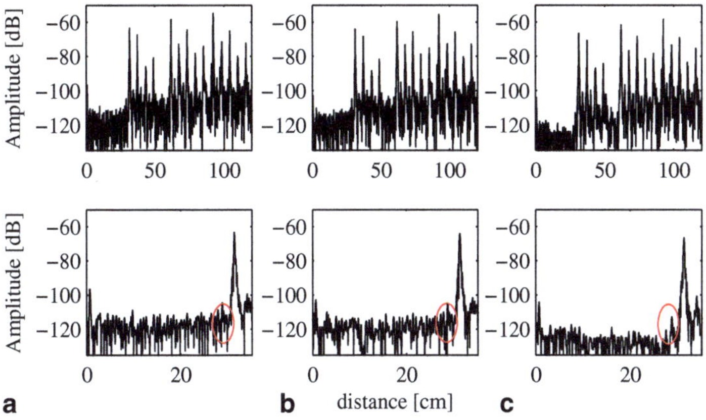

Fig. 4.22 Typical $h_u(n)$ envelopes reconstructed with PuC for the steel SUT used for Example 4.6 and illustrated in Fig. 4.21. The plots in the *top row* refer to the received signals, while those in the *bottom row* illustrate the signals after the compression (matched) filter. In all the plots, the abscissa is given in centimeter due to the fact that range can be evaluated by knowledge of the ultrasonic wave propagation speed in steel. The plots in column **a** refer to the case of an *unwindowed* LChirp; the plots in column **b** refer to the case of an *unwindowed* LChirp in excitation filtered by a matched filter consisting of a *windowed* LChirp. Finally, plots in column **c** refer to the case of *windowed* LChirp used both in excitation and in detection. The adopted window is the Tukey-Elliptical one. The main peak in the plots in the *bottom row* is given by the reflection on the SUT boundary opposite to the probes. The *smaller peak*, identified by the *red circle*, is the one relative to reflection on the defect. The figure clearly illustrates that this peak is only distinguishable in case (c), where the windowing provides a sufficient attenuation of the sidelobes

In practice, whenever two paths have time-of-flights (ToFs) that are sufficiently similar, there can be a self-interference issue in their detection. Since different excitation signals can provide different resolutions and can lead to self interference to different extents, it is important to be able to quantify these differences. Having appropriate *merit factors* is the key to be able to *design* good excitations and to make an informed choice between them when dealing with the requirements of specific applications.

4.6.1 Performance Quantification Issues

Quantifying the quality of an excitation is not an easy task. As just noticed, the main way in which the excitation affects system performance (as long as the matched filter and PuC are involved) is via its autocorrelation function. This is a mathematical object with *infinitely many dimensions*. Summarizing them in a few scalars is obviously a challenge.

Second, important as it is, autocorrelation cannot still be representative of the whole picture. Other aspects necessarily exist. For instance, some hints have already been given to the fact that excitations should be tuned to the transducers properties, in order to avoid wasting part of their energy (or even overstressing the TX). This matter will be specifically addressed in Sect. 4.7.

Finally, in defining merit factors, one would like to abstract as much as possible from specific applications. While it may be relatively easy to define a performance index capable of expressing the excitation quality with respect to a single ultrasound NDT problem and setup, the value of such index would be modest if merely changing some detail could make it invalid or misleading. Conversely, introducing performance indexes that can retain validity across a wide range of applications is generally a complex operation. Ideally, it should involve the definition of application independent quantities that can be transformed into application-specific ones by combining them with indicators summarizing the application traits and features.

With these premises, it now possible to underline that the current state of matters with respect to excitation related merit factors is in some senses suboptimal. Currently, two main options are available. One of them is to rely on so to say *classic* quantities that have a clear and rather direct physical meaning. In many cases, these are either derived from or common to sibling disciplines, such as *radar*. As further positive points, they tend to be well established, well understood, and sufficiently easy to derive. On the negative side, these indicators may fail at fully capturing all the aspects of excitation quality and may be better suited for some measurement setups than for others. One of the reasons why it is so is that these classic merit factors tend to be ultimately based on *local* measures. For instance, they may rely on the value of a certain feature point in the excitation autocorrelation function, or on the distance between two feature points. Conversely, the qualities that make one excitation better or worse than another often appear *spread in time* in the autocorrelation function and particularly along its tails.

The second option is the proposal of new figures of merit. While these may be more information-rich than classical ones, they may have disadvantages of their own. First of all, not being standard nor established, they may complicate comparisons to other researchers results. Second, they may appear much more *indirect* than classic merit factors. In other words, they may be more complex to derive and require a more sophisticated mathematical framework for their computation. Most important, their relationship with physical features of the signals or systems under exam may be less evident. Consequently, working with them may end up being more difficult, as the designer can rely less on his expertise and sensitivity for their evaluation.

In the following, both options are illustrated. In Sect. 4.6.2, some conventional merit factors are reviewed. Then, in Sect. 4.6.3, a very recent proposal is described. This will be particularly useful in view of next Chap. 5 where systems deploying multiple TXs are described. In fact, the proposal is principally meant at providing a unified excitation assessment framework for single and multi-TX ultrasound NDT systems.

4.6.2 Conventional Merit Factors

In order to quantify the quality of a PuC set-up, the first step is to verify the *compression* performance of the chosen excitation signal, taking also into account the characteristics of the TX and RX. Therefore, the attention can be focused on the quantity:

$$\delta_{ex} = [u * \Psi * h_T * h_R](t) = [\hat{\delta} * h_T * h_R]. \tag{4.27}$$

If δ_{ex} can maintain a good fidelity with respect to the ideal pulse, then it is possible to estimate with a good approximation the SUT impulse response h_u. Four main quantities are often computed in order to characterize δ_{ex} with respect to detection and resolution. Clearly, to gather information applicable to real world problems, in the definition of these quantities, the presence of noise must be considered. The above mentioned merit factors are:

(a) the main lobe width (MLW);
(b) the SNR at the main lobe;
(c) the near sidelobe level (NSL);
(d) the far sidelobe level (FSL).

Clearly, MLW is related to the range resolution, while the other three quantities are related to the ability to detect defects, in the presence of either inherent system noise (quantified via SNR) or noise that is self-generated due to the presence of multiple propagation paths (indirectly quantified via NLS and FLS).

Furthermore, note that each quantity is computed with reference to a distinct region of δ_{ex}. To gain deeper insight, consider the three regions labeled in Fig. 4.23 by the capital letters A,B,C:

A indicates the main lobe, that is the region outside of which the envelope of the δ_{ex} is 20 dB below its maximum;
B indicates the near sidelobes, i.e., two regions at the sides of the main lobe, each with the same width as the main lobe;
C indicates the far sidelobes, i.e., the outer regions beyond the near sidelobes.

As mentioned, MLW provides a measure of the resolution of the PuC procedure, and can be formally defined as:

$$\mathrm{MLW} = 2\min\left\{ \bar{n} \in \mathbb{N} \text{ s.t. } \sup_{\substack{n > n_p + \bar{n} \vee \\ n < n_p - \bar{n}}} \delta_{ex}(n) < 0.1\, \delta_{ex}(n_p) \right\} \tag{4.28}$$

where n_p is the position where $\delta_{ex}(n)$ peaks, namely

$$n_p = \arg\max_{n \in \mathbb{N}} \delta_{ex}(n). \tag{4.29}$$

Obviously, the smaller MLW, the better the temporal-spatial resolution of the measurement. Consequently, this parameter is of particular interest in ToF measurement or in range detection.

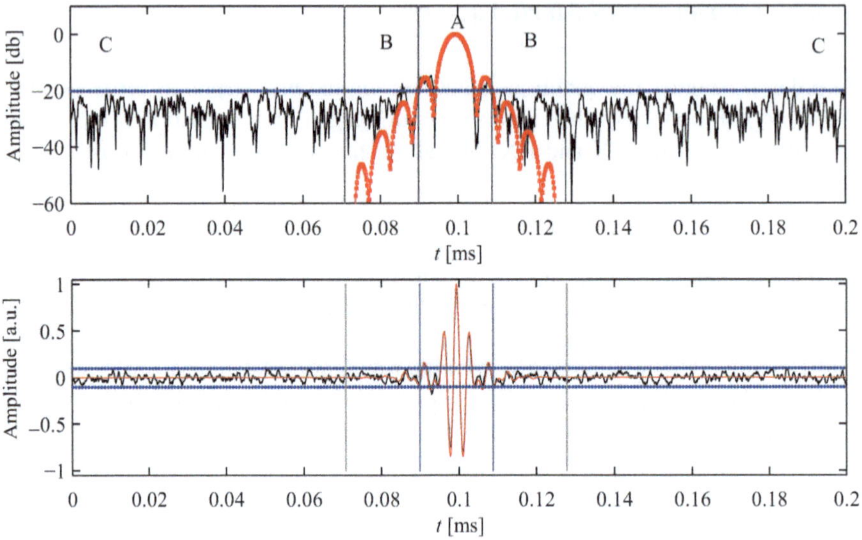

Fig. 4.23 Separation of feature-regions in δ_{ex}. *Top plot*: the δ_{ex} envelope in dB. *Bottom plot*: the actual δ_{ex}. The envelope view makes the regions containing the signal main lobe *A*, near side lobes *B* and far sidelobes *C* quite evident to spot. In the plots, a measured δ_{ex} obtained with a chirp excitation (*red*) is compared with one obtained in a very noisy environment to illustrate the importance of very low sidelobes (*black*)

SNR related to the main lobe expresses the capability of detecting a signal in presence of noise and is defined as the ratio between the energy of the expected impulse response after PuC without noise, to the energy of the noise inside the main lobe. The numerator of this ratio is in fact the energy of $\delta_{ex}(n)$ itself. Thus:

$$\mathrm{SNR_{ML}} = \frac{\sum\limits_{\substack{n\in\\ \text{MainLobe}}} (\delta_{ex}(n))^2}{\sum\limits_{\substack{n\in\\ \text{MainLobe}}} (\delta_{ex}(n) - \delta_m(n))^2} \tag{4.30}$$

where $\delta_m[n]$ is the measured impulse response of the TX–RX system in presence of a noise component $e_n(n)$.

NSL is defined as the ratio (in dB) between the maximum absolute amplitude of $\delta_{ex}(n)$ and the mean value of the envelope of $\delta_m(n)$ in the near sidelobe region, given by

$$\mathrm{NSL} = \frac{\max\limits_{n\in\mathbb{N}}(\delta_{ex}(n))}{\operatorname*{mean}\limits_{\substack{n\in\\ \text{NearSidelobes}}}(\mathrm{env}[\delta_m](n))} \tag{4.31}$$

where the operator "env" extracts the signal envelope.

Analogously, FSL is defined as the ratio (in dB) between the maximum absolute amplitude of $\delta_{ex}(n)$ and the maximum value of the envelope of $\delta_m(n)$ in the far

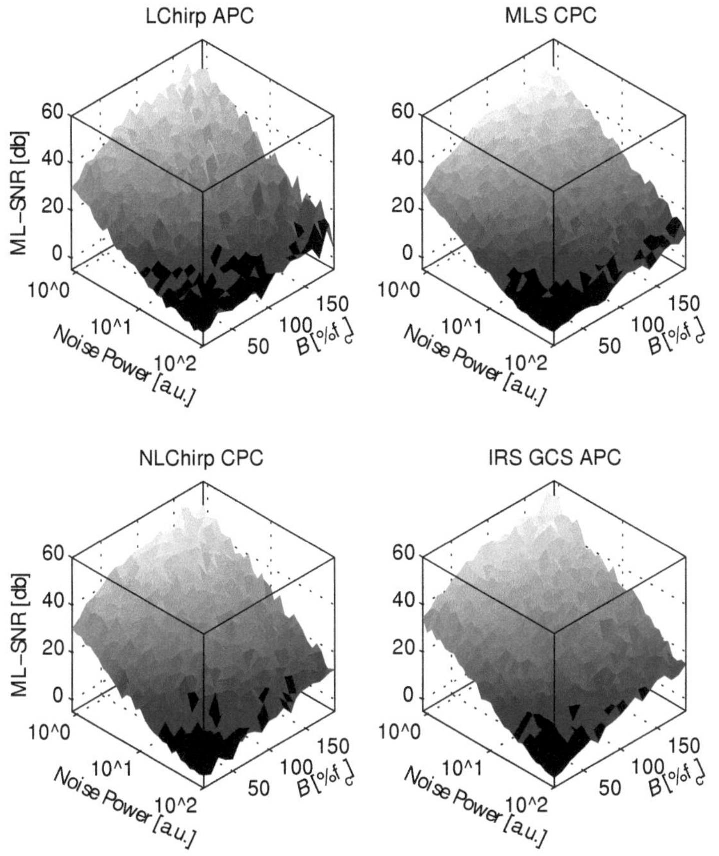

Fig. 4.24 Trends of the main lobe SNR for various excitation waveforms and for different noise energy and TX–RX system bandwidth (expressed in percentage of f_c). The noise is normalized to the energy of the signal and the LChirp signal bandwidth is assumed to match that of the transducers

side-lobes region. Thus

$$\mathrm{FSL} = \frac{\max\limits_{n \in \mathbb{N}}(\delta_{ex}(n))}{\operatorname*{mean}\limits_{\substack{n \in \\ \mathrm{FarSidelobes}}}(\mathrm{env}[\delta_m](n))}. \tag{4.32}$$

NSL and FSL quantify the level of noise and at the same time express a limit value for secondary reflected signals to be detectable. These quantities are of utmost importance when complex impulse responses have to be measured, as in the case of multipath reflection, multilayered structures, etc.

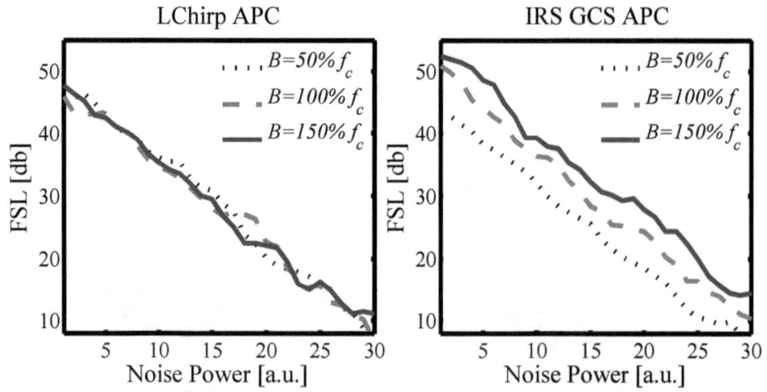

Fig. 4.25 Comparison of the FSL values for an LChirp matching the bandwidth of the TX–RX system and for an IRS-based excitation derived from GCSs for different TX–RX system bandwidth as a function of the noise level. If the TX–RX system bandwidth is large enough ($B > f_c$), the IRS excitation assures the best performances. Otherwise, for a narrowband system, the optimal choice is a narrowband chirp signal

Example 4.7 *(Choice of the Optimal Waveform)*
Once the above merit factors are defined, one can use them to choose the optimal excitation waveform given a certain set-up, i.e., given the transducers transfer function, and some hypothesis on the measurement noise level. Of course no waveform can be a *universal* optimum applicable to all the cases. Conversely, the optimal choice depends on the transducers features and on the noise level. To illustrate this key point, Figs. 4.24 and 4.25 show the value of the main lobe SNR and the FSL, as some system parameters (including the noise level) are varied for a few different excitations. While the trends are similar, the punctual values of the SNR and FSL at different conditions are different for the excitations under exam, with none of them beating all the others for all the merit factors and through the whole of the parameter-region under exam.

4.6.3 *Merit Factors Based on a Probabilistic System View*

In order to introduce an alternative approach to the identification of merit figures, let the expression of signal delivered by the matched filter be recalled

$$y_f(t) = [u * h_T * h_u * h_R * h_f](t) + [e_n * h_f](t). \tag{4.33}$$

Since the aim here is to focus just on the excitation features, it is convenient to remove the effects of $H_T(s)$, $H_R(s)$. This gives

$$y_f(t) = [u * h_u * h_f](t) = \left[u * \left(\sum_{i=1}^{N} h_{u_i} \right) * h_f \right](t) + [e_n * h_f](t) . \qquad (4.34)$$

The previous expression can be rearranged with the aim to focus on the estimation of the ith path. This gives

$$y_f(t) = [u * h_{u_i} * h_f](t) + \left[u * \left(\sum_{\substack{j \in \{1,\ldots,N\} \\ j \neq i}} h_{u_j} \right) * h_f \right](t) + [e_n * h_f](t) . \quad (4.35)$$

Here, the first term is the useful one for the estimation, the last term represents the filtered noise components, while the central term accounts for self interference. In fact, at $t = \Delta t_i$, where one should ideally see just a peak as high as $\alpha_i R_{uu}(0)$, one actually gets

$$y_f(\Delta t_i) = \alpha_i R_{uu}(0) + \sum_{\substack{j \in \{1,\ldots,N\} \\ j \neq i}} \alpha_j R_{uu}(\Delta t_j - \Delta t_i) + e_f(\Delta t_i) . \qquad (4.36)$$

In plain words, there is an interfering component from the generic path j onto path i given by $\alpha_j R_{uu}(\Delta t_j - \Delta t_i)$. If $|\Delta t_j - \Delta t_i| > T$ this component is null, but otherwise it is not. In order to better perceive the amount of interference, it is convenient to reason in terms of signal power. The useful ith component has power

$$\alpha_i^2 R_{uu}^2(0) \qquad (4.37)$$

while the jth interfering component has power

$$\alpha_j^2 R_{uu}^2(\Delta t_j - \Delta t_i) . \qquad (4.38)$$

Being an autocorrelation, $R_{uu}(\tau)$ must clearly have maximum modulus at $\tau = 0$. Then, its modulus must necessarily decay as $|\tau|$ grows larger. In order to minimize the interference, one evidently wants the autocorrelation modulus to decay *rapidly*. Incidentally, this is the reason why conventional merit factors include both NLS and FLS, since jointly they provide an idea of the autocorrelation decay profile.

In this section, another route is taken to evaluate the decay. Temporarily ignoring he power of $e_f(t)$, the SNR due to self interference from path j to path i is evidently

$$\mathrm{SNR}_{i,j} = \frac{\alpha_i^2}{\alpha_j^2} \frac{R_{uu}^2(0)}{R_{uu}^2(\Delta t_j - \Delta t_i)} . \qquad (4.39)$$

If this quantity falls below 1 (0 dB), the detection of path i can be completely hindered by the interference of path j. Unfortunately, such event may happen. In fact, even if $R_{uu}^2(\Delta t_j - \Delta t_i)$ is certainly smaller than $R_{uu}^2(0)$, also α_i can be smaller than α_j.

To be sure that path i is not hidden by the interference from path j, one needs

$$\frac{\alpha_i}{\alpha_j} \frac{R_{uu}(0)}{\left|R_{uu}(\Delta t_j - \Delta t_i)\right|} > 1 . \tag{4.40}$$

In order to actually use the previous inequality, one should preliminary consider what are the actual values that α_i/α_j can take. In fact, it makes sense to assume that the specific ultrasound NDT system can set some bounds for the range of interest of α_i/α_j. On one hand, there will certainly be a minimum level of attenuation such that no path can be attenuated less than that. If the notation α_M is introduced for it, one gets $\alpha_i < \alpha_M, \alpha_j < \alpha_M$. Second, it has already been noticed that there must be an α_t such that all paths that get an attenuation worse than α_t can be ignored. This gives $\alpha_i > \alpha_t, \alpha_j > \alpha_t$. Altogether, one has

$$\frac{\alpha_t}{\alpha_M} < \frac{\alpha_i}{\alpha_j} < \frac{\alpha_M}{\alpha_t} . \tag{4.41}$$

Since the inequality in (4.40) is certainly satisfied for $\alpha_i/\alpha_j > 1$, the values to consider are those in

$$\frac{\alpha_t}{\alpha_M} < \frac{\alpha_i}{\alpha_j} < 1 . \tag{4.42}$$

Let the lower bound α_t/α_M be indicated as θ for brevity. With this, if

$$\theta \frac{R_{uu}(0)}{\left|R_{uu}(\Delta t_j - \Delta t_i)\right|} > 1 \tag{4.43}$$

one can be sure that path i is not hidden by interference from path j, *regardless of the attenuations that the two paths may have*.

The issue is that the inequality may hold for some $\Delta t_j - \Delta t_i$, but not for some other. How can this situation be interpreted? When an ultrasound NDT system is set up, one cannot know the times-of-flight associated to the various paths in advance. For any abstract evaluation of the excitation quality, they need to be considered as random variables. Furthermore, it is quite difficult to make any hypothesis on the statistical distribution of $\Delta t_j - \Delta t_i$. However, it is known already that $R_{uu}(\tau) = 0$ for $|\tau| > T$. Hence, for the assessment of the excitation quality one only needs to consider the distribution of $\Delta t_j - \Delta t_i$ when $|\Delta t_j - \Delta t_i| < T$. Let this latter conditioned distribution be described by a probability density function (PDF) $\rho_\Delta(t)$ and suppose that it is known. Also, build the indicator function

$$\xi(t) = \begin{cases} 1 & \text{if } \theta \frac{R_{uu}(0)}{|R_{uu}(t)|} > 1 \\ 0 & \text{otherwise} . \end{cases} \tag{4.44}$$

With this, the integral

$$P(\theta) = \frac{1}{2T} \int_{-T}^{T} \xi(t) \rho(t) \mathrm{d}t \tag{4.45}$$

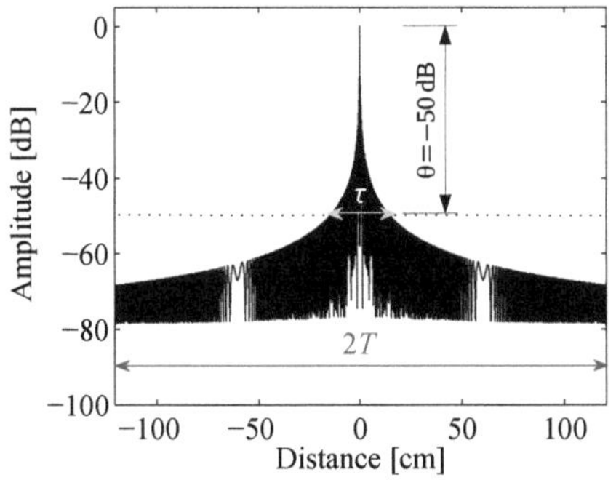

Fig. 4.26 Graphical illustration of the merit factor $P(\theta)$: τ is the measurement of the support of the normalized $\hat{\delta}(t)$ for which $\hat{\delta}(t) > \theta$. In the example $\theta = -50$ dB and the autocorrelation function derives from an LChirp signal. The function $\hat{\delta}(t)$ is defined in the interval $t \in [-T, T]$. The resulting probability $P(\theta)$ is equal to $\tau/(2T)$

returns the probability that *for the worst attenuation ratio* θ *that needs consideration and limiting to those times-of-flight that can create interference* there is no issue in the detection of path i due to the interference of path j. Clearly, one would like this probability to be as large as possible.

At this point, one may conjecture that $\rho(t)$ is uniform in $[-T, T]$. With this, the integral above can be computed as the fraction of the $[-T, T]$ interval where the indicator function $\xi(t)$ is 1. This quantity is quite easy to derive. It is sufficient to introduce a version of $R_{uu}(t)$ normalized over $R_{uu}(0)$, namely $\hat{R}_{uu}(t) = R_{uu}(t)/R_{uu}(0)$, and to take the relative measure of the subset of $[-T, T]$ where $|\hat{R}_{uu}(t)| < \theta$. Graphically, one needs to intersect $\hat{R}_{uu}(t)$ with an horizontal line at ordinate θ and measure the fraction of the $[-T, T]$ interval where the function is below the line. The whole $P(\theta)$ profile can be obtained by scanning the $\hat{R}_{uu}(t)$ function with an horizontal line that is progressively lowered from ordinate 1. The process is illustrated in Fig. 4.26 for a particularly simple autocorrelation profile deriving from an LChirp excitation. For more complex excitations, the horizontal line may get multiple intersections with the autocorrelation curve.

But does the conjecture make sense? It is reasonable to say it does, at least in some approximate way. In fact, the unconditioned distribution of $t = \Delta t_j - \Delta t_i$ will likely span a relatively large interval with respect to T. When the conditioning is introduced, a very little interval of the original unconditioned PDF of t is isolated. In this little interval one may expect the fluctuation of the PDF to be sufficiently small to be ignored.

With the tool introduced so far, one can evaluate the quality of different excitations through the corresponding functions $P(\theta)$. The advantage over classic merit factors comes from the integral that is present in Eq. (4.45). Due to it, the whole of the self-correlation function enters the quality evaluation, and not just a punctual feature. Unfortunately, a function such as $P(\theta)$ is not exactly a figure of merit: one would expect a scalar. Yet, it is certainly possible to say that if an excitation A gives a $P_A(\theta)$ that is consistently higher than the $P_B(\theta)$ given by excitation B, then A is better than B. Furthermore, it is possible to relate $P(\theta)$ with the classic merit factors. The

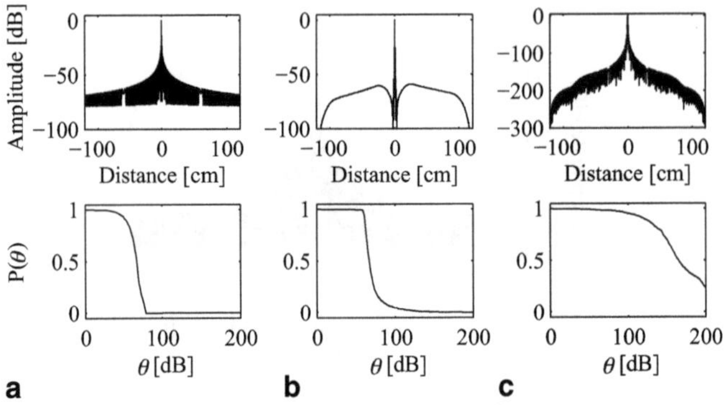

Fig. 4.27 Profiles of $\hat{\delta}(t)$ (*top*) and $P(\theta)$ (*bottom*) corresponding to the excitations used in Fig. 4.22: in **a**, $u(t)$ is an *unwindowed* LChirp and $\Psi(t) = \text{Rev}[u](t)$; in **b** $u(t)$ is an *unwindowed* LChirp and $\Psi(t)$ is a *Tuckey-windowed* reversed version of $u(t)$, namely $w_{\text{Tukey}}(t) \times \text{Rev}[u](t)$; in **c** $u(t)$ is a *Tukey-windowed* LChirp and $\Psi(t) = \text{Rev}[u](t)$

portion of $P(\theta)$ with θ values very close to 1 (0 dB) is connected with MLW. The portion of $P(\theta)$ relative to very small θ values is related to FSL. Finally, the portion of $P(\theta)$ for intermediate θ values is related to NSL.

Figure 4.27 shows the $P(\theta)$ profiles of the $\hat{\delta}(t)$ curves corresponding to the excitations used in Fig. 4.22. The better profiles are those which stay close to 1 for a wider span of θ values. From these plots, one can see how the higher $P(\theta)$ exhibited by scheme in subplots c, namely the one employing windowing techniques, for all θ values is in agreement with the fact that this excitation was the best in the practical detection of the small defect in the steel sample.

Furthermore, one reason why using a curve such as $P(\theta)$ can be an interesting way to assess excitations is the flexibility that can be obtained. In fact, the function $P(\theta)$ can be used to obtain scalar merit factors when it is "combined" with numbers that qualify the specific application. Basically, there are two possible approaches to do so.

In the first approach, the application problems sets a target probability $\hat{P}$, say 50 %, and one derives the $\hat{\theta}$ such that $P(\hat{\theta}) = \hat{P}$. The lower this $\hat{\theta}$, the better. Graphically, this is like fixing an ordinate on the $P(\theta)$ plot and taking the corresponding abscissa as a merit figure. This is like saying that when there are two paths that (for sure) interfere with each other, one wants anyway to ensure a certain probability $\hat{P}$ that the paths can be detected correctly even if there is a very bad attenuation ratio between them. From this premise, the worst possible attenuation ratio that can be tolerated is found. A word of caution is necessary. The quantity $\hat{P}$ is not an *absolute* probability of detecting a path here. It is an abstract quantity to which the significance of a detection probability can be given in extremely bad conditions (restricting to cases where interference certainly occurs and to paths that have very different attenuations). The unconditioned detection probability for a generic path would be much higher than $\hat{P}$. This means that even rather low $\hat{P}$ values make sense applicationwise. The problem with this approach is that identifying which $\hat{P}$ is good for a given application is nontrivial.

The second approach consists in noticing that it is generally possible to quantify α_t starting from the average power of $e_f(t)$, i.e., the noise floor at the matched filter output. In fact, it is not sensible to try to estimate paths for which the matched filter output is already know to peak below the noise floor. If α_t and α_M are known from the application, also θ is known. Hence, in this case one can directly read the $P(\theta)$ value of interest, the higher the better. Graphically, this is like fixing an abscissa on the $P(\theta)$ plot and taking the corresponding ordinate as a merit figure.

As a conclusion, one cannot help noticing that this *statistically inspired* approach, that was originally introduced in [7] may appear more complicated than the use of conventional merit factors. However, the extra complexity is justified by the fact that differently from conventional merit factors the approach can be straightforwardly extended to deal with the multi-TX set-up that shall be introduced in the next Chap. 5. Furthermore, this approach is also suitable for excitations where the autocorrelation tails do not decay monotonically, which is precisely the situation where conventional merit factors may not work so well.

4.7 System Adaptation to the Probe Features

The previously introduced merit factors seem to evaluate excitations regardless of the transducers' transfer functions. Indeed, in the previous sections, sentences such as "removing the effects of $H_T(s)$ and $H_R(s)$" where frequent. This was necessary for the sake of simplicity, yet such effect cannot be unconditionally forgotten and shall now be given some attention.

Since the very definition of the channel model (Sect. 4.2), it has been observed that the role of $H_T(s)$ and $H_R(s)$ can be considered marginal only if one has an excitation $u(t)$ that can pass through them almost undistorted. In order to provide some feeling of what this means, Fig. 4.28 provides a couple of examples of actual transfer functions that can be encountered. These refer to matched TX and RX couples rated 4for 5 MHz (in the plot panes a and b) and 2.25 MHz (in the plot panes c and d). The examples condense $H_T(s)$ and $H_R(s)$ together, since they always appear cascaded in the channel. Furthermore, real world curves, such as the shown ones, are generally experimentally obtained by recording the channel response in test setups where the TX and the RX are directly coupled, with no SUT between them. This arrangement precisely records $H_T(s)H_R(s)$ as an ensemble.

As evident from the examples, the transducers tend to have band-pass (BP) transfer functions, possibly with a nonflat in-band magnitude. The bandwidth restriction causes a spreading in time of the impulse response, as evident from plots 4.28b and 4.28c. For the 5 MHz probes, the bandwidth is approximately 6 MHz and the ideal pulse becomes as large as 160 ns (which is coherent with the bandwidth). For the 2.25 MHz probes, the bandwidth is approximately 2.4 MHz, and the impulse gets spread as much as a whole µs (a number not totally coherent with the inverse of the bandwidth, that may denote a rich phase response from the probes). This spreading is important, because it already places a bound on the system time-resolution. Another

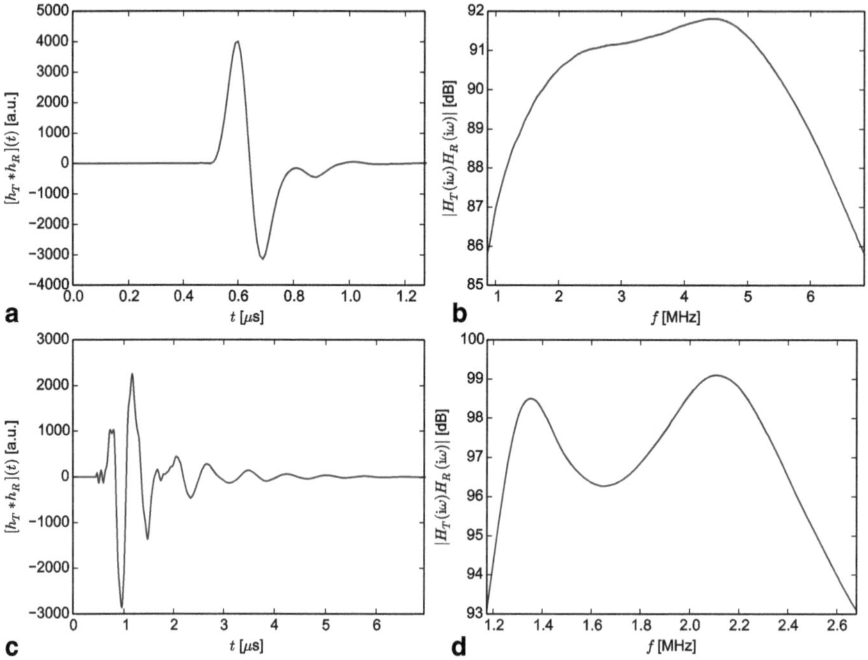

Fig. 4.28 Behavior of some real probes. **a** and **b** Impulse response and magnitude response of a matched TX and RX couple rated for 5 MHz. **c** and **d** Impulse response and magnitude response of a matched TX and RX couple rated for 2.25 MHz

aspect that can be noticed is that in plots 4.28a and 4.28c the "pulses" arrive with some delay with respect to the time origin. This indicates a rich dynamics inside the probe, where part of the behavior resembles that of a nonlumped system. The delay is approximately 500 ns for the 5 MHz probes and just slightly more for the 2.25 MHz probes. Such delays may need to be compensated in estimating the paths' ToFs when the probes are used for ultrasound NDT.

4.7.1 Matching Signal and System Features

The probe behaviors that have just been evidenced can clearly have impacts that depend on the specific excitations that are being used. If one employs an excitation that can *pass through* the transducers' frequency response, the impact of $H_T(s)$ and $H_R(s)$ can be modest. Otherwise, the impact may be relevant.

A first aspect that need to be noticed is *distortion*. As an example, Fig. 4.29 shows how some excitations are passed through the 5 MHz probes illustrated in Fig. 4.28.

From the figure it is quite evident that with the chirp it is easy to achieve modest distortion. Conversely, with the PN code excitation, the distortion is more evident, even if IRS can help its containment.

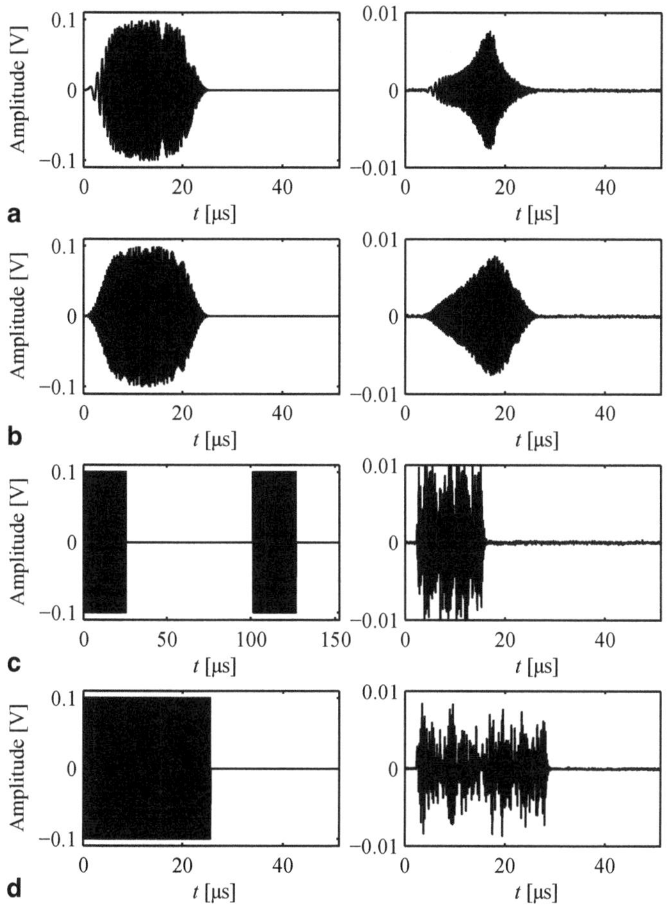

Fig. 4.29 **a** LChirp defined in the bandwidth 2–6 MHz: the exciting signal at the TX is on the *left*, the collected signal at the RX on the *right*. **b** The same arrangement as in **a**, this time considering an LChirp defined in the 0–8 MHz frequency range. **c** An MLS signal with $f_c = 4$MHz. **d** An IRS derived from a GCSs with $f_c = 4$MHz

The aspect of distortion is quite important. In fact, the matched filter is known to be optimal when the filter design is based on the actual template that needs to be isolated from noise. Hence, one should use $[u * h_T * h_R](-t)$ as the $H_f(s)$ impulse response. However, in practice, one always prefers to base the matched filter on the excitation alone. Namely, H_f is typically designed with an impulse response given by $\mathrm{Rev}[u](t)$. This is particularly true when PN codes are adopted, since fast convolution methods exist based on the fact that the matched filter impulse response takes discrete levels. As a consequence, if the distortion is significant, namely if $\mathrm{Rev}[u * h_T * h_R](t)$ is rather different from $\mathrm{Rev}[u](t)$, the matched filter becomes sub-optimal. This fact alone can have an impact on SNR. Furthermore, since $H_f(s)$

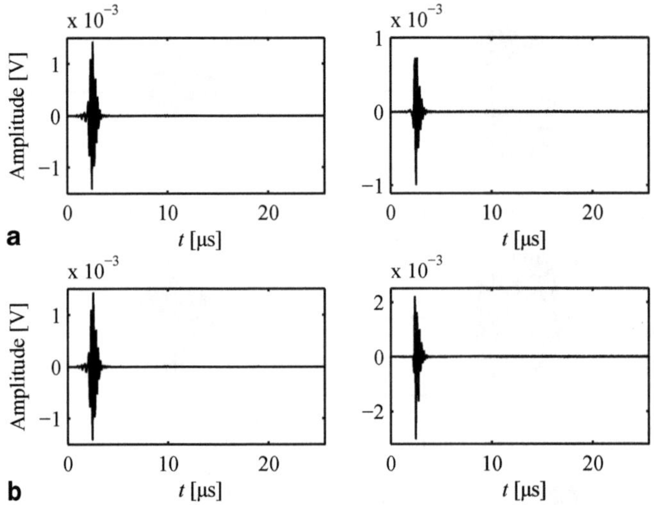

Fig. 4.30 **a** Comparison between the δ_{ex} curves obtained with an LChirp defined in the 2–6 MHz frequency range (*left*) and an LChirp defined in the 0–8 MHz frequency range. **b** Comparison between the δ_{ex} curves for an LChirp defined in the 2–6 MHz frequency range (*left*) and an IRS excitation with $f_c = 4\text{MHz}$

plays the dual role of the SNR enhancing filter and the 'pulse compressor' distortion can also have an impact on the PuC quality and thus on time resolution.

As an example, Fig. 4.30 shows some actual examples of the compression quality that can be achieved in a realistic setup. Enlarging the signal bandwidth can somehow improve the resolution, but may imply a reduction in the signal level since less energy gets properly converted by the transducers. This phenomenon is well depicted in Fig. 4.29a, b that reports the input and output signals for an LChirp in the range 2–6 and 0–8 MHz, respectively. It can be clearly seen that in the latter case the output signal at the beginning and at the end, i.e., at the extremal frequency values, exhibits a lower amplitude than in the former one. Moving to a coded excitation can also improve resolution due to their broadband nature, but still can have an impact on the collected signal energy especially when the standard PN waveforms are used. On contrary, by employing IRS excitation with a proper central frequency, or in general by applying spectral shaping to binary codes, the percentage of the energy transferred to the RX can be very similar to that of a chirp signal. In this case, the PN approach can be more effective than the chirp-based one since for any given duration the energy delivered to the system doubles that provided by a FM excitation, due the constant power of any binary and bipolar waveform. After PuC, the higher energy of the received signal translates into a higher amplitude of the experimentally measured $\delta_{ex}(t)$ and then to a better SNR for a fixed noise level.

This latter example introduces another important matter, namely the ability to pass the excitation power through the transducers, that can be even more important than distortion. That part of the excitation that falls outside of the transducers' pass-band (or on the transition zones at its boundaries) cannot be pushed into the SUT and results in an energy loss. The resulting extra attenuation directly goes to the detriment of the system SNR. It may not be very important in setups where there is spare SNR. However, in arrangements where the SNR is already critical (e.g., because there are air-couplings that introduce a large attenuation themselves), the extra attenuation due to excitations with PSDs mismatched to the probes magnitude response may make the difference.

With chirp excitations, it can be quite easy to match the transducers' bandwidth, since this is just a matter of choosing the right initial and final frequencies (and possibly of using appropriate time-windows to avoid artifacts at the PSD boundaries). Conversely, with PN codes, matching the transducers' bandwidth is necessarily harder (even if IRS can help). For this reason, PN codes can be fairly more problematic than chirps in SNR-critical setups.

Another aspect that is worth mentioning is the *quasi-instantaneous* signal power. This is the instantaneous power *averaged in the very short term* and for cyclic or quasicyclic signals is proportional to the squared envelope level. If two signals with identical peak levels of quasi-instantaneous power are considered, one with a constant evolution of its quasi-instantaneous power and another one with an irregular power evolution, the former will certainly deliver more energy per time unit. This consideration is relevant because transducers typically place a limit on the quasi-instantaneous power level. Thus, to exploit at its best the power conversion capacity of the TX, it is convenient to use excitation signals with constant or nearly constant quasi-instantaneous power (or with constant envelope). Unfortunately, in non-IRS PN codes the PSD extends toward DC. This can cause a significant fluctuation of the quasi-instantaneous power.

After having established that excitations should have a bandwidth reflecting the probe bandwidth and an envelope as constant as possible, one may consider if further optimizations are possible. Indeed, this is the case in all those occasions where the transducers magnitude response is nonflat in their pass-band. In fact, energy transfer is maximized if most of the energy is concentrated where the transducers' transfer function peaks. If energy and SNR were the only goals, it would be convenient to put *all* the excitation energy at the single frequency where the transducers' magnitude response is largest. For the transducers used as examples, this would be at about 4.5 MHz for the 5 MHz rated probes and at about 2.2 MHz for the 2.25 MHz rated probes. Doing so, rather than using excitations with PSDs uniformly spread across the whole of the transducers' bandwidth would reduce the attenuation (and improve the SNR) by several dBs Unfortunately, this approach cannot be followed. In fact, it has already been established that excitations need to be *wide band* to assure a good time-resolution. Yet, it is possible to *tradeoff* between the wide-band requirement and the desire to concentrate energy where the transducers' have a lower attenuation.

An excellent compromise can be represented by using excitations whose PSD mirrors the shape of the transducers' magnitude response. If such a choice is made, the optimal tradeoff between SNR and resolution should be reached [37]. Clearly, this consideration leaves open the issue of synthesizing excitations capable of both

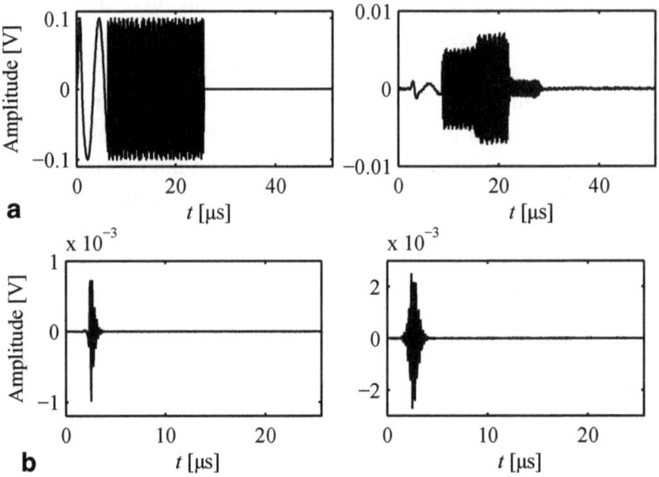

Fig. 4.31 **a** An NLChirp defined in the 0–8 MHz frequency range coping with the transducer transfer function. The transmitted and received signals are plotted on the *left* and on the *right* respectively. Although not perfect, the nonlinear frequency sweep provides some equalization of the received signal amplitude. **b** A comparison among the δ_{ex} curves of the LChirp and the NLChirp (*left and right*, respectively). As expected in the linear case, δ_{ex} is shorter but in the nonlinear case, the energy that can reach the receiver is larger

having a pre-assigned spectrum and a nice autocorrelation function. Fortunately, it has already been shown that NLChirps offer a rather easy way to practice spectrum shaping. As an example, Fig. 4.31 illustrates how an NLChirp can help increasing the amount of energy reaching the RX when tuned to deliver an excitation spectrum seconding the combined TX–RX transfer function profile. This goes at a slight detriment of resolution, but for some applications the tradeoff can be convenient.

Obviously, none of the considerations about SNR can make sense without taking into account that the system under exam is not just affected by inherent noise, but also by "self-inflicted" interference due to the tails of δ_{ex}. A priority is thus to contain them by choosing schemes that are appropriate for the application at hand. As an example, Fig. 4.32 shows by experiments on a practical setup how much self interference can appear by using an acyclic scheme with MLS excitations. This can be tolerable in some applications and less tolerable in others.

4.7.2 Merit Factors and the Matching of Excitations and System Features

Before moving to further topics with other chapters of the book, there is one last observation that is worth making. All the latter considerations on the ability of the excitation signals to work accordingly to the system features, passing most of their

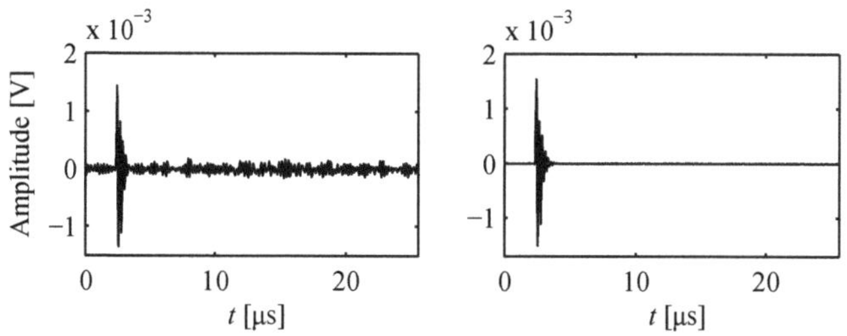

Fig. 4.32 A comparison of δ_{ex} curves obtained with MLS excitations, in the acyclic (*left*) and cyclic (*right*) case

energy through the transducers, may appear somehow unrelated with the merit factors that have been introduced in Sects. 4.6.2 and 4.6.3.

In fact, it is not completely the case, particularly when the statistically inspired merit factors are taken into account. In fact, as it should be clear from the end of Sect. 4.6.3, the channel attenuation has a direct impact on the choice of the quantity θ used to evaluate performance through the $P(\theta)$ curves. This aspect will be better detailed shortly. Before that, in view of the large use that conventional merit factors enjoy, it can be appropriate to augment that set of merit figures with a new one more explicitly tied to the adaptation of the excitation to the probe features. This can be done by defining the quantity

$$\mathrm{EEE} = A_S \cdot A_E \tag{4.46}$$

where A_S is the spectral efficiency, namely the overall power attenuation (actually the "less than one" gain) that the excitation encounters when passing through $H_T(s)$ and $H_R(s)$, and A_E is the envelope efficiency, namely the ratio between the average power and the quasi-instantaneous power. For a constant envelope signal, A_E is 1. When excitations with different EEE values are compared, the EEE difference can reflect almost directly on SNR.

With this, it is now possible to go back to the statistically inspired merit evaluation. In fact, EEE is directly related to both α_M and α_t, and thus, to the θ that is used to convert the $P(\theta)$ curve in a single system dependent merit figure.

4.8 Conclusion

In this chapter the role of the excitation choice and of the signal processing methods involved in ultrasonic NDT has been introduced, by means of both theory and examples. Specifically, the way in which proper excitation and signal processing methods allow one to escape from certain resolution-SNR tradeoffs by means of techniques such as PuC has been detailed. Some alternatives approaches for PuC have been extensively described (including some notable variants) and compared. Because comparison should be as much as possible quantitative, the most important merit factors currently used in the field have been introduced and some new ones have been proposed and motivated. All the concepts proposed in the chapter have been presented trying to distill the most important aspects and trends by first simplifying as much as possible the models and then by reading one by one some relevant effects present in real world systems. Furthermore, the presentation has been structured in view of providing a toolbox suitable for further extensions, such as those necessary to deal with the multitransducer setups that will be the object of the next chapter.

References

1. Adams JW (1991) A new optimal window. IEEE Trans Signal Process 39(8):1753–1769
2. Betta G, Burrascano P, Ferrigno L, Laracca M, Ricci M, Silipigni G (2013) On the use of complex excitation sequences for eddy current testing. 19th symposium IMEKO TC4 and 17th IWADC workshop on advances in instrumentation and sensors interoperability, Barcelona, Spain
3. Blitz J, Simpson G (1996) Ultrasonic methods of non-destructive testing, non-destructive evaluation series, vol 2. Springer, New York
4. Bordier J, Fink M, Le Brun A, Cohen-Tenoudji F (1991) The influence of multiple scattering in incoherent ultrasonic inspection of coarse grain stainless steel. In: Proceedings of the 1991 IEEE ultrasonics symposium, IEEE, pp 803–808
5. Budišin SZ (1991) Efficient pulse compressor for Golay complementary sequences. Electron Lett 27(3):219–220
6. Burrus CSS, Parks TW (1991) DFT/FFT and convolution algorithms: theory and implementation. Wiley, New York
7. Caporale S, Callegari S, Ricci M, Burrascano P (2013) Constant envelope pseudo orthogonal excitations for ultrasound testing. In: Proceedings of the 18th international conference on Digital Signal Processing (DSP), IEEE, pp 1–8
8. Cau F, Fanni A, Montisci A, Testoni P, Usai M (2006) A signal-processing tool for non-destructive testing of inaccessible pipes. Eng Appl Artif Intell 19(7):753–760
9. Challis RE, Ivchenko VG (2011) Sub-threshold sampling in a correlation-based ultrasonic spectrometer. Meas Sci Technol 22(2):1–12
10. Choma M, Sarunic M, Yang C, Izatt J (2003) Sensitivity advantage of swept source and Fourier domain optical coherence tomography. Opt Express 11(18):2183–2189
11. Cohen FS (1992) Modeling of ultrasound speckle with application in flaw detection in metals. IEEE Trans Signal Process 40(3):624–632
12. Cohn M, Lempel A (1977) On fast M-sequence transforms (corresp.). IEEE Trans Inf Theory 23(1):135–137
13. Cook C (2012) Radar signals: an introduction to theory and application. Elsevier, New York
14. Farina A (2000) Simultaneous measurement of impulse response and distortion with a swept-sine technique. Audio Engineering Society Convention 108, Audio Engineering Society
15. Gan TH, Hutchins DA, Billson DR, Schindel DW (2001) The use of broadband acoustic transducers and pulse-compression techniques for air-coupled ultrasonic imaging. Ultrasonics 39(3):181–194
16. Garcia-Rodriguez M, Yañez Y, Garcia-Hernandez MJ, Salazar J, Turo A, Chavez JA (2010) Application of golay codes to improve the dynamic range in ultrasonic lamb waves air-coupled systems. NDT E Int 43(8):677–686
17. Gardner WG (1994) Efficient convolution without input/output delay. Audio Engineering Society Convention '97, Audio Engineering Society
18. Golay MJE (1961) Complementary series. IRE Trans Inf Theory 7(2):82–87

19. Golomb SW, Welch LR, Goldstein RM, Hales AW (1982) Shift register sequences, vol 78. Aegean Park Press, Laguna Hills
20. Harris FJ (1978) On the use of windows for harmonic analysis with the discrete fourier transform. Proc IEEE 66(1):51–83
21. Hernández A, Ureña J, Hernanz D, García JJ, Mazo M, Dérutin JP, Serot J, Palazuelos SE (2003) Real-time implementation of an efficient golay correlator (EGC) applied to ultrasonic sensorial systems. Microprocess Microsyst 27(8):397–406
22. Hunt B (1971) A matrix theory proof of the discrete convolution theorem. IEEE Trans Audio Electroacoust 19(4):285–288
23. Hutchins D, Burrascano P, Davis L, Laureti S, Ricci M (2014) Coded waveforms for optimised air-coupled ultrasonic nondestructive evaluation. Ultrasonics 54(7):1745–1759
24. Klauder JR, Price AC, Darlington S, Albersheim WJ (1960) The theory and design of chirp radars. Bell Syst Tech J 39(4):745–808
25. Lazaro JC, San Emeterio JL, Ramos A, Fernandez-Marron JL (2002) Influence of thresholding procedures in ultrasonic grain noise reduction using wavelets. Ultrasonics 40(1):263–267
26. Lee BB, Furgason ES (1983) High-speed digital Golay code flaw detection system. Ultrasonics 21(4):153–161
27. Legendre S, Massicotte D, Goyette J, Bose TK (2000) Wavelet-transform-based method of analysis for lamb-wave ultrasonic NDE signals. IEEE Trans Instrum Meas 49(3):524–530
28. Lu Y, Michaels JE (2008) Numerical implementation of matching pursuit for the analysis of complex ultrasonic signals. IEEE Trans Ultrason Ferroelectr Freq Control 55(1):173–182
29. Luke HD (1988) Sequences and arrays with perfect periodic correlation. IEEE Trans Aerosp Electron Syst 24(3):287–294
30. Milleit RE (1970) A matched-filter pulse-compression system using a nonlinear FM waveform. IEEE Trans Aerosp Electron Syst 6(1):73–78.
31. Morozov M, Tian GY, Edgar D (2009) Comparison of pec and sfec nde techniques. Nondestruct Test Eval 24(1–2):153–164
32. Mulaveesala R, Tuli S (2006) Theory of frequency modulated thermal wave imaging for nondestructive subsurface defect detection. Appl Phys Lett 89(19):191913
33. Nowicki A, Trots I, Lewin PA, Secomski W, Tymkiewicz R (2007) Influence of the ultrasound transducer bandwidth on selection of the complementary Golay bit code length. Ultrasonics 47(1):64–73
34. Oppenheim AV, George VC (2010) 6.011 introduction to communication, control, and signal processing. MIT OpenCourseWare
35. Pallav P, Gan TH, Hutchins DA (2007) Elliptical-Tukey chirp signal for high-resolution, air-coupled ultrasonic imaging. IEEE Trans Ultrason Ferroelectr Freq Control 54(8):1530–1540
36. Phang APY, Challis RE, Ivchenko VG, Kalashnikov AN (2008) A field programmable gate array-based ultrasonic spectrometer. Meas Sci Technol 19(4):1–13
37. Pollakowski M, Ermert H (1994) Chirp signal matching and signal power optimization in pulse-echo mode ultrasonic nondestructive testing. IEEE Trans Ultrason Ferroelectr Freq Control 41(5):655–659
38. Prada C, Kerbrat E, Cassereau D, Fink M (2002) Time reversal techniques in ultrasonic nondestructive testing of scattering media. Inverse Probl 18(6):1761–1773
39. Ricci M, Senni L, Burrascano P (2012) Exploiting pseudorandom sequences to enhance noise immunity for air-coupled ultrasonic nondestructive testing. IEEE Trans Instrum Meas 61(11):2905–2915
40. Rodríguez MA, San Emeterio JL, Lazaro JC, Ramos A (2004) Ultrasonic flaw detection in NDE of highly scattering materials using wavelet and Wigner–Ville transform processing. Ultrasonics 42(1):847–851
41. Sachs J, Peyerl P, Wöckel S, Kmec M, Herrmann R, Zetik R (2007) Liquid and moisture sensing by ultra-wideband pseudo-noise sequence signals. Meas Sci Technol 18(4):1074–1087
42. Sarwate DV, Pursley MB (1980) Crosscorrelation properties of pseudorandom and related sequences. Proc IEEE 68(5):593–619

43. Schmerr L, Song JS (2007) Ultrasonic nondestructive evaluation systems: models and measurements. Springer, New York
44. Schroeder MR (1979) Integrated-impulse method measuring sound decay without using impulses. J Acoust Soc Am 66(2):497–500
45. Stan GB, Embrechts JJ, Archambeau D (2002) Comparison of different impulse response measurement techniques. J Audio Eng Soc 50(4):249–262
46. Strickland D, Mourou G (1985) Compression of amplified chirped optical pulses. Opt Commun 55(6):447–449
47. Svilainis L, Lukoseviciute K, Dumbrava V, Chaziachmetovas A (2013) Subsample interpolation bias error in time of flight estimation by direct correlation in digital domain. Measurement 46(10):3950–3958
48. Turin G (1960) An introduction to matched filters. IRE Trans Inf Theory 6(3):311–329.
49. Turin GL (1976) An introduction to digital matched filters. Proc IEEE 64(7):1092–1112
50. Wei L, Huang Zy, Que Pw (2009) Sparse deconvolution method for improving the time-resolution of ultrasonic NDE signals. NDT E Int 42(5):430–434
51. White JDH, Challis RE (1992) A Golay sequencer based NDT system for highly attenuating materials. In: IEE colloquium on non-contacting and remote NDT, IET, pp 1–7
52. Wiener N (1949) Extrapolation, interpolation, and smoothing of stationary time series, vol 2. MIT Press, Cambridge
53. Xiang N (1992) Using m-sequences for determining the impulse responses of LTI-systems. Signal Process 28(2):139–152
54. Xiang N, Schroeder MR (2003) Reciprocal maximum-length sequence pairs for acoustical dual source measurements. J Acoust Soc Am 113(5):2754–2761
55. Zhu Y, Weight JP (1994) Ultrasonic nondestructive evaluation of highly scattering materials using adaptive filtering and detection. IEEE Trans Ultrason Ferroelectr Freq Control 41(1): 26–33

Chapter 5
Excitations and Signal Processing for Multiprobe Setups

Salvatore Caporale, Sergio Callegari, Marco Ricci and Pietro Burrascano

Abstract The possibility of taking advantage of multiprobe setups for ultrasound non-destructive testing (NDT) is introduced with reference to its physical and signal processing aspects. Preliminarily, a major distinction is made between phased-arrays and actual multi-transducer configurations. Then, most of the chapter is devoted to the latter type of setup. In this framework, the previously introduced channel model developed for single-probe operation, is extended to setups featuring multiple transmitting transducers (TXs) and receiving transducers (RXs) meant to be employed simultaneously in a so-called multiple input, multiple output (MIMO) configuration. It is also illustrated how this configuration relates to multi-user communication networks. Proper merit factors suitable for characterizing the performances of multiprobe systems are presented by taking advantage of the probabilistic system view introduced for single-probe systems. Finally, various classes of signals that can be adopted in MIMO NDT systems are described. These excitations may be designed according to random or deterministic approaches. In better detail, special techniques based on tessellation of the time-frequency (TF) plane are developed to design excitation signals capable of joining assuring contextually the good degree of appropriateness for MIMO setups and the good signal-to-noise (SNR) and resolution properties of chirps.

S. Caporale (✉)
Advanced Research Center on Electronic Systems for Information and Communication Technologies "E. De Castro" (ARCES), University of Bologna, Bologna, Italy
e-mail: salvatore.caporale@unibo.it

S. Callegari
Department of Electrical, Electronic and Information Engineering "G. Marconi" (DEI) and Advanced Research Center on Electronic Systems for Information and Communication Technologies "E. De Castro" (ARCES), University of Bologna, Bologna, Italy
e-mail: sergio.callegari@unibo.it

M. Ricci · P. Burrascano
Department of Engineering, Polo Scientifico Didattico di Terni, University of Perugia, Terni, Italy
e-mail: marco.ricci@unipg.it

P. Burrascano
e-mail: pietro.burrascano@unipg.it

P. Burrascano et al. (eds.), *Ultrasonic Nondestructive Evaluation Systems*,
DOI 10.1007/978-3-319-10566-6_5

5.1 Introduction

In order to enhance the inspection cability of an ultrasound system, it can be convenient to increase the number of probes used for the monitoring. However, the way in which one can take advantage of multiple probes is not obvious and different approaches exist. To some degree, different approaches may be even combined.

A most notable distinguishing element is the level of coordination required among the probes. On one extreme, one has highly coordinated elementary probes organized in arrays, where array elements must be accurately located on with respect to the other and are required to emit (or receive) slightly different version of the same excitation. The most common embodiment of strategy is the *phased array* approach [9, 14]. On the other extreme, one finds uncoordinated or loosely coordinated elements, which basically return information in a way similar to what would be achieved in a single transmitting transducer (TX) single receiving transducer (RX) system, by taking multiple measurements, each with a different placement of the probes [5]. These opposite strategies are better discussed in Sect. 5.2. In this book, emphasis is placed on the loosely coordinated setup, which is regarded as a *proper* multiprobe system. Conversely, abstracting from its internal architecture, a phased array can in many senses be interpreted as single *macro-probe* with a particularly strict and configurable aperture angle and directionality. In principle, a proper multiprobe setup could even be built on top of phased-arrays probes.

Multiprobe arrangements have a great advantage over single-TX, single-RX systems in the reduction of testing time. Consider a setup with a single TX and n RXs. Since the latter can be operated simultaneously, in a single shot one can gather an equivalent amount of information as in n subsequent measurements practiced by a single-TX, single-RX system by moving the RX. Thus, there is an n-fold reduction in testing times even before considering the overhead of rearranging the probes. This advantage would be increased if one could also rely on multiple simultaneously operated TXs, as m TXs would make the testing time reduction $n \times m$ fold.

Note that the single-TX, multiple-RX case can be analyzed and designed using the same principles required by a single-TX, single-RX system. The same holds for a multiple-TX system where the TX operation is serialized in time (loosing most of the time saving advantage). Conversely, the presence of multiple simultaneously operated TXs makes the system radically different, introducing the possibility of interference among the multiple excitations [5]. The treatment and containment of such interference is the core topic of this chapter. The corresponding issues are somehow similar to those encountered in communication systems where multiple "speakers" need to share the same communication resource. In that context, sharing can be achieved in different ways. For instance, one can rely on time division. In a multiprobe ultrasound system this is the same as serializing the operation of the TXs. Alternatively, one may use frequency division. In ultrasound non-destructive testing (NDT), this is generally not possible because the transducers bandwidth is typically modest and a large bandwidth in the excitations is indispensable for successfully practicing pulse-compression (PuC) [11] and for the achievement of a good time resolution (cfr. Chap. 4). Finally, one can practice code division [19]. By exclusion, this emerges as the most promising technique for ultrasonic systems. Notwithstanding this analogy, one has to take into account that a communication system and a multiprobe ultrasound system are remarkably different as far as the way *information* is managed. In fact, in the first system the aim is to send information through the channel by means of distributed transmitters and receivers, while in the second case the purpose is to collect information about the channel. Hence, the separability issue about waveforms has to be reviewed according to this paradigm, i.e., the fact that the channel itself creates delayed and attenuated copies of the waveforms coming from the TXs cannot be ignored.

In this respect, a few points are worth underlining. First of all, as it can intuitively be expected, interference can be contained but not fully eliminated. This means that there is a trade off between the time saving achievable by using a simultaneously operated multiprobe setup and the measurement quality. In some contexts where quality requirements prevail, serializing the TX operation may be the best strategy. However, in other contexts timing constraints may dominate. For instance, this can be the case in industrial environments where testing must be practiced on an moving assembly line. Secondly, industrial environments may impose other trade-offs. For instance, interference containment may not justify a complete disruption of hardware architecture with respect to conventional single-TX systems. The subtleties of these issues and the fact that proper multiprobe systems are way less established than other approaches make this a challenging area in ultrasonic NDT.

5.2 A Quick Review of Phased Arrays

As mentioned in the introduction, the traditional ways of taking advantage of multiple elementary transmitting and receiving elements consists in placing them close to each other in a fixed geometry that is typically linear or matrix shaped. These elements constitute an *array* [9, 14]. Within an array, the transmission elements are operated simultaneously, using the same fundamental excitation that is replicated among them with a phase delay. The delays influence the way in which the individual ultrasound waves produced by the elementary transmitters interfere with each other in the sample under test (SUT). Hence, by proper calibration, one can reinforce the overall wave in a desired direction and suppress it in undesired ones. Similarly, in receiving arrays, the individual signal collected by the array elements are merged by superposition after the application of suitable delays. With this, one can reinforce sensitivity in a desired direction and suppress it in undesired ones. Since operation is based on a strict control of the relative phases of the array components, this structure is commonly indicated as a *phased array*. The result of obtaining favored axes, directions, and depths where ultrasound waves are best propagated or sensed is indicated as *beam forming*. Figure 5.1 illustrates a very schematic view of a phased array probe, applied on a SUT.

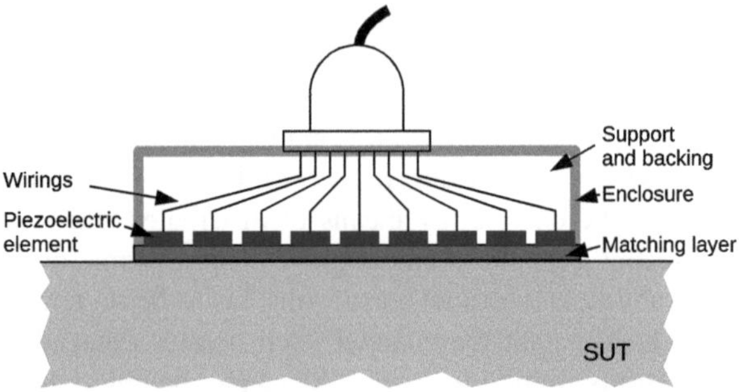

Fig. 5.1 Schematic view of a phased array probe, applied onto a SUT. The representation is highly schematic and missing many elements, being mostly aimed at comparison to setups where individual transducers are individually and arbitrarily placed on the SUT, as in Fig. 5.2

An interesting aspect of phased arrays is that direction control is obtained by signal processing techniques that can be modified over time. This means that contrarily to a transducer where precise focusing is achieved by a fixed geometries, here the preferred direction and focus can be altered electronically. This is known as *beam steering* and represents a very attractive feature that allows a single probe to be used for accurately scanning different regions of a SUT. Depending on the array construction and operation, different types of scanning are possible. For instance, the scanning can be practiced by changing the beam direction, achieving a so-called angular steering (or sectorial scanning also called S-scan). Alternatively, by altering not just the relative phases, but also the actuated power on the individual transducers (or merely their activation state), it is possible to perform linear scans. In this case the beam angle stays fixed, while the beam is "rigidly" translated across the SUT. As a result, multiple "apertures" are, so to say, stacked one to the other until a cross-sectional view of the SUT is obtained. This latter way of using phased arrays typically requires a larger number of individual elements than the former. For instance, having a 64-element array, one may think of using them 16 at a time. In each instant, their phases can be controlled to achieve a thin beam. Over time, the 16 elements in use are progressively updated by shifting by one position at each step the starting one. Up to $64 - 16 + 1 = 49$ different views of the SUT can so be obtained. By both using electronically controlled linear scans and also physically stepping the relative position of the probe and the SUT on an axis orthogonal to that used for the electronic scan, bidimensional views of a SUT can also be achieved. This is often indicated as a C-scan. In the most general cases, arrays can be found ranging from 16 to 200 and more individual elementary transducers.

From what has been said so far, the benefits of phased array ultrasonic techniques should be well clear. The ability to perform scanning operation can be a great aid in the inspection of SUTs characterized by complex geometries. In spite of the higher cost of an array in comparison to a traditional probe, phased arrays may also help making the testing activity more affordable, by reducing or removing altogether the need to actuate the probe or the SUT in order to alter their relative positions when multiple views are required. Furthermore, an aspect that has a great impact on test costs is test time. The possibility to do scanning without physical movement can reduce the latter quite significantly (up to ten times in certain setups).

Typically, the ability to perform scans also has a positive influence in reducing the number of defects that may go unnoticed during testing. A similar result comes from the electronic focusing that generally guarantees a very good beam helping the detection of small defects. An optimized beam shape, together with the ability to focus at multiple depths, also means that defect localization and sizing can be practiced with better accuracy. An aspect that is particularly interesting, in relation to all the matters considered in the previous chapter, is that focusing can significantly improve SNR and the self-interference issues due to the multiple propagation paths.

That said, phased arrays are not further discussed in this book, and the interested reader is invited to check the relevant literature on the topic. The reasons for this choice are threefold. First of all, the features of a phased array are strictly related to construction geometry and to technological aspects that are often out of the control of the user. Conversely, the choice in this book is to focus on aspects that can be easily experimented upon, even in an education context. Secondly, even if the widespread use of phased-array testing is only recently entering the industrial NDT field, commercial phased array systems for human body diagnostic have existed since the 1970s. Hence, the technique can be considered in many senses established. It is true that industrial NDT poses specific challenges of its own (most important, there is a much wider spread of the acoustic properties of materials, and in the geometries under investigation), so that the application of phased arrays cannot be a mere transposition of techniques already developed elsewhere. Still, the choice made in this book is to prefer the discussion of less established approaches when these are related to aspects where the reader can experiment directly. Finally, a phased array, when one abstracts from the internal architecture, can in many regards be interpretable as a single probe with a particularly strict and configurable beam aperture (even if this may not be exactly the case).

5.3 Multiprobe Ultrasound NDT Setups as MIMO Systems

After the previous illustration of multiple TX ultrasound NDT setups using coordinated arrays of transmitters and receivers, it is now possible to concentrate on sensing systems where the TXs are not strictly organized. This means that the relative probe position needs not to be accurately set for the probe system to operate correctly. Conversely, the positions can be arranged to provide the most thorough internal scanning of the SUT. Furthermore, the TXs do not need to operate in a coherent way. Conversely, they can be totally asynchronous and only the RXs need to know their actual timing to estimate the *times of flight* of the sound waves. In this arrangement, the only constraint is that the excitations used at the TXs are designed together in order to minimize the cross interference [5].

Figure 5.2 shows a sample probe arrangement for a 2D circular object. The arrangement makes any defect in the SUT visible from many RXs. Ideally, this should enable a thorough inspection of the SUT and correct identification and location of defects even if the dimensions are large and the individual probes have relatively

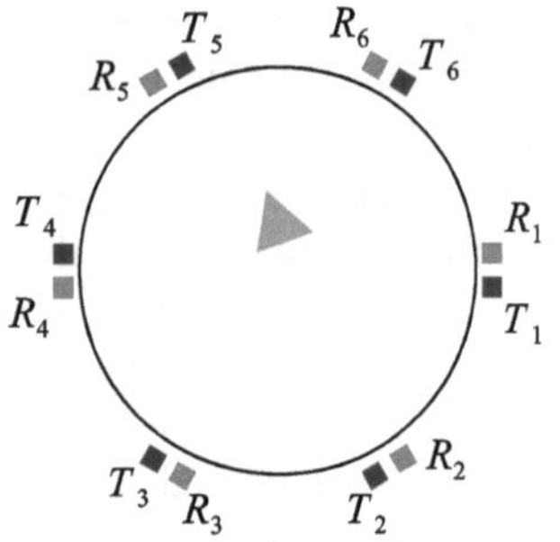

Fig. 5.2 Scheme of principle of a multiprobe setup. In this idealized illustration, the SUT geometry is 2D circular and a single defect is evidenced with a triangular marker inside the SUT

large apertures. Strategies for the joint processing of the information collected at many RXs are detailed in Part III.

As mentioned before, this setup is in many ways similar to that of a communication system, where multiple speakers and listeners need to share a single communication resource, without any separation in time or in frequency [19]. The main difference between the communication system and the current one is that in the former the listeners do not know what message the speakers are passing and need to estimate it accurately, notwithstanding the attenuation and distortion introduced by the channel, the self-interference due to multiple propagation and the cross-interference due to the sharing of the channel. Here, the message (or better the excitation) is known and what needs to be estimated is the channel between each TX and RX pair. Apart from this fundamental "swap" in the system goals, many of the cross-interference issues may have similar traits.

Although one aims to estimate the channel between each TX and RX pair, in a subsequent analysis and processing phase one wants to collect the information about all these individual channels and extract some information about the SUT considered as an overall channel between the sets of TXs and RXs, which may themselves be regarded as a single compound TX-RX pair. In fact, the SUT is shared between each TX and RX pair, hence its overall behavior can be considered as the actual information which is revealed in many different ways at the RXs, thanks to the excitations emitted by the TXs. This scenario is more similar to a multiple input, multiple output (MIMO) communication system [15] where multiple antennas are employed for receiving and transmitting the same information with high efficiency rather than a multiuser communication system where the fundamental issue is multiplexing the streams without interference.

In other words, from a practical point of view, the estimation of a channel, i.e., the collection of propagation paths between a TX and RX pair, requires the ability of multiplexing the excitations coming from each TX so that they can be separated and correctly identified at the RXs, hence the technique to be adopted for this aim make us look at the system as a multiuser communication system. Conversely, from an abstract point of view, the employment of a nonsequential multiple inspection of the SUT makes the system resemble a proper MIMO communication system. Since the final aim is the global inspection of the SUT, the approach will always be referred to as ultrasonic MIMO although this chapter will be mainly dedicated to the description of multiplexing techniques. For the illustration of cross interference issues, an indispensable preliminary step is a formalization of the MIMO model.

5.3.1 Extension of Channel Model to Multiprobe Setups

In order to identify the optimal waveforms to be employed in a multiprobe setup, the channel model presented for the single-probe case has to be properly extended. In better detail, at this stage we are more interested in characterizing the propagation on multiple channels rather than in giving a global description taking into account the effects of the transducers and the presence of different kind of noise. In fact, one may assume that the channels are all equivalent as far as the TXs and RXs are concerned also in view of the noise model. On the other hand, highlighting the difference between the single-probe and the multiple-probe setups is certainly required. As far as the signal processing operated at the generic receiver is concerned, even if the global system consists of a MIMO setup, it can be now thought as a multiple input, single output (MISO) system, since the estimation of information on the SUT as a whole by the assembly of data coming from each RX is not performed at this stage. To further justify this assumption, recall that in the positioning of the TXs and RXs of a MIMO system which has to maximally and jointly exploit the presence of multiple paths it is sensible that there is no privileged TX and RX pair. Hence, we can refer to a generic RX without any need to specify the presence of other RXs. All the transducers can be considered identical, so there is no need to distinguish between different TX and RX impulse responses. Hence, at a generic RX, with respect to the kth TX alone and temporarily ignoring the electric noise and the analog to digital converter (ADC) noise, one has

$$y^{(k)}(t) = [([u^{(k)} * h_T * h_u^{(k)}] + e_p) * h_R](t) = \left[h_R * \left(\sum_{i=1}^{N_k} [[u^{(k)} * h_T] * h_{u_i}^{(k)}] + e_p \right) \right](t) \tag{5.1}$$

where N_k paths have been considered for the kth TX and each variable referred to the kth TX is represented by superscript (k). Note that, as for the previous Chap. 4, also here the excitations are indicated as $u(t)$ in order to avoid confusion with the Laplace variable s. By collecting the contributions coming from all the TXs and adding electric and ADC noise we get

$$y(t) = \sum_{k=1}^{K} y^{(k)}(t) + e_n(t) + e_q(t) = \left[h_R * \left(\sum_{k=1}^{K} \sum_{i=1}^{N_k} [[u^{(k)} * h_T] * h_{u_i}^{(k)}] + e_p \right) \right](t) + e_n(t) + e_q(t) \tag{5.2}$$

Each channel linking the kth TX and the generic RX is composed by N_K propagation paths which can be roughly described via attenuations and delays as in $h_{u_i}^{(k)} = \alpha_i^{(k)} \delta(t - \Delta t_i^{(k)})$.

When considering single-probe setups, the problem of choosing the excitation is related to a compromise between resolution and noise, together with the need of facing the limits of the transducers. In case a multiprobe system is considered, there are intrinsically more degrees of freedom since each TX can employ a different excitation. In fact, this is needed in order to allow the RXs to correctly identify the contributions of the individual transmitters to the received signals. Apart from this, the issues which have been posed when designing optimal excitation for single-probe setups have still to be taken into account, as the MISO system can still be considered as a collection of single input, single output (SISO) systems. From this point of view, the criteria and merit factors that have been identified for single-probe setups are still valid and have to be integrated with new merit factors capable of taking into account the effect of cross-channel interference. Hence, the paradigm of noise filtering by means of the matched filter together with the matter of increasing the transferred energy by means of PuC have still to be considered as a starting point for identifying excitations that can be transmitted simultaneously from different TXs and received on a single RX [11, 13, 18].

Expectably, dealing with a multiprobe system is a quite challenging operation. In fact, the optimal excitation of a MISO system considered as a collection of SISO systems would be a chirp function, as it has been already illustrated in Chap. 4. Nevertheless, the paradigm of employing multiple TXs imposes to differentiate the excitations used in each TX so that the information about the SUT collected by each channel, i.e., the impulse responses $h_{u_i}^{(k)}$, can still be separated. As mentioned before, this requirement makes the system similar to a multiuser communication system where the simultaneous messages transmitted by various users have to be separated at the receiver without interference. In this field, the main property allowing for the separation of the received signals is their orthogonality. In order to understand the applicability of this paradigm, one needs first to characterize the system in terms of number of *users* involved in the *network* and number of *messages* sent by each *user*.

The kth TX sends to the RX N_k messages about its propagation paths consisting of the amplitude $\alpha_i^{(k)}$ and the delay $\Delta t_i^{(k)}$. These messages can be detected and distinguished in case their delays are sufficiently different and their amplitude are larger than noise. On the contrary, they might be self-interfered in case some delays have similar values. These issues have been already illustrated with reference to the resolution issue for single-probe setups. For the generic kth *user* which *modulates* its messages by means of the excitation $u^{(k)}$, by defining the self-correlation as

$$R_{uu}^{(k,k)}(\tau) = \int_{\mathbb{R}} u^{(k)}(t-\tau)u^{(k)}(t)\,\mathrm{d}t \quad k = 1, \dots, K \tag{5.3}$$

the following property would be desirable

$$R_{uu}^{(k,k)}(\tau) = \delta(\tau) \quad k = 1, \dots, K \tag{5.4}$$

which may be referred to as *orthogonality by shift*, representing a nonnumerable infinite number of conditions. When $u^{(k)}(t)$ is a time-limited and band-limited function, it is expandable in finite number of orthogonal functions, hence the above equation can be only approximately satisfied. So, the orthogonality condition has to be better interpreted in a probabilistic sense, as it has been previously illustrated for the single-probe setup.

When considering multiple TXs, one has to deal with a system having K *users* transmitting simultaneous $\sum_{k=1}^{K} N_k$ *messages*. If a single excitation was used for all the TXs, the probability that the whole set of $\Delta t_i^{(k)}$ are sufficiently different would be much decreased. Moreover one would need to have a methodology to refer each path to its TX, i.e., refer each *message* to its *user*. Hence, the excitations at the many TXs must be jointly designed so that it is possible, according to a proper probabilistic criteria, to identify the TX. It is worth noting that, by considering K different TXs synchronous with the RX, it is relatively easy to design a set of K orthogonal excitation if a sufficient time bandwidth product (TBP) is available. Unfortunately, even if in our setup the TXs and the RX are synchronized, when it comes to the *messages*, i.e., the reflectors generating multiple copies of the same excitation along each channel, there cannot be any synchronization, hence getting orthogonal waveforms at the receiver is not feasible. Instead, by taking also into account the *orthogonality by shift* property, the ideal specification would be to design a set of K waveforms, such that every pair is orthogonal for every possible shift. By posing

$$R_{uu}^{(n,m)}(\tau) = \int_{\mathbb{R}} u^{(n)}(t-\tau)u^{(m)}(t)\,\mathrm{d}t \quad k = 1,\dots,K \tag{5.5}$$

one asks

$$R_{uu}^{(n,m)}(\tau) = \delta(\tau)\delta_{n-m} \quad n,m = 1,\dots,K \tag{5.6}$$

where δ_n represents the Kronecker symbol. The term δ_{n-m} guarantees that, when $\tau = 0$ and $n \neq m$, the integral is null. Of course, being an extension of the requirement given for a SISO system, the above property cannot be realized in practice. Again, approximate solutions need to be sought.

To summarize, the aim is to design a set of excitations such that each one behaves like a chirp when taken as a single function out of the set, and such that each pair of functions, under arbitrary shifts, has nearly null scalar product, (ideally, the difference from zero in the scalar product should decrease as the shift is increased). The orthogonality condition has to be better interpreted in a probabilistic sense, as it has been previously illustrated for the single-probe setup.

5.3.2 Extension of Merit Factors to Multiprobe Setups

Clearly, merit factors applicable to SISO systems can still be applied to MIMO systems and in particular to MISO systems viewed as a collection of SISO systems. In more detail, both the conventional and probabilistic merit factors considered in Chap. 4 aim to measure how the self-correlation $R_{uu}^{(k,k)}(\tau)$ of the employed excitation

$u^{(k)}(t)$ is good at approximating the *Dirac* function, i.e., how $u^{(k)}(t)$ is close to satisfy the orthogonality by shift condition. Consequently, when moving from a SISO to a MISO setup, merit factors have to be modified in order to measure how much the correlation between each excitation pair, $u^{(n)}(t)$ and $u^{(m)}(t)$ with $n, m = 1, \dots, K$, namely $R_{uu}^{(n,m)}(\tau)$, is close to $\delta(\tau)\delta_{n,m}$ [5]. From this point of view, one has to derive merit factors starting from more than a single self-correlation function. Hence, the number of function to be considered is K^2, being split in the K self-correlations $R_{uu}^{(k,k)}$, $k = 1, \dots, K$, and $K^2 - K$ cross-correlations $R_{uu}^{(n,m)}$ (actually adding up to $(K^2 - K)/2$ unique pairs).

So, at the generic RX we consider the signal coming out of the matched filter relative to the nth TX by ignoring noise and the effect of both the TX and RX

$$y_f^{(n)}(t) = \sum_{k=1}^{K} \left[u^{(k)} * h_u^{(k)} * h_f^{(n)}\right](t) = \left[R_{uu}^{(n,n)} * h_u^{(n)}\right](t) + \sum_{\substack{k=1,\dots,K \\ k \neq n}} \left[R_{uu}^{(k,n)} * h_u^{(k)}\right](t) \tag{5.7}$$

which has been split in a term relative to the nth channel, being the useful term to be identified, and in a sum of contributions being the cross-interference of the other channels. The signal $r_f^{(k)}$ would ideally result in $h_u^{(n)}$ if the condition $R_{uu}^{(k,n)}(\tau) = \delta(\tau)\delta_{k-n}$ was satisfied. Now, one can assume that the delay of the ith path of the nth channel, namely $\Delta t_i^{(n)}$, is known and evaluate the capability of the considered MISO system to estimate the coefficient $\alpha_i^{(n)}$ by measuring the amount of self-interference and cross-interference. The previous expression can be expanded as follows

$$y_f^{(n)}(t) = \left[R_{uu}^{(n,n)} * h_{u_i}^{(n)}\right](t) + \left[R_{uu}^{(n,n)} * \sum_{\substack{j=1,\dots,N_n \\ j \neq i}} h_{u_j}^{(n)}\right](t) + \sum_{\substack{k=1,\dots,K \\ k \neq n}} \left[R_{uu}^{(k,n)} * \sum_{j=1}^{N_k} h_{u_j}^{(k)}\right](t) \tag{5.8}$$

where we highlighted the ith path of the kth channel. So, for $y_f^{(n)}(\Delta t_i^{(n)})$ one gets

$$y_f^{(n)}(\Delta t_i^{(n)}) = \alpha_i^{(n)} R_{uu}^{(n,n)}(0) + \sum_{\substack{j=1,\dots,N_n \\ j \neq i}} \alpha_j^{(n)} R_{uu}^{(n,n)}\left(\Delta t_j^{(n)} - \Delta t_i^{(n)}\right) + \sum_{\substack{k=1,\dots,K \\ k \neq n}} \alpha_j^{(k)} R_{uu}^{(k,n)}\left(\Delta t_j^{(k)} - \Delta t_i^{(n)}\right). \tag{5.9}$$

Basically, the evaluation of the MIMO setup can be done by measuring how much the information about the ith path, namely $\alpha_i^{(n)} R_{uu}^{(n,n)}(0)$, is hidden by either by another path belonging to the same channel or by a path derived from cross-channels. More precisely, merit factors do not have to measure the actual amount of interference generated by all possible interfering paths, rather they should provide a general metrics for a proper comparison of different designs. Conventional merit factors would not apply to summarize the behavior of the cross-correlations $R_{uu}^{(k,n)}$, since these are not supposed to be decaying functions like self-correlations. Instead, the merit factors based on a probabilistic point view which have been already introduced in Chap. 4 for single-probe setups do not suffer from this limitation. Hence, we focus on the *probability* that

$$\frac{\left|R_{uu}^{(k,n)}\left(\Delta t_j^{(k)} - \Delta t_i^{(n)}\right)\right|}{R_{uu}^{(n,n)}(0)} < \frac{\alpha_i^{(n)}}{\alpha_j^{(k)}} \tag{5.10}$$

for all possible values of $\Delta t_j^{(k)}$ and $\alpha_j^{(k)}$ when $\Delta t_i^{(n)}$ and $\alpha_i^{(n)}$ are fixed. It is worth noting that the ratio on the left is not exactly a correlation normalized to its peak. Anyway, it is reasonable to assume that each excitation $u^{(k)}$ has the same energy as long as system specifications prescribe a constant envelope and the same duration for all the excitations. Moreover, the above relation depends on the difference $\Delta t_j^{(k)} - \Delta t_i^{(n)}$ only rather than on the values $\Delta t_j^{(k)}$ and $\Delta t_i^{(n)}$ separately. Hence, from a practical point of view, one can write

$$\frac{\left|R_{uu}^{(k,n)}(t)\right|}{R_{uu}^{(k,n)}(0)} < \frac{\alpha_i^{(n)}}{\alpha_j^{(k)}} . \tag{5.11}$$

Clearly, the right term just represents a threshold to decide how it is probable that the considered correlation does not cause interference. The term probability has been employed rather than the term *occurrence* since some values of t might have a larger weight than others. As done in the previous chapter, we introduce the threshold θ, such that

$$\frac{|R_{uu}^{(k,n)}(t)|}{R_{uu}^{(k,n)}(0)} < \theta \quad \theta \in [0,1] \tag{5.12}$$

We point out that limiting θ in the interval $[0, 1]$ is a worst case assumption, since it may also happen that $\alpha_i^{(n)} \gg \alpha_j^{(k)}$, i.e., the upper limit for θ would be ∞. Nevertheless, one might give another interpretation to the above probability condition. Let us suppose that all paths have comparable or even equal coefficients, that is the same as assuming that $\alpha_i^{(k)} = 1, \forall i, k$. One may also need some margin in order to identify a peak with respect to self and cross interferences. Then, the above condition measures the capability of identifying the ith path of the kth channel with respect to the margin θ (which may potentially be expressed in dB for better visualization, as in the following figures).

Finally, as it has been done for single-probe systems, we define

$$\xi^{(k,n)}(t) = \begin{cases} 1 & \text{if } \frac{|R_{uu}^{(k,n)}(t)|}{R_{uu}^{(k,n)}(0)} < \theta \\ 0 & \text{otherwise} \end{cases} \tag{5.13}$$

such that the normalized weighted measure of the set where $\xi^{(k,n)}(t) = 1$ represents the wanted probability

$$P^{(k,n)}(\theta) = \frac{1}{2T}\int_{-T}^{T} \xi^{(k,n)}(t)\rho(t)\mathrm{d}t. \tag{5.14}$$

The typical behavior of such probabilities is to be monotonically increasing functions obviously stating from 0 when $\theta = 0$ and being equal to 1 when $\theta = 1$.

With respect to merit factors and desirable features, two final aspects are worth remarking. First of all, when employing a family of excitations, one must typically accept a degradation of self-correlation properties with respect to a single-probe system (that can be fully optimized for self-correlation only). Secondly, the way in which $P(\theta)$ decreases as θ is reduced has a different impact in the single and multi TX case. In the single TX case, one can only operate for θ values where $P(\theta)$ is really close to 1. The overall number of propagation paths associated to each defect is low, thus missing the detection of some paths can immediately hinder defect revelation and localization. Conversely, in the multi-TX case, a defect can produce many propagation paths associated to different TX–RX couples. Even if one of them goes undetected due to self or cross interference, the other ones may still get detected, permitting the revelation of the defect. Thus, values of $P^{(k,n)}(\theta)$ that are somehow less than 1 can be tolerated, as long as they are large enough, this feature clearly deriving from the *diversity* among the many receivers and transmitters. As a consequence, it may be convenient to design excitations where the decay of $P^{(k,n)}(1/\vartheta)$, $\vartheta \in [1, \infty)$, is made less steep, even at the cost reducing the plateau a little or of compromising a bit on cross-correlation.

5.4 Excitations for Multiprobe Setups

In the previous sections the main issues involved in the introduction of MIMO systems have been illustrated and analyzed. Focus has been put on outlining the analogy between a multiprobe setups where TXs operate simultaneously and multi-user communication networks. From this point of view, the properties allowing for the actual functioning of such a system have been identified together with some metrics quantifying the quality of a potential design. Although this preliminary analysis gives useful guidelines for the identification of suitable excitations, nothing has been said about how to effectively synthesize these excitations. Nevertheless, merit factors have been introduced as an extension of merit factors employed for single-probe setups, hence, at first attempt, it comes quite sensible to design excitations for MIMO systems by modifying excitations designed for SISO systems [11, 13, 18].

Some main aspects of the design are worth mentioning before some techniques are derived, namely:

- *Random versus Deterministic Design.* In designing excitation for SISO setups two fundamental strategies have been exploited: (i) employ deterministic functions such as chirps; (ii) employ random sequences with special self-correlation properties. Clearly, these approaches represent countertrend solutions. On one hand, chirps are intrinsically satisfying the requirements of excitations suitable for SISO setups, on the other hand, random sequences need to be properly defined in order to get the required properties but they present some degrees of freedom which may be exploited for application purposes.

- *Joint design.* One might think that excitations for MIMO systems can be designed as a sequential process. According to the previous considerations, such sequences must have mutual properties that has to be intrinsically satisfied. Hence, the problem is to identify a class of signals which can be employed as excitations. Nevertheless, when a random approach is adopted, it is also possible that excitations are selected among a set of feasible signals according to proper criteria.
- *Degrees of freedom.* The number of channels, i.e., the number of TXs, imposes some constraints on the number of degrees of freedom which have to be available in the design process. In more detail, the main resource determining the amount of flexibility in finding suitable excitations is the TBP. Nevertheless, it has already been outlined that the matter of finding excitations which minimize the cross-channel interference cannot be approached as the identification of an orthogonal set. Because of the orthogonality by shift property, a larger number of degrees of freedom with respect to the number of TXs might be necessary.
- *Orthogonality versus Resolution.* When considering a SISO system, the merit factor quantifying the orthogonality by shift property is essentially measuring the system resolution. In a MIMO setup, although the concept is similar, the issue is to keep the channels independent, hence the actual aim is to design excitations being orthogonal. So, these two aspects may end up being conflicting: when aiming for a small cross-channel interference one might have to give up to a self-correlation accurately approximating the Dirac function; on the contrary, good self-correlations might result in poor performances in terms of cross-interference.

The above issues give a simplified perspective on the problem of excitations design, since in practice things are more complicated and many other factors contribute to determine the quality of a certain solution.

5.4.1 Inadequacy of Chirps for Multiprobe Setups

The excitations of a MIMO system, which is supposed to be an upgrade of a SISO system, should preserve the good quality of chirp excitations. Hence, one might ignore the cross-channel interference and consider a system where all the TXs employ simultaneously the same chirp excitation [11]. From a theoretical point of view, this system can be modeled as a single channel where multiple messages consisting of the amplitude coefficient, the time of flight, and the belonging class are sent. In case the minimum distance between time of flights is larger than a certain threshold, coefficients and time of flights are easily estimated. Conversely, the belonging class is not explicitly transmitted and might be extracted by combining information about coefficients and time of flights at the information processing level. This operation is potentially troublesome since there is no a priori information about the channel structure, that is mainly the number of reflectors.

Apart from the above considerations, a chirp excitation still has a degree of freedom which can be employed in order to realize a simple system consisting of two probes. In fact, given a certain bandwidth $[f_0, f_M]$ one can define:

$$u^{(1)}(t) = \sin\left(2\pi f_0 t + \pi\kappa t^2\right) \quad (5.15)$$

$$u^{(2)}(t) = \sin\left(2\pi f_M t - \pi\kappa t^2\right) \quad (5.16)$$

where $\kappa = (f_M - f_0)/T$, being $[0, T]$ the time window, so that $u^{(1)}$ linearly sweeps the frequency interval from f_0 to f_M while $u^{(2)}$ behaves in the opposite way. In order to evaluate if this pair can be actually employed for a simple two-probe setup, one has to estimate the behavior of the cross-correlation $R_{uu}^{(2,1)}(t)$. An explicit calculation cannot be performed, hence we provide a qualitative analysis which will be useful for the understanding of further developments. By considering its definition one gets

$$R_{uu}^{(2,1)}(\tau) = \int_{\max(0,\tau)}^{\min(T,T+\tau)} u^{(1)}(T - t + \tau)u^{(1)}(t)\,\mathrm{d}t \quad (5.17)$$

being nonnull for $\tau \in [-T, T]$. It is worth noting that if the integral is performed by adding up individual integrals referred to a suitable partitioning of the interval $[\max(0, \tau), \min(T, T + \tau)]$, the sub-interval mainly contributing to the integral is the one around $t = (T + \tau)/2$. This can be written as follows

$$R_{uu}^{(2,1)}(\tau) \simeq \int_{\frac{T+\tau}{2}-\varepsilon}^{\frac{T+\tau}{2}+\varepsilon} u^{(1)}(T - t + \tau)u^{(1)}(t)\,\mathrm{d}t \quad (5.18)$$

where ε is a suitable value. In fact, the representation of $u^{(1)}(t)$ and $u^{(2)}(T-t+\tau)$ with respect to the time–frequency rectangle $[-T, T] \times [f_0, f_M]$ are lines with opposite slopes which cross each other for $t = (T + \tau)/2$. Hence, inside the restricted integration interval, $u^{(1)}(t)$ and $u^{(2)}(T - t + \tau)$ are approximately sine functions at the same frequency $f_0 + \kappa(T + \tau)/2$. In case they are in phase, the result of the integration is about constant not depending on the frequency, otherwise it is null. By a further simplification, the integral above behaves as

$$R_{uu}^{(2,1)}(\tau) \simeq \gamma \int_{\mathbb{R}} \delta\left(t - \frac{T+\tau}{2}\right) u^{(1)}(t)\mathrm{d}t \quad (5.19)$$

where γ is a suitable constant, and finally

$$R_{uu}^{(2,1)}(t) \simeq \gamma \sin\left(2\pi f_M \frac{T+t}{2} - \pi\kappa\left(\frac{T+t}{2}\right)^2\right) \quad (5.20)$$

with $t \in [-T, T]$. The fact that the cross-correlation is shaped as a chirp is not relevant from an application point of view. More interestingly, the cross-correlation has a constant envelope. This qualitative analysis has two interesting points to be highlighted:

- *Time–Frequency dissimilarity.* In this simple two-probe setup, the strong difference between $u^{(1)}(t)$ and $u^{(2)}(t)$ in terms of time–frequency trajectory has been exploited. This can be considered a general guideline for the design of a class of functions suitable for multiprobe systems. Nevertheless, the kind of dissimilarity which has been employed here, being actually the sign of the slope on the time–frequency plane, cannot be systematically used for system with number of TXs larger than 2. Furthermore, it is worth highlighting that these two functions always have a crossing point in the time–frequency plane, thus making the cross-correlation lower bounded by γ.

- *Cross-correlation decay.* The self-correlation of a chirp has been shown to decay according to its TBP. Instead, the cross-correlation of $u^{(1)}(t)$ and $u^2(t)$ has a constant envelope, i.e., the amount of interference coming from a cross-path is not depending on its time of flight. Without defining a specific technique at the information processing level to gather the information about each channel in view of accurately characterizing the SUT it is not possible to assert which features are most desirable in self and cross correlations. At this point, we can just remark that being able to realize excitations with decaying cross-correlation might be a requirement for certain applications and frameworks.

Finally, according to the proposed analysis, we have illustrated how simple chirps are not suited for multiprobe system. Notwithstanding this, some interesting points have been highlighted which may be useful for the introduction and understanding of more effective approaches.

5.4.2 PN Sequences in Multiprobe Setups

The employment of randomness in order to get a class of excitations having δ-like self-correlations and sufficiently small cross-correlations has been already experimented in digital communications and is known as the framework of code division multiple access [19]. In multiuser networks, a very efficient exploitation of the available bandwidth is reached by means of assigning to each user a spreading sequence, so that each transmission always uses the whole spectrum and variable fading effects typical of wireless communications are averaged. In such wireless networks, two kinds of transmission are possible, namely synchronous and asynchronous to the receiver. The synchronous mode requires that all the transmissions from various users arrive at the receiver at the same time, hence the transmitters have to know their time-of-flight. Of course this approach cannot be pursued in ultrasound systems, since the time-of-flight is itself a value to be estimated through the transmission. Conversely, the asynchronous mode employs pseudo noise (PN) spreading sequences which are designed so that a certain amount of tolerable cross-channel interference is always present but no knowledge about the time-of-flight is required [7, 10, 12].

The paradigm of code division multiple access might be exploitable in ultrasound multiprobe setups. Nevertheless, it is worth remarking some points:

- *Constant envelope.* In communications, PN sequences are upconverted to the high frequency where the wireless system operates. In ultrasound system this full upconversion is not necessary and the sequences are typically fed into the transmitter power amplifier by merely applying a proper upsampling, or at most by using inverse repeated sequences (IRS). Hence, the transducer works with a signal that is never perfectly matching its bandwidth and whose envelope may not be constant, thus potentially encountering troubles due to its nonideality.
- *Correlations.* The quality of correlations of PN sequences is important as long as one has the possibility of getting back from the SUT signals of sufficient quality. Conversely, because of the modulation technique which does not take into account the band-pass characteristic of ultrasound systems, signals might get corrupted (particularly when IRS is not used) and the capability of building a multiprobe system with multiple simultaneous excitations might be compromised.

According to the above mentioned issues, in order to take advantage of randomness for generating excitations suitable for MIMO setups we refer to specific techniques taking a constant envelope specification as strict requirement.

5.5 Random-FM Excitations

In Chap. 4, PuC has been practiced in ultrasound NDT systems using chirps. As illustrated in this chapter, chirps are inadequate for setups where multiple probes need to operate concurrently in MIMO arrangements. Conversely, systems based on coded excitations like PN sequences miss some chirp advantages (constant envelope excitation, easiness of bandwidth control, etc.) and may not be easily implemented on hardware originally conceived for chirp excitations. So, systems based on random-FM excitations have been proposed [4], capable of enabling MIMO with minimal changes with respect to a chirp-based setup. Random-FM excitations are shown to retain many advantages of chirps and provide the ability to frequency-shape the excitations matching the transducers features.

It has been already illustrated how wide-band excitations such as those needed for PuC may offer the opportunity to create families of excitations where the members are approximately orthogonal and thus suited for MIMO operation.

The random-FM approach may be seen as a bridge between chirp-based excitations and PN excitations. It is based on the FM-modulation of pulse amplitude modulation (PAM) sequences so that hardware changes with respect to chirp based systems can be minimal and constant envelope operation can be guaranteed. It offers sufficiently good cross-correlation properties for MIMO. It can be extended to differentiate excitations based on discrete codes. It provides an easy adjustment to the probe bandwidth. Furthermore, it has advantages of its own, like the possibility to carefully shape the power spectral density (PSD) of the excitations, following [3]. Similarly to nonlinear chirps [17] this lets one equalize the probe response, to enhance the actual bandwidth (and thus resolution) or transferred power [16].

5.5.1 Review of Random-FM

A chirp can be also written as a function of a normalized instantaneous frequency $x(t)$ in the following way

$$u(t) = \mathrm{Re}\left(e^{2\pi \mathrm{i}(f_c t + \Delta f \int_{-\infty}^{t} x(\tau)\,\mathrm{d}\tau)}\right) \tag{5.21}$$

where f_c is the central frequency, Δf is the maximum frequency deviation from f_c, and $x(t)$ is a monotonically increasing, smooth modulating signal taking values in $[-1, 1]$. The simplest case is evidently that of a linear chirp in which $x(t) = -1 + 2t/T$ where T is the chirp length.

Random-FM excitations can be generated with the same modulator as chirps, given in (5.21), yet using a more articulated PAM modulating waveform

$$x(t) = \sum_{k=-\infty}^{+\infty} x_k g(t - kT_c) \tag{5.22}$$

where $g(t)$ is a unit pulse of duration T_c and the values $x_k \in [-1, +1]$ make up a random modulating sequence. By Fig. 5.4 it is possible to see that arbitrary couples of random-FM excitations have relatively low cross-correlations, since the PAM sequence acts as a *signature* differentiating them. This makes random-FM suited for MIMO.

A thorough theory of random-FM signals has been developed in [3]. Particularly, it is proved that the power spectral density of an (infinitely long) modulated signal depends on the probability density function (PDF) of the modulating sequence according to

$$\Phi_{uu}(f) = \int_{-1}^{1} K_1(x, f - f_c)\rho(x)\,\mathrm{d}x + \left(\frac{\left(\int_{-1}^{1} K_2(x, f - f_c)\rho(x)\,\mathrm{d}x\right)^2}{1 - \int_{-1}^{1} K_3(x, f - f_c)\rho(x)\,\mathrm{d}x}\right) \tag{5.23}$$

where $K_1(x, f)$, $K_2(x, f)$, $K_3(x, f)$ are integration kernels, $\Phi_{uu}(f)$ is the PSD, the modulating sequence is assumed to be made of independent samples, and $\rho(x)$ is its PDF. The relationship can be interpreted as a (nonlinear) *smoothing and leaking* operator so that the $\Phi_{uu}(f)$ tends to be shaped as $\rho(x)$ with some distortion. According to relationship (5.23) one can *design* a modulating sequence PDF capable of producing a desired PSD by means of iterative methods [2] or optimization techniques [1].

Furthermore, for *large* T_c the above equation can be greatly simplified according to the following approximation:

$$\lim_{T_c \to \infty} \Phi_{uu}(f) = \frac{1}{2\Delta f}\,\rho\left(\frac{f - f_c}{\Delta f}\right) \tag{5.24}$$

where the PSD accurately *copies* the PDF. Clearly, T_c cannot be considered large in an absolute way, instead, one needs to refer it to the system bandwidth $2\Delta f$. Thus, it is convenient to define a *modulation index* $m = T_c \Delta f$. As long as m is a few units or more, (5.24) provides a satisfactory approximation and the modulation will be referred to as *slow*. Otherwise, the modulation will be referred to as *fast* and the reference equation is (5.23). Interestingly, in the *slow* modulation case, a desired PSD can be designed by means of (5.24) with direct inversion.

5.5.2 Application of Random-FM to Multiprobe Setups

The application of random-FM excitation to ultrasound systems must be supported by some considerations about pulse-compression and MIMO.

- *Self-correlation away from the peak.* These correlation entries contribute to the noise floor at the receiving probes as self-interference, which may hide the detection of secondary paths. By comparing Figs. 5.3 and 5.4, it is evident that the self-correlation is higher for the random-FM excitation than for the chirp, which may appear problematic. However, a few aspects are worth considering. First, the self-correlation of the chirp keeps decreasing at a significant rate even at large lags. Conversely, the random-FM one only decreases rapidly for small lags, then it flattens. Thus, the best self-correlation entries (large lags) are certainly much better for the chirp, yet this does not necessarily mean that the random-FM excitation is much worse in the worst case (small lags). Secondly, the higher self-correlation profile is the price that one pays for the MIMO abilities. Indeed, other coded excitations suitable for MIMO also pay a price in self-correlation with respect to the chirp. Furthermore, in a MIMO setup, there is not just self-interference, but also mutual-interference that can be visualized through the cross-correlation entries. Having a self-interference much lower than the mutual-interference would only bring marginal advantages since the latter would dominate. Consequently, a fair evaluation of the random-FM self-correlation curve also requires a comparison to the cross-correlation curves. Since the self-correlation floor of the random-FM excitation is no worse than its cross-correlation floor or the cross-correlation floor of the chirp, one may conclude that the self-correlation floor is actually *good enough* since any improvement would only bring marginal advantages in a MIMO setup.
- *Correlation dependency on system parameters.* Even if a complete discussion of this aspect is out of the scope of this chapter, it is worth noticing that (as expectable) the correlation floors of random-FM modulations scale (in amplitude) with $1/\sqrt{TB}$ where TB is the TBP of the excitation. Incidentally, the same relationship holds for the cross-correlation floor of PN sequences such as Gold or Kasami. This provides some ability to tune the noise levels due to self- and mutual-interference to the application needs.
- *Secondary peaks in the correlation curves.* In Fig. 5.4, the correlation plots of the random-FM excitation appear irregular and peaky compared to those of the chirp in Fig. 5.3. Secondary peaks are dangerous as they might be misinterpreted for weak

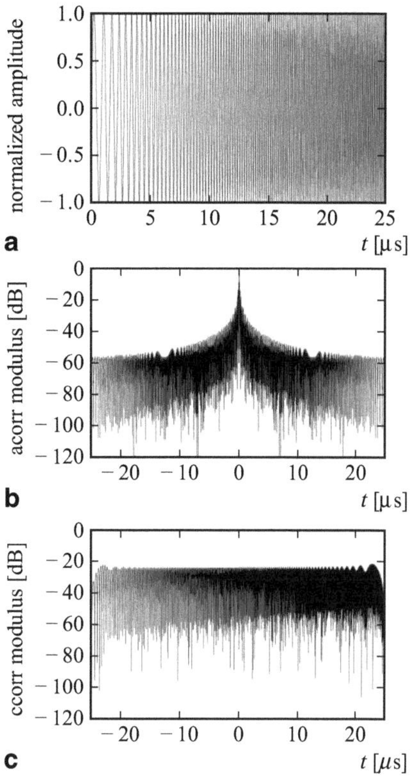

Fig. 5.3 A linear chirp **a**, together with the modulus of its acyclic auto-correlation **b**, that illustrates the suitability for matched filter schemes, having a well-localized central pulse and being rapidly vanished elsewhere. Clearly, with this excitation it is impossible to have many arbitrary transmitters simultaneously operated. The only family of quasi-orthogonal functions that can be derived from it is made just of the chirp and its reversed waveform. The modulus of their cross-correlation (accounting for the mutual interference in a matched filter-based receiver) is shown at the **c**

paths from other reflectors at the receiving probes. Focusing on the self-correlation curve, an intuitive investigation of their origin is possible. Self-correlation is tied to PSD by the Fourier transform. The PSD of random-FM signals, as returned by Eq. (5.23) is, by the very properties of the involved operators, quite regular. The corresponding self-correlation should necessarily be smooth. Thus, the peakiness in Fig. 5.4 (middle) is not intrinsic in the random-FM waveforms. Its origin gets evident considering that Eq. (5.23) holds for infinitely long signals. Peakiness emerges as a short-length effect since the number of PAM pulses used to build the excitation is limited, so that the value distribution of the modulating sequence can significantly differ from the prescribed PDF. Similar considerations could hold for the peaks in the cross-correlations. Thus, the secondary peaks can be reduced by enlarging T/T_c. Being $T_c = m/\Delta f$, peaks can also be seen as a consequence of low $(T\Delta f)/m$. Therefore, large m values, which may appear convenient for spectral control, can actually be undesirable. Note that in Fig. 5.4 the peakiness is accentuated by the choice of very short excitations (25 μs) and relatively large m (1), used for representation purposes.

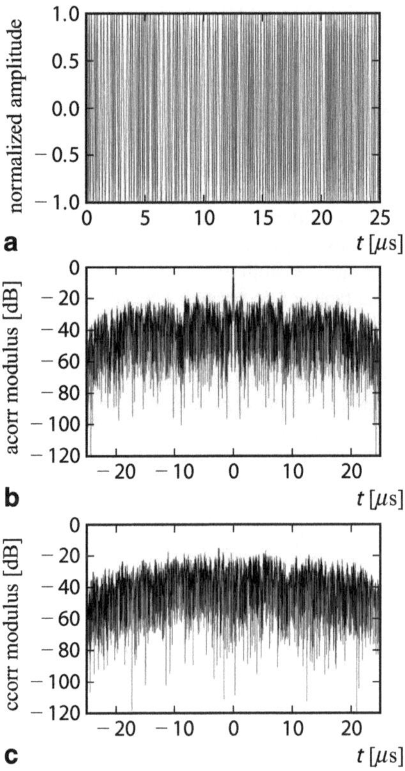

Fig. 5.4 A random-FM excitation **a**, together with the modulus of its acyclic auto-correlation **b**, which exhibits a central peak being remarkably larger than the *floor* due to self-interference. The decay of the self-correlation is not as good as in the chirp case, nevertheless the cross-correlation **c**, resembles the self-correlation apart from the peak and many excitations keeping this behavior can be generated and employed simultaneously

5.5.3 Flexibility and Advantages of the Random-FM Approach

The intrinsic advantage of random-FM with respect to other code-division based approaches is the possibility of employing the same hardware used for chirps. Furthermore, random-FM offers other advantages, mainly related to the possibility of performing spectral shaping with flexibility without compromising the pulse-compression property and the MIMO applicability.

To appreciate this point, we review the issue of the system adaptation to the probe features, i.e., the excitations are passed to/from the SUT through probes and amplifiers that are typically characterized by nonflat frequency responses, hence the excitations should be designed in order to compensate this fact and to potentially take advantage of it. Typically, in common probes a remarkable part of the excitation bandwidth and of the excitation power is lost. Spectrum shaping can be theoretically employed in order to compensate either the loss of power or the loss of bandwidth, hence, depending on the particular application setup, a suitable tradeoff can be identified.

1. If the presence of noise is critical and the dominant noise components do not scale with the excitation power (i.e., dominant components are not self/mutual interference due to the transmitters, nor dispersion noise due to the intrinsic structure of the material, rather effects like thermal or quantization noise), reducing the power loss on the probes may be beneficial. Hence, it is useful to make the excitation spectrum second the probe response according to the paradigm of matched filtering. As an effect, the power loss is reduced while the loss in bandwidth is tolerable.
2. In case the noise does not remarkably affect performances, spectral shaping of the excitation can be exploited in order to maximize the bandwidth. This can be done by picking a PSD equalizing the probe response.

It is worth recalling that a spectral shaping cannot be practiced starting with a uniform-PSD excitation and applying a linear filter, because the filter would make the envelope of the excitation nonconstant. Conversely, random-FM signals can obtain the desired power distribution quite simply by picking a modulating PAM sequence with a PDF shaped as the desired spectrum. Indeed, being the required spectrum (be it seconding or equalizing the probe response) typically smooth, either slow or fast modulations can generally achieve it reasonably well.

5.6 Excitations Based on Tessellations of the Time–Frequency Plane

In the investigation of techniques for generating pseudo-orthogonal excitations, it is not trivial to foresee the self-correlation (SC) and cross-correlation (CC) behaviors by simply studying their properties in the time domain. Instead, a better insight into the pseudo-orthogonality of an arbitrary family of excitations can be gathered by visualizing the corresponding waveforms in the time–frequency (TF) domain [6]. This task might be approached by evaluating spectrograms as well as other TF transforms or, since the signals to be analyzed are analytically known and have constant envelopes, by considering their frequency trajectories in the TF plane.

This point of view has already been hinted in the previous sections, especially about chirps. The frequency trajectory of a linear chirp is a diagonal line crossing the rectangular area identified by the intervals $[0, T]$ and $[f_c - \Delta f, f_c + \Delta f]$. Conversely, in random-FM excitations the frequency trajectory is represented by the modulating PAM sequence $x(t)$ after the application of a proper scaling factor and offset. Obviously, having to visit all the available frequency intervals, random-FM signals approximately span the same region of the TF plane. So, the *coverage* of the TF plane is not by itself a sufficient criteria to characterize the properties of an excitations family. Actually, it is the specific *way* in which the *TB* region is spanned that counts and an empirical approach can be set up to gather hints at how this influences SC and CC profiles.

5.6.1 Fundamentals of Tessellation Based Approaches

When analyzing a signal according to its TF structure, since time and frequency features are not independent, it is convenient to assume a tiling of the TF plane in basic blocks, which may possibly be a basis, according to rectangular scheme, such that all the possible excitations can be represented with respect to same set of functions. Tessellation schemes other than the rectangular one may be considered. Nevertheless, having excitations with constant envelope is a major specification, and a rectangular tiling is well suited to constant envelope signals.

By means of this approach, it comes intuitive that to achieve a rapid decay of the $R_{uu}^{(k,k)}(\tau)$ envelope for large $|\tau|$, one should assure that a single instantaneous frequency is never repeated. Hence, with respect to a grid where each block is identified by a (t, f) couple in which t is the central time and f the central frequency, to impose that in a single excitation each f is employed only once might be a feasible strategy to control the SC envelope. With this, to span all the available bandwidth, it can be convenient to assume the same number of divisions of T and B when creating the grid. Thus, given a desired area of the basic tile Δ_{TB}, the number of divisions can be determined as

$$N \simeq \sqrt{\frac{(TB)}{\Delta_{TB}}} \tag{5.25}$$

where some adjustment is applied in order to make $N \in Z$. For instance, attempting to take $\Delta_{TB} = 1$, one might want to keep the area covered by the resulting tiles strictly inside $[0, T] \times [f_c - \Delta f, f_c + \Delta f]$, so that N can be conveniently determined by floor approximation in (5.25).

After this procedure, a square $N \times N$ matrix can be associated to a generic FM-excitation such that the column index corresponds to frequencies and the row index corresponds to time instants. The matrix is necessarily *sparse*, i.e., most entries must be 0, while nonnull entries are equal to 1 and identify the active tiles in the corresponding excitation. For instance, a linear chirp excitation is associated to the diagonal matrix while other signals such as random-FM excitations, are associated to generic sparse matrices. In better detail, to avoid repetition of any specific frequency and to guarantee the complete span of all the available frequencies, each column and row must sum to 1. This condition is satisfied by all matrices $\boldsymbol{P}$ associated to permutations of the N-dimensional space

$$\{\boldsymbol{P} \in \{0,1\}^{N\times N} : \boldsymbol{P}\boldsymbol{P}^{\mathrm{T}} = \boldsymbol{P}^{\mathrm{T}}\boldsymbol{P} = \boldsymbol{I}\}. \tag{5.26}$$

It is worth noticing that excitations based on permutations like the one above do not allow any sort of spectrum shaping to be practiced. However, in some cases it is possible to recover some spectrum shaping abilities by a minor relaxation of some constraints.

For a N-dimensional space, the set of orthogonal permutation matrices has $N!$ elements. Even for low N, this number is large, so that an exhaustive search is unfeasible and other techniques are required in order to identify the permutations suited to design specifications. So, we review two different strategies for generating pseudo-orthogonal excitations and base them on the exploration of a significant subset of the permutation space. The first one may be seen as a variant of the random-FM approach, while the other attempts at recovering some desirable chirp features by sticking to permutations where only a limited number of diagonals around the main one can contain nonnull entries.

5.6.2 Design of the TF Tessellation

Although the problem of designing excitations with prescribed SC might be tackled in many ways, taking advantage of the TF representation has been outlined as an effective strategy for controlling the global envelope of correlations [5]. In order to define random energy distributions being capable of generating excitations suitable for multiple TX and potentially preserving some advantages of chirps, a description of the TF in terms of available degrees of freedom is needed. For this reason, a continuous-wise design procedure would be unfeasible. Instead, splitting the TF plane in discrete atoms allows for tackling the problem with a combinatory approach. So, the basic idea is to decompose the available time–frequency area in basis tiles of area equal to 1, then a custom procedure is applied in order to identify permutations to address some specifications. The choice of $\Delta_{TB} = 1$ allows for considering an orthogonal basis as well, which might be convenient for guaranteeing null SC or CC for certain lags. Yet, this is not mandatory.

Hence, the time and frequency steps are referred to as δt and δf respectively with $\delta t = T/N$, where N is somehow related to T and B, but is not necessarily $N = \sqrt{TB}$, since one might want either to take into account the effective bandwidth occupied by a sinusoidal burst or to maximize the exploitation of the available area on the TF plane so enlarging N. A generic expression of an excitation is given by

$$u(t) = \sum_{n=1}^{N} \cos\left(2\pi\phi(p(n))t + \gamma_{p(n)}\right) \chi_{[0,1]}\left(\frac{t}{\delta t} - n + 1\right) \tag{5.27}$$

where $p : \{1, \ldots, N\} \rightarrow \{1, \ldots, N\}$ is a permutation function, ϕ is a function selecting a frequency over a set of allowed tones, $\chi_{\ldots}$ is the characteristic function of the set in the subscript, and $\gamma_{p(n)}$ is a suitable phase making the phase of $u(t)$ continuous. Hence, the excitation is expressed as the concatenation of subsequent sinusoidal bursts at different frequencies. In order to emulate the SC behavior of a chirp by means of a nonlinearly varying frequency modulation, having a continuous phase excitation is a necessary requirement to avoid sudden variations which might cause unwanted peaks. As it has already been said, the choice $\Delta_{TB} = 1$ is considered since $\Delta_{TB} > 1$ would produce underdetermined excitations not taking advantage of all the available degrees of freedom, while $\Delta_{TB} < 1$ would produce overdetermined excitations where the redundancy might cause some issues when generating multiple sequences with requirements on CC properties. Hence, it follows $\delta f = 1/\delta t$, while the set of frequencies is given by

$$\phi(n) = f_0 + \frac{n-1}{\delta t} \qquad n = 1, \ldots, N. \tag{5.28}$$

being f_0 the minimum available in-band frequency. With this choice, it can be easily verified that imposing the continuous phase for $t = k\delta t, k = 0, \ldots, N-1$, i.e., $2\pi\phi(p(k))k\delta t + \gamma_{p(k)} - 2\pi\phi(p(k+1))k\delta t + \gamma_{p(k+1)} = 2\pi m, m \in \mathbf{Z}$ results in $\gamma_{p(k)} = 0$. In particular, if f_0 was chosen to be an integer multiple of δf, the tessellation would result to be a nonlapped short time fourier transform basis, hence $[u * \mathrm{Rev}[u]](k\delta t) = 0,\ k \neq 0$.

In case multiple excitations $u^{(k)}$, $k = 1, \ldots, K$, are considered, for evaluating performances, we refer to the merit factors which have been introduced in Sect. 5.3.2, namely $P^{(k,n)}(\theta)$. In order to better display these functions, we introduce

$$\nu_S(\vartheta) = P^{(k,k)}(1/\vartheta) \tag{5.29}$$

$$\nu_C(\vartheta) = P^{(k,n)}(1/\vartheta) \tag{5.30}$$

where $\vartheta \in [1, \infty)$ represents an attenuation and the reference to the specific excitations, i.e., (k,k) or (k,n), has been omitted since these parameters are meant to represent the generic performances according to the specific class of excitations which is evaluated.

5.6.3 Radar Derived Approaches

According to the point of view introduced so far, the task of designing excitations for ultrasound systems can be approached as the identification of a suitable permutation matrix. When the size of the permutation space, namely N, is large, finding these matrices may be a challenging task. From number theory, it is known that permutations carrying special properties useful for radar applications can be obtained in many ways, even for large N. For instance, given a prime number p, excitations with $N = p - 1$ are obtained by taking the primitive roots g's of p and defining the sequences of the powers of the primitive roots modulo p, $\boldsymbol{P}_i[n] = (g_i)^n \bmod p$. The number of primitive roots n_g is equal to the Totient Euler function of $p : n_g = \phi(p)$. Starting from the permutations generated by the primitive roots, other permutations can be retrieved by exploiting particular properties [8].

Permutation families suitable for designing excitations for radar applications have been extensively studied in literature, although often with additional requirements to those discussed here, since radar poses challenges of its own, like dealing with moving targets and the Doppler effect. The latter imposes to evaluate features from the self- and cross *ambiguity* functions [8] rather than SC and CC. However, since none of the requirements described here is removed, solutions designed for radar can be regarded as acceptable for NDT, although the extra constraints may affect

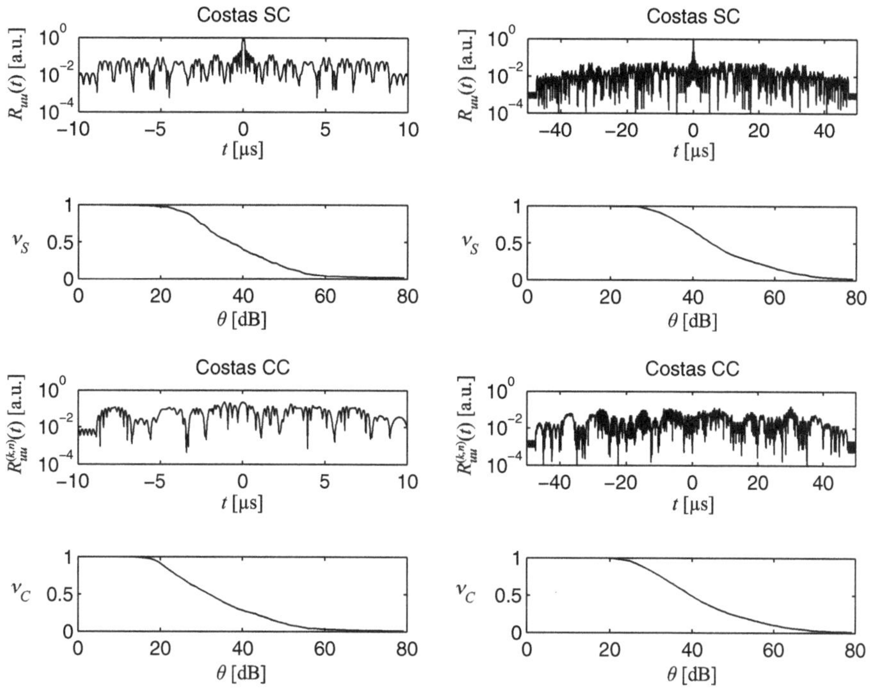

Fig. 5.5 SC, CC, and merit factor plots for excitations based on Costas arrays. *Left*: $T = 10\ \mu$s; *right*: $T = 50\ \mu$s. Other test conditions include $B \approx 2\Delta f = 8$ MHz with an $f_0 = 5$MHz central frequency

the quest for optimality according to the introduced merit factors, namely $P^{(k,k)}(\theta)$. A remarkable approach devised for radar is given by the so-called Costas Arrays [10] that are precisely based on the primitive root generation method illustrated above. These sequences are characterized by an ideal self-ambiguity function (i.e., thumbtack) and good cross-ambiguity functions.

From the thumbtack self-ambiguity function, one gets an SC that is characterized by an almost constant envelope at large lags. Consequently, these codes can marginally improve over random-FM, but the advantage is minor. In Fig. 5.5 the behavior in terms of ν_S and ν_C is illustrated.

5.6.4 Diagonal-wise Excitations

The excitations introduced in Chap. 4 and in the present chapter exhibit different behaviors in terms of SC decay at increasing lags. Ultrasonic systems which employ single frequency bursts feature a linearly decaying SC, where the decay is uniquely due to the increased shift between the bursts. With regard to chirps, a faster SC decay has been obtained, thanks to the employment of waveforms spanning the whole available bandwidth, such that the decrease in correlation due to shift between the bursts is also increased by the decrease in correlation between different local frequencies. This results in a progressive decay with a rate that becomes approximately constant as soon as the modulus of the lag is sufficiently large. On the other hand, random-FM and Costas excitations employ all the available bandwidth in an *unstructured* way, so that the correlation for increasing lag is further decreased by the resulting incoherence but shows an approximately constant floor out of the main lobe. According to this review, an optimal excitation should be designed to match both the behavior of PuC excitations for large lags and the behavior of random-FM and Costas excitations for small lags. It it is worth recalling that Costas and random-FM excitations have been mainly introduced in compliance with MIMO paradigm whereas decaying properties have been only qualitatively considered as a performance parameter but no design criteria have been set.

In order to establish a design criteria for matching the fast decay property for both small and large lags at the same time, the mechanism of decay has to be modeled in a single manner. The main idea is to detect a domain where these two properties can be easily described and potentially merged. Since it has been already outlined that the decrease in correlation can be obtained by either the decrease in superimpositions between bursts or by their intrinsic incoherence due to different frequency components, it comes natural to evaluate the behavior of correlation in the TF domain. Both the desired behaviors can be effectively characterized by focusing on the energy distribution on the TF plane. In fact, the scalar product between two shifted versions of the same chirp decreases as the distance between their TF representations, which are ideally two lines, increases. Conversely, the TF representation of a random-FM excitation is a collection of spots being spread on the whole TF plane, so that the scalar product between a pair of shifted versions of the same sequence for short lags can result in a small correlation thanks to the exploitation of a large TF area, while for large lags random coverages of the TF are still superimposed resulting in an almost stationary behavior of correlation.

Hence, excitations matching the good property of random-FM and chirps can be considered in order to be able to deal with various system design requirements and specifications, while keeping the MIMO paradigm [5]. These excitations are here referred to as *diagonal-wise* since the basic idea behind their construction is to identify multiple energy distributions on the TF plane being concentrated along the diagonal as for the ideal TF distribution of a chirp. The result is a set of excitations bridging the gap between linear chirps and random-FM sequences. i.e., approximating the decay of linear chirps for large lags and resembling the noise floor of the random-FM approach for small lags.

The framework might be summarized as follows. Given certain duration and bandwidth, a chirp exhibits the most favorable performances in terms of $\nu_S(\vartheta)$, i.e., it starts decaying from 1 to 0 for a larger value of ϑ and the whole $\nu_S(\vartheta)$ curve represents an upper bound with respect to all other excitations. The performances of random-FM can be equivalently obtained according to the above mentioned TF tessellations by considering permutations guaranteeing a sufficiently random coverage of the TF plane. The resulting $\nu_S(\vartheta)$ can be considered a lower bound, meaning that a specific choice of permutations might bring better performances potentially approaching the performances of chirps.

Hence, the idea is to look for permutations being similar to the identity, so that the corresponding performances will approximate the one of chirps for large lags, but still being able of producing CC suitable for MIMO setups. So, one may define the concept of maximum distance of a permutation from the identity and identify a combinatory strategy to describe all the possible permutations with fixed maximum distance from the identity one. The resulting combinatory problem is mathematically challenging since the number of total permutations increases with $N!$, hence a complete scan cannot be performed. Moreover, one might want to characterize the *whole* space of desired permutations so that optimization procedures can possibly be pursued.

With these premises, let us suppose that the grid size is $N \times N$, the vector to be permuted is $\boldsymbol{v} = [1, \dots, N]$, the permuted vector is referred to as $\boldsymbol{p}$, and the maximum distance from the diagonal is D, i.e., max $|\boldsymbol{v} - \boldsymbol{p}| = D$. A generic permutation can be represented as a $N \times N$ permutation matrix $\boldsymbol{P}$. In order to accomplish the specification on D, it comes natural to cover the main diagonal with smaller combinational blocks. Yet, the coverage of the entire space of permutations can not be performed by splitting the permutation matrix in subblocks, so a different strategy is required. The basic idea is to split the permuted sequence in subsequences, such that each one can be chosen among a set of feasible subsequences and has to be compatible only with its contiguous ones, i.e., the *memory* is equal to 1. With this, the entire sequence can be described as a collection of *states* which are chosen according to a state diagram.

Let these considerations be formalized. We define as *state* an ordered collection of R items with $2D \leq R < N$. As it will be immediately clear, it is convenient to take R as small as possible, namely $2D$. For the sake of simplicity, N is assumed to be an integer multiple of R, i.e., $N = LR$, $L \in Z$. The overall permutation $\boldsymbol{P}$ is built from L sub-permutations each expressed by a state. These are in fact partial R-permutations of $R + 2D$ elements represented by $R \times (R + 2D)$ matrices $\boldsymbol{P}_s^{(i)}$, where the additional $2D$ items take into account the distance from the diagonal. The following constraints are in place for all $1 \leq i \leq L$

$$\begin{aligned} \forall j \, \boldsymbol{P}_s^{(i)}(j, j+k) &= 0 \quad k \geq 1 \\ \forall j \, \boldsymbol{P}_s^{(i)}(j+2D, j-k) &= 0 \quad k \geq 1. \end{aligned} \tag{5.31}$$

Furthermore, the initial and final state also require some additional conditions since the overall permutation $\boldsymbol{P}$ is defined on a limited set. Accordingly, a generic input vector is defined for the partial permutations as

$$\boldsymbol{v}_s = [1, \dots, R + 2D]. \tag{5.32}$$

With this, the i-th state is $\boldsymbol{p}_s^{(i)} = \boldsymbol{v}_s \boldsymbol{P}_s^{(i)}$ and the overall sequence $\boldsymbol{p}$ is

$$\boldsymbol{p} = [\boldsymbol{p}_s^{(1)}, \, \boldsymbol{p}_s^{(2)} + R, \, \dots, \, \boldsymbol{p}_s^{(i)} + (i-1)R, \, \dots, \boldsymbol{p}_s^{(L-1)} + (L-2)R, \, \boldsymbol{p}_s^{(L)} + (L-1)R] \tag{5.33}$$

where the comma indicates the stacking of subvectors and the addition of a scalar to a vector is applied componentwise.

From a practical point of view, the partial permutations $\boldsymbol{P}_s^{(i)}$ can be generated by first considering the combinations of R items over $R + 2D$, then generating all the possible permutations from them and finally sieving away all those not respecting the constraints. Keeping R small assures that the partial permutations are manageable and practically enumerable.

In the mechanism presented so far, one constraint still needs to be applied, namely that there is no item repetition in $\boldsymbol{p}$. For the sake of simplicity, one wants to be able to enforce this constraint by only looking at the relationships that each state has with its *contiguous* ones. In other words, the relationship should be expressible by a memory-1 constraint on the sequence of states $\boldsymbol{p}_s^{(i)}$. To see how this can happen let $\mathcal{S}[\boldsymbol{v}]$ indicate the set of the entries of vector $\boldsymbol{v}$. Now, the condition can only be satisfied if for all possible states

$$\mathcal{S}[\boldsymbol{p}_s^{(i-1)} - R] \cap \mathcal{S}[\boldsymbol{p}_s^{(i+1)} + R] = \emptyset \tag{5.34}$$

which in turn requires $R \geq 2D$. This justifies the lower bound placed on R at the beginning.

Now, let the memory-1 constraint be finally expressed by building a state transition matrix describing which configurations are allowed for state $\boldsymbol{p}_s^{(i+1)}$ assuming known the configuration at state $p_s^{(i)}$. The allowed configurations must clearly satisfy

$$\mathcal{S}[\boldsymbol{p}_s^{(i)}, \boldsymbol{p}_s^{(i+1)}] \cap \{D+1, \ldots, 2R-D\} = \{D+1, \ldots, 2R-D\}. \tag{5.35}$$

To exemplify this rather long procedure, we report the case $D = 1$ with $R = 2D$. The total number of partial permutations referred to each state $\binom{4D}{2D}(2D)! = 12$, but only 7 configurations are compliant with the condition (5.31) on the maximum distance from the diagonal. Furthermore, by border conditions one can identify only 3 configurations that are suitable as an initial state and only 3 as a final state. Applying the transition constraints, the transition diagram in Fig. 5.6a is obtained. Figure 5.6b eventually illustrates the construction of a frequency hopping sequence from the transition diagram. Note that by construction, the diagram must always incorporate 2 isolated states, that—not being suitable as initial or final states—are completely useless in building hopping sequences.

As far as the detection of states and the building of a transition matrix is concerned, enumerating and sieving partial permutations is currently feasible only for D no larger than 3. This can also be inferred by looking at Table 5.1, which summarizes the number of useful (nonisolated) states and of useful state connections. Furthermore, the table estimates the number of deliverable hopping sequences (ignoring the constraints on initial and final states). Clearly, all numbers increase very rapidly with D. For larger distances, the complete listing can be substituted by a random choice of permutations up to a certain number of states. In this approach, states which are not forwardly connected to other states must clearly be discarded. It is worth remarking that, as D increases, the number of permutations to be collected has to grow, since the probability of having nonconnected states also increases.

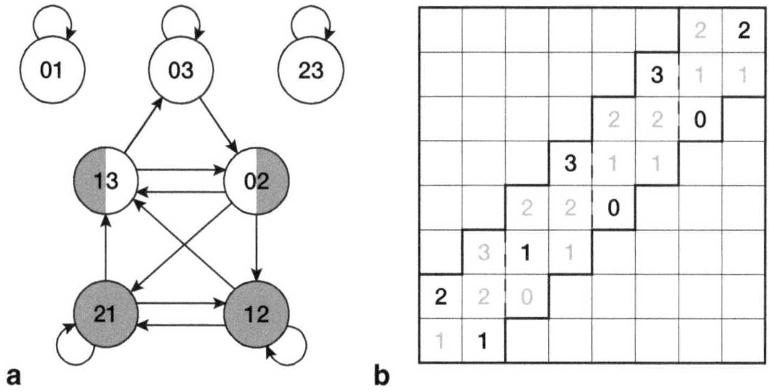

Fig. 5.6 State diagram corresponding to the construction of diagonal-wise permutations with maximum distance from the main diagonal $D = 1$ (**a**) and a sample sequence of states built on a 8×8 grid (**b**). In (**a**) states *grayed* on the *left side* are suitable as initial states, states *grayed* on the *right* are suitable as final states, and *fully grayed* states are suitable for both roles

Table 5.1 Number of states, connections, and sequences for increasing D values

D	No. of states	No. of connections	Connection density	No. of sequences
1	5	13	0.52	1.9^N
2	244	5616	0.094	2.94^N
3	21,650	13,403,090	0.029	3.93^N

As a last remark, it is worth hinting at a possible exploitation of the proposed permutation-based model. Since the excitations are built by an orthogonal basis, for $n \in N$, one has $R_{uu}(n\,\delta t) = 1$ for $n = 0$ and is null otherwise. Thus, the sum of the SC function sampled at lags multiple of δt is 1, and a similar argument also holds for the CC. Hence, the shape of $R_{uu}^{(k,n)}(n\delta t)$, which is nonnull for $n = -2D, \ldots 2D$, might be optimized according to some specifications. For instance, one might want to make this shape as flat as possible, or to annihilate the cross correlation for a specific n, etc. Moreover, it follows that, as D is increased, the peaks can be reduced to approximately $1/(2D + 1)$.

Behaviors corresponding to excitations based on diagonal-wise permutations are illustrated, for the usual test conditions, in Fig. 5.7. The distance 2 and distance 4 cases have been evaluated.

As evident from the plots, the approach lets the SC and CC envelope profiles resemble that of random-FM and Costas array-based excitations for short lags, while converging to the chirp behavior at larger lags. In ν_S and ν_C this allows more gentle decay profiles to be obtained.

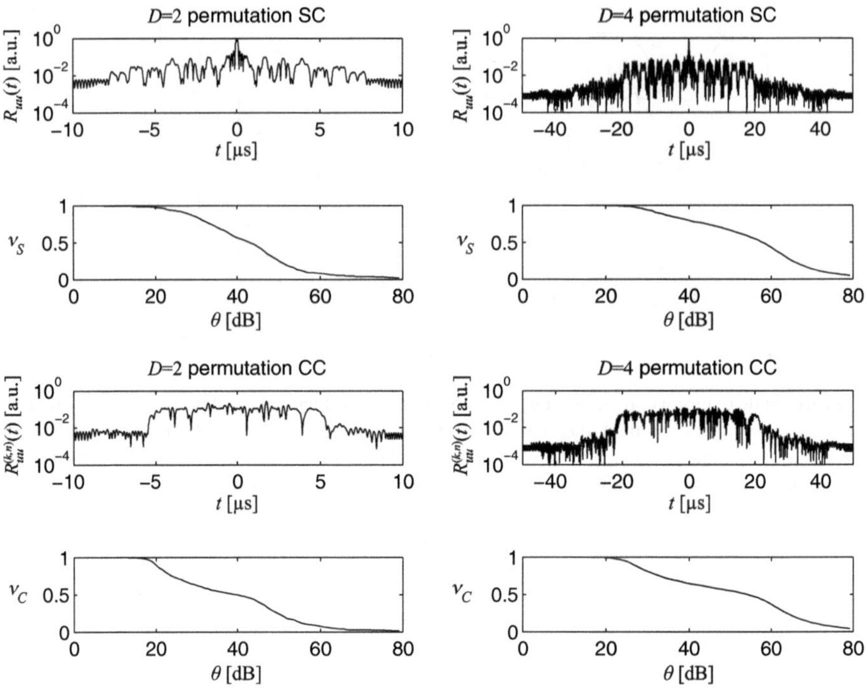

Fig. 5.7 SC, CC, and merit factors for diagonal-wise permuted excitations. *Left*: T equal to 10 μs; *right*: T = 50 μs. Other test conditions include $B \approx 2\Delta f = 8$ MHz with an $f_0 = 5$ MHz central frequency

5.7 Conclusion

In this chapter, main issues concerning the extension of single-probe NDT setups to multiprobe setups have been analyzed and discussed. The system model has been properly redefined by taking into account the need of employing multiple excitations, thus requiring to evaluate performances in terms of both SC and CC. Furthermore, different strategies to synthesize excitations suitable for multiprobe systems and corresponding to different design specifications have been presented together with their advantages, drawbacks, and overall performances. In particular, a major attention has been devoted to random-FM excitations and diagonal-wise excitations for their good features such as constant envelope and continuous phase.

References

1. Callegari S (2014) Achievement of preassigned spectra in the synthesis of band pass constant envelope signals by rapidly hopping through discrete tones. Proceedings of ISCAS 2014, pp 2776–2779
2. Callegari S, Rovatti R, Setti G (2003) Chaos based FM signals: applications and implementation issues. IEEE Trans Circuits Syst I: Regular Papers 50(8):1141–1147. doi:10.1109/TCSI.2003.815222
3. Callegari S, Rovatti R, Setti G (2003) Spectral properties of chaos-based FM signals: theory and simulation results. IEEE Trans Circuits Syst I: Regular Papers 50(1):3–15. doi:10.1109/TCSI.2002.807510
4. Callegari S, Ricci M, Caporale S, Monticelli M, Eroli M, Senni L, Rovatti R, Setti G, Burrascano P (2012) From chirps to random-FM excitations in pulse compression ultrasound systems. In: 2012 IEEE international ultrasonics symposium (IUS), pp 471–474. doi:10.1109/ULTSYM.2012.0117
5. Caporale S, Callegari S, Ricci M, Burrascano P (2013) Constant envelope pseudo orthogonal excitations for ultrasound testing. In: Proceedings of the 18th international conference on digital signal processing (DSP), IEEE, pp 1–8
6. Cohen L (1995) Time–frequency analysis: theory and applications. Prentice-Hall, Upper Saddle River
7. Cohn M, Lempel A (1977) On fast M-sequence transforms (corresp.). IEEE Trans Inf Theory 23(1):135–137
8. Cook C (2012) Radar signals: an introduction to theory and application. Elsevier, Amsterdam
9. Erhard A, Schenk G, Möhrle W, Montag HJ (2000) Ultrasonic phased array technique for austenitic weld inspection. In: 15th world conference on non-destructive testing, Rome, pp 674–677
10. Golomb SW, Welch LR, Goldstein RM, Hales AW (1982) Shift register sequences, vol 78. Aegean Park Press, Laguna Hills
11. Lam F, Szilard J (1976) Pulse compression techniques in ultrasonic non-destructive testing. Ultrasonics 14(3):111–114
12. Lee BB, Furgason ES (1983) High-speed digital Golay code flaw detection system. Ultrasonics 21(4):153–161
13. Milleit RE (1970) A matched-filter pulse-compression system using a nonlinear FM waveform. IEEE Trans Aerosp Electron Syst AES-6(1):73–78. doi:10.1109/TAES.1970.310012
14. O'Donnell M (1992) Coded excitation system for improving the penetration of real-time phased-array imaging systems. IEEE Trans Ultrason Ferroelectr Freq Control 39(3):341–351
15. Paulraj A, Gore D, Nabar R, Bolcskei H (2004) An overview of MIMO communications—a key to gigabit wireless. Proc IEEE 92(2):198–218. doi:10.1109/JPROC.2003.821915
16. Pollakowski M, Ermert H (1994) Chirp signal matching and signal power optimization in pulse-echo mode ultrasonic nondestructive testing. IEEE Trans Ultrason Ferroelectr Freq Control 41(5):655–659
17. Ricci M, Callegari S, Caporale S, Monticelli M, Battaglini L, Eroli M, Senni L, Rovatti R, Setti G, Burrascano P (2012) Exploiting non-linear chirp and sparse deconvolution to enhance the performance of pulse-compression ultrasonic NDT. In: IEEE international ultrasonics symposium (IUS 2012), IEEE, pp 1489–1492
18. Turin G (1960) An introduction to matched filters. IRE Trans Inf Theory 6(3):311–329. doi:10.1109/TIT.1960.1057571
19. Viterbi AJ (1995) CDMA: principles of spread spectrum communication. Addison Wesley, Redwood City

Part III
Dealing with the Collected Data: Information Processing

The ability to create monitoring systems for the quality control of industrial processes capable of taking into account the vagueness of the data is still an open problem. Since the assessment of the quality of an industrial production process encompasses the characterization of the defects (in terms of detection and classification) that may be present, it needs to exploit efficient data analysis techniques that can cope with vagueness and that imply a low computational load, is particularly attractive, if not a true necessity, for real-time applications. Starting from the idea that soft computing techniques, and in particular fuzzy techniques, can be considered as good candidates for the formulation and resolution of the problem, this Part of the book is devoted to the development and design of information processing systems based on fuzzy approaches. Particularly, Chap. 6 structures the data analysis process by means of banks of rules managed by inference techniques operating on principles of linguistic logic introducing the reader to the design of fuzzy and neuro-fuzzy systems for the localization and classification of defects for ultrasonic non-destructive evaluation (NDE) application. Finally, Chap. 7 presents two innovative investigative approaches for the characterization of defects in ultrasonic NDE. The first one exploits the concept of "Computing with Words" where a word is considered as a label of a fuzzy set of points put together per similarity (granule) which leads to a bank of fuzzy rules structured per classes. The second one starts from the idea that a defect (often not visible at the naked eye) can be considered as the cause of reduction of the similarity of its signal with one free of defects for which, taking into account the vagueness in a signal, it is desirable the use of Fuzzy Similarities techniques.

Chapter 6
Standard Soft Computing Techniques for Characterization of Defects in Nondestructive Evaluation

Mario Versaci, Salvatore Calcagno, Matteo Cacciola, Francesco Carlo Morabito, Isabella Palamara and Diego Pellicanò

Abstract Nondestructive evaluation (NDE)/nondestructive testing (NDT) of industrial manufactured items holds a strategic importance both for the quality assessment of the productive process and for taking monitoring actions during the relative life cycle. Since the quality assessment passes through the characterization of the defects that can be generated, it is imperative to exploit inspection techniques able to produce signals characterizing the defects themselves. Current inspection techniques do not provide, for several reasons, signals free from errors, inaccuracies and imprecision, so, in the information-processing step, it is necessary to face the problem by approaches capable of taking into account the inherent vagueness. While the form reconstruction of a defect is still an open problem, its localization and classification has been carried out with excellent results by the scientific community with the development of efficient and accurate methodologies also in terms of vagueness management. Considering the technological transfer point of view, the approaches developed so far are burdened by a less than desirable computational complexity that translates into expensive hardware requirements. For this reason, it is necessary to elaborate alternative methodologies capable of combining high-quality results and low-computational complexity. Specifically, among the many possible computing

M. Versaci (✉) · S. Calcagno · M. Cacciola · F. C. Morabito · I. Palamara · D. Pellicanò
DICEAM Department, Mediterranea University, Via Graziella Feo di Vito,
Reggio Calabria, Italy
e-mail: mario.versaci@unirc.it

S. Calcagno
e-mail: calcagno@unirc.it

M. Cacciola
e-mail: matteo.cacciola@unirc.it

F. C. Morabito
e-mail: morabito@unirc.it

I. Palamara
e-mail: isabella.palamara@unirc.it

D. Pellicanò
e-mail: diego.pellicano@unirc.it

P. Burrascano et al. (eds.), *Ultrasonic Nondestructive Evaluation Systems*,
DOI 10.1007/978-3-319-10566-6_6

techniques, attention is given to the soft techniques, and in particular, to the fuzzy logic (FL). The latter, generalizing dichotomic logic and taking into account the vagueness of signals, can deliver results comparable as a whole to those obtainable by more sophisticated techniques, but with a reduced computational charge. Moreover, the formalization in terms of natural language (NL) leads to the structuring of systems managed by legible linguistic rules even by nonexperts in the field and, at the same time, easily revisable by the expert. The present chapter is completely dedicated to introduce the reader to the basic principles of the logic of fuzzy inference system (FIS), applied functionally within NDE/NDT, presenting an example of study of the inverse problem.

6.1 Introduction

The ability to create monitoring systems for the quality control of industrial processes, taking into account the vagueness of the data, is still an open problem. Since the assessment of the quality of an industrial production process encompasses the defects characterization that may be present, it needs to exploit efficient data analysis techniques that can cope with vagueness and imply a low-computational load. The latter is a particularly attractive feature, if not a true necessity, for real-time applications [1, 2, 11, 12]. In this context, soft computing techniques, and in particular, fuzzy ones can be considered as good candidates for the formulation and resolution of the problem because they structure the data analysis process by means of banks of rules managed by inference techniques operating on the principles of linguistic logic. In this section, the reader is introduced to the design of fuzzy and neuro-fuzzy systems (NFSs) for localizing and classifying defectiveness for ultrasonic nondestructive testing (NDT) applications.

6.2 Nomenclature and Basic Principles of Fuzzy Logic and Systems

All the scientific constructs tend to include idealizations, such as dot-like bodies in physics, perfect gases in thermodynamics, and so on. They represent useful abstractions for the theory but impose approximations when faced with reality. According to Black (1937)

> It is a paradox that scientific theories, highly developed and useful, are expressed extensively in terms of objects never met in the real experience.

The purpose of Black's writing is to point out the need of an appropriate symbolism for the vagueness of real experience, where the standard interpretations can appear as particular cases. The main problems that a theory about formal languages has to face are the vagueness and the nondenoting terms. The birth of fuzzy logic (FL) was determined by the need of treating problems where full precision is relatively unimportant or, anyway, very difficult or expensive to achieve. Reality and precision are two concepts that may come in conflict—one comes at the detriment of the other. Furthermore, when speaking about products that need to be put on the market, a factor of great importance is cost cutting, which is desired by both the productions firms and the consumers. So, it is obvious that if there exists a technology that guarantees a good quality with a low price, this is preferable to a technology of greater quality but coming at a very high cost. Even if the current technology can provide more precise and powerful algorithms, often being less meticulous is a necessity.

Vagueness can be exploited to study problems that are also apparently solvable by traditional approaches, obtaining cheaper solutions with a better adherence to the needs of the market. FL and computing with words (CW) represent notable examples of technology getting close to the artificial intelligence (AI) constantly sought by all programmers [3, 4]. Allowing a calculator to act as a human being is not a concept of science fiction anymore (excluding, of course, the emotional involvement) [10]. In CW, items such as words, adjectives, and concepts are employed as expressions in the usual ways by the human reasoning. Starting from premises stated in the same forms used by humans and by applying semantic rules, computation is no more a mere manipulation of numbers, but instead, it can reach conclusions by a natural language (NL) framework based on expressions and words. In this way, the similarity with human reasoning becomes evident [3, 4, 10]. The logical criteria that are typically at the scholar's disposal are the result of a bivalent logic. At the same time, real problems, common situations, as well as our way of thinking and facing them, are not actually bivalent. However, thinking through the principles of bivalent logic means not contradicting the fundamental principles of such a logic. Conversely, FL is a polyvalent logic, namely an extension of the Boolean one, where each proposition can be given a level of truth between zero and one. It is strongly linked to the theory of *blurred groups*. While already informally conceived by Cartesio, Russel, Einstein and others, it was formalized by Lotfi Zadeh in 1965 [3–5].

There are two key aspects of FL: capacity to *treat uncertainty* and *emulation of the human reasoning*. Even if the matching of the noun "logic" by the adjective "fuzzy" may sound paradoxical, FL has conversely been introduced precisely to solve the paradoxes already known to be present in traditional logic. Being bivalent, the latter does not allow intermediate grades of truth between *true* and *false*. At the basis of FL, one has the concept of fuzzy set constituting a natural extension of the concept of set according to Cantor [10]. In particular, the Aristotelian principles of noncontradiction and excluded middle are not valid for it. The consequence is twofold. On one hand, an element can belong to different sets at the same time, but at a different degree. On the other hand, a set and its complement may produce a nonvoid intersection. Fuzzy values can vary from zero to one as probability values [3–5]. Probability may indicate the frequency of an event but does not say if such event occurs, and in case it occurs, in which measure it occurs. Conversely, fuzziness is meant for events that occur anyway . . . but at a certain degree! The relations between FL and probability are extremely controversial and, in many occasions, have provoked hard and often nonconstructive debates among their followers. On one hand, *probabilists*, strong of a century-old tradition and a consolidated position, have

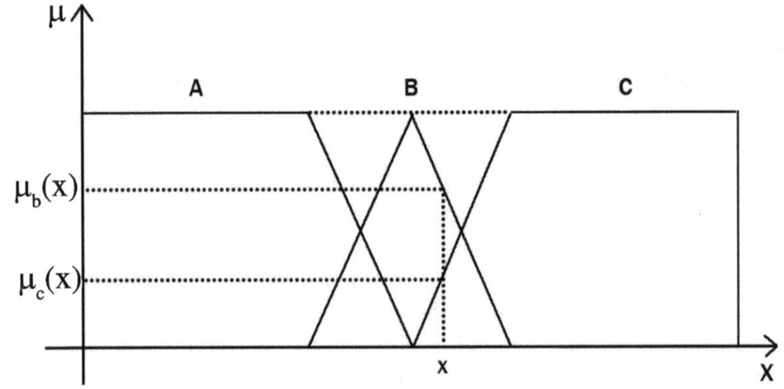

Fig. 6.1 Representation of a family of fuzzy sets on an universe of discourse X. The image illustrates the partial membership of a generic element x to multiple fuzzy sets

been defending a monopoly, historically held in matter of casualness and uncertainty. Their point is that FL is the probability in disguise supported by the purely fortuitous circumstances that measures of probability are expressed in the interval [0, 1]. On the other hand, *fuzzy scholars* have been showing that also probabilistic theory, in its different formulations, is ultimately a theory of the case tightly linked to a dichotomy defining probability as *the whole in the part*, namely as the measure of how much the part contains the whole. The part, in fact, can contain the whole in the measure in which its extension can superimpose to that of the universal set. This conception implies that a part can fully contain the whole only in the trivial case in which the part coincides with the whole. The containing, in fact, is no more bivalent but is fuzzy itself, and so it can get any value between zero (noncontaining) and one (containing and, at the limit, being coincident) [3–5].

6.2.1 The Theory of Fuzzy Sets: An Overview

To express the membership of an element x to a crisp set $\overline{A}$, it is enough to consider a membership function μ defined on the set of possible values X (universe of the discourse) ranging on $\{0, 1\}$: if $\mu_{\overline{A}}(x) = 1$, $x \in \overline{A}$; if $\mu_{\overline{A}}(x) = 0$, $x \notin \overline{A}$. So, any element x can completely belong or not belong to $\overline{A}$.

A fuzzy set A is characterized by the fact that the membership degree of each element x to the set A can be any real number from zero to "unit." For this reason, A is completely defined once a membership function $\mu_A : X \to [0, 1]$ is introduced, where X is the universe of the discourse, grading and shading in this way the membership concept. In particular, an element x can belong to several fuzzy sets even with different degrees. To indicate the membership function, it is common to use the name of the corresponding set which typically is a linguistic label, for instance, simply writing $A : X \to [0, 1]$. As an example, Fig. 6.1 represents three fuzzy

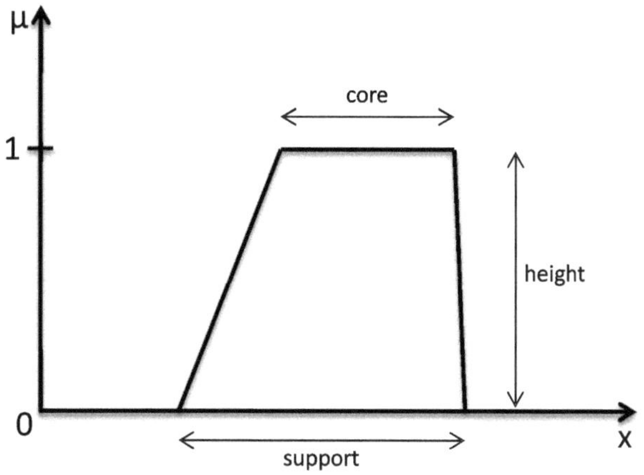

Fig. 6.2 Representation of a fuzzy set marking the intrinsic features: support, core, and height

sets, labeled respectively as A, B, and C, and partially overlapping. From such an arrangement, it comes out that a generic numeric value x_1 belongs to multiple fuzzy sets, but at different degrees. On the contrary, there are membership values to which multiple numeric values can correspond. This is typical of the shaded transition between adjacent fuzzy sets. This fact underlines that particular elements can belong to multiple sets, at the same degree [3–5].

6.2.1.1 Notations and Possible Representations of Fuzzy Sets

Due to the fact that in the application field a major goal is to express fuzzy sets in ways manageable by a computer, it is appropriate to open the discussion by providing those notations which can be representative of the quantities characterizing a fuzzy set. If the universe of the discourse is fairly adequate, a fuzzy set can be enumerated by the following notation:

$$A = \frac{\mu_1}{x_1} + \frac{\mu_2}{x_2} + \cdots + \frac{\mu_N}{x_N} = \sum_X \frac{\mu}{x} \tag{6.1}$$

where N is the number of elements of A. For continuous domains, the formulation in Eq. (6.1) can be translated into:

$$A = \int \frac{\mu_A(x)}{x} \mathrm{d}x \,. \tag{6.2}$$

In either case, both the summation and the integral represent the set operation of *union*. Moreover, the fuzzy sets theory, besides relying on new operations among sets, also needs new feature extracting operators among which the *support*, *core*, and *height* operators certainly stand out (Fig. 6.2). Mathematically:

$$support(A) = \{x \in X \mid \mu_A(x) > 0\} \tag{6.3}$$

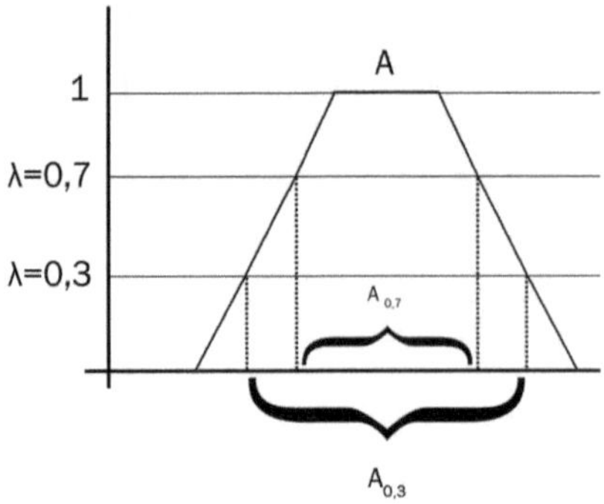

Fig. 6.3 Representation of a fuzzy set by the α-cuts procedure

$$core(A) = \{x \in X \mid \mu_A(x) = 1\} \tag{6.4}$$

$$height(A) = \sup_{x \in X} \mu_A(x)\,. \tag{6.5}$$

If $support(A)$ is a single point, A is denominated *fuzzy singleton.*

There are several ways of representing fuzzy sets in a numeric computation procedure, yet it is convenient to limit our attention to two procedures that are relatively simple to carry out. The first one is indicated as the *L–R procedure*, while the other (that shall shortly be shown in many ways more attractive) is known as *α-cuts (or α-levels) procedure* [3–5]. L–R method considers any fuzzy set as a sequence of four numeric values and a couple of shape functions. The numeric quantities represent the extremes of the core and the support, while the shape functions make explicit the type of link (linear, parabolic, ...) between the right (left) extremes of the core and the support. This procedure, notwithstanding its theoretic interest, has limited applications because it requires setting the functions in advance (here they are indicated with an "L" suffix for the left part and an "R" for the right one). Thus, it does not deliver the required flexibility with respect to the membership values. Conversely, the procedure based on α-cuts considers a fuzzy set A and an $\alpha \in [0, 1]$. An α-cut is nothing more than a set $A_\alpha = \{x \mid \mu_A(x) \geq \alpha\}$ so that for $\forall\lambda \leq \alpha,\ 0 \leq \alpha \leq 1$, one has $A_\alpha \subseteq A_\lambda$ where $A_0 = X$ (Fig. 6.3). Such a representation is, with no doubt, more attractive than the L–R representation because a fuzzy set can be implemented by two arrays as large as two times the number of α-cuts: starting from the extremes, they comprise the extremes of each cut in the first array and the corresponding cut values in the second array [3–5]. For example, if in a fuzzy set A, we operate three cuts (0, 0.5, and 1), the first array gets the form:

$$\begin{bmatrix} \mu_{0\,left} & \mu_{0.5\,left} & \mu_{1\,left} & \mu_{1\,right} & \mu_{0.5\,right} & \mu_{0\,right} \end{bmatrix} \tag{6.6}$$

while the second array can be written as

$$\begin{bmatrix} 0 & 0.5 & 1 & 1 & 0.5 & 0 \end{bmatrix} \tag{6.7}$$

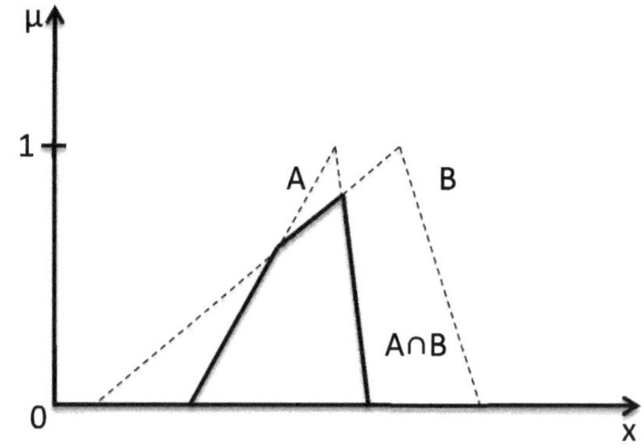

Fig. 6.4 Intersection of two fuzzy sets A and B

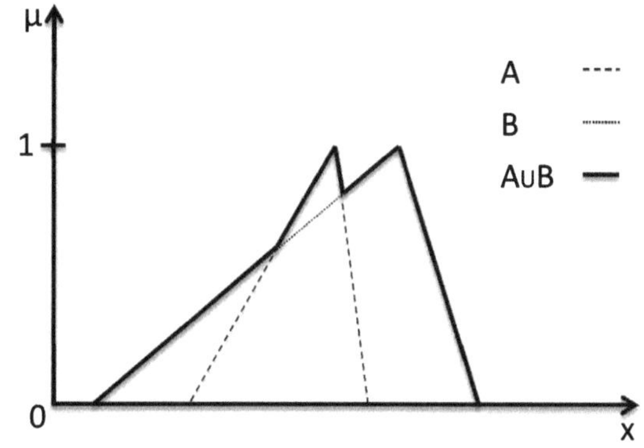

Fig. 6.5 Union of two fuzzy sets A and B

Operations Among Fuzzy Sets As in theory of crisp sets, also in FL, it is possible to define the elementary set operations as:

- Intersection $\mu_{A\cap B}(x) = \min\{\mu_A(x),\ \mu_B(x)\}$ (Fig. 6.4);
- Union $\mu_{A\cup B}(x) = \max\{\mu_A(x),\ \mu_B(x)\}$ (Fig. 6.5), and
- Complement $\mu_{A^c}(x) = 1 - \mu_A(x)$ (Fig. 6.6).

It can be demonstrated that the main laws of set theory hold. This is the case for the law of the excluded middle $A \cup A^c \neq X$ (Fig. 6.7) and the law of noncontradiction $A \cap A^c \neq \emptyset$ (Fig. 6.8). From the observation of the operations defined above, it can be easily realized that the key aspect is not the membership of an element to the set, but the shape of the fuzzy membership function (FMF). The choice of such FMF and the optimization of its characterizing parameters is a design problem where continuity, nonmonotonicity, and differentability should be provided. Generally speaking, the main shapes being adopted are the triangular, trapezoidal, gaussian, and sigmoidal. However, the criteria of choice for the shape of a membership function are substantially linked to the problem under study and to the subjectivity of the operator's experience [3–5].

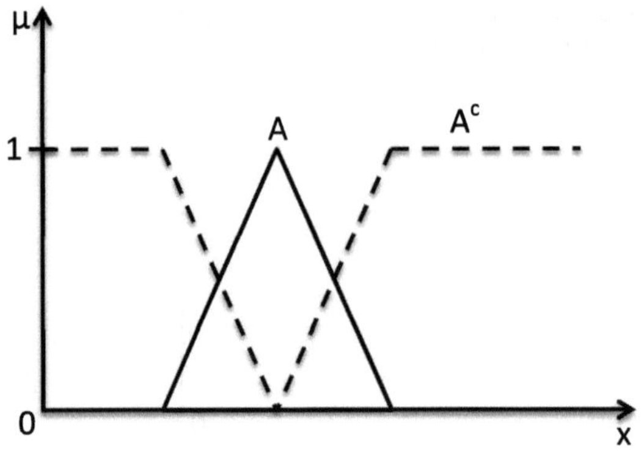

Fig. 6.6 Complement of a fuzzy set *A*

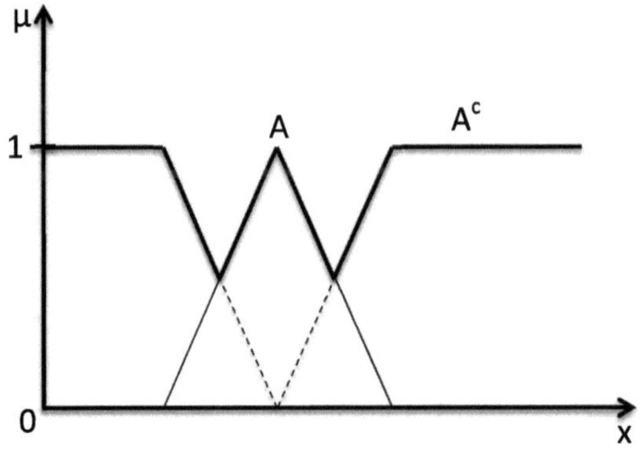

Fig. 6.7 Law of the excluded middle

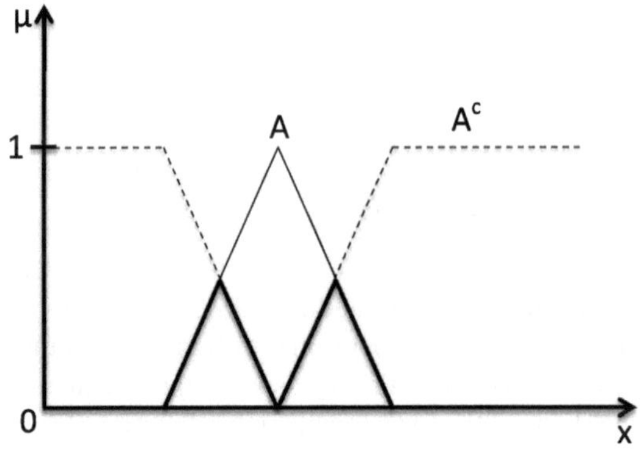

Fig. 6.8 Law of noncontradiction

Linguistic Variables The approach starts from the observation that the key elements are not numbers, but labels of fuzzy sets which are classes of objects in which the transition between belonging and not belonging is fuzzy [6]. Fuzziness suggests us that the logic is made up by fuzzy statements, fuzzy connectors, . . . , and so on, able to support partial truths, imprecision, and subjectiveness. If we refer to one of the peculiarities of the human thought, namely the capacity to synthesize information, i.e., the capacity to extract from sets or big masses of data only those subsets that are relevant for a certain purpose, we realize that it deals with an approximation problem by the codification of the most relevant information via grouping in fuzzy sets and labeling [6, 9, 8, 13, 14]. This approach presents three main features:

1. Use of linguistic variables whose values are not numbers but terms of a natural or artificial language;
2. Individuation of simple relations among variables by means of conditional fuzzy statements;
3. Characterization of complex relations among variables by means of fuzzy algorithms.

A *linguistic variable* is defined in terms of a base variable (in a classic sense) whose values are real numbers inside a given interval. The *linguistic terms*, which represent approximated values of the base variable, are interpreted as *fuzzy numbers*. To define a linguistic variable, we need a quintuple [8, 9]:

$$(v, T, X, g, m) \tag{6.8}$$

where v is the name of the variable; T is the set of linguistic terms of v; X is the universe of discourse; g is a syntactic rule used to generate linguistic terms (for example "very low," "not very low," ...); m is a semantic rule assigning to each linguistic term $t \in T$ its meaning $m(t)$, which is a fuzzy set on X. The atomic terms such as "middle," "high," "good," ... of the NL can be modified by means of adjectives or adverbs such as "very," "almost," ... . The effect of a linguistic modifier is to alter the FMF of an atomic term. For example, an expression like "very" is a *concentration* because it reduces the membership degree of all the elements (reduction of the uncertainty). The lower is the membership degree of an element with respect to a fuzzy set, the more reduced is its membership by concentration. An expression like "lightly," instead, involves some *dilatation* because it increases the membership values of elements which are partially in the set (increase of the uncertainty) [13, 14]. Finally, *intensification* increases the membership degree of the elements whose original membership degree is larger than 0.5 and decreases the membership degree when the original membership degree is smaller than 0.5. Possible mathematical translations of linguistic modifiers, for an atomic term $\alpha = \int_X \mu_\alpha(x)/X$, can be:

$$\text{very } \alpha = \alpha^2 = \int_X \frac{(\mu_\alpha(x))^2}{X} \tag{6.9}$$

or

$$\text{a little } \alpha = \alpha^{1/2} = \int_X \frac{(\mu_\alpha(x))^{1/2}}{X} . \tag{6.10}$$

Cartesian Product, Fuzzy Relations According to the dichotomic logic, a Cartesian product of two universes of discourse X and Y is the set of the ordered pairs (x, y) with x belonging to X and y to Y. Formally:

$$X \times Y = \{(x, y) : x \in X \wedge y \in Y\}. \tag{6.11}$$

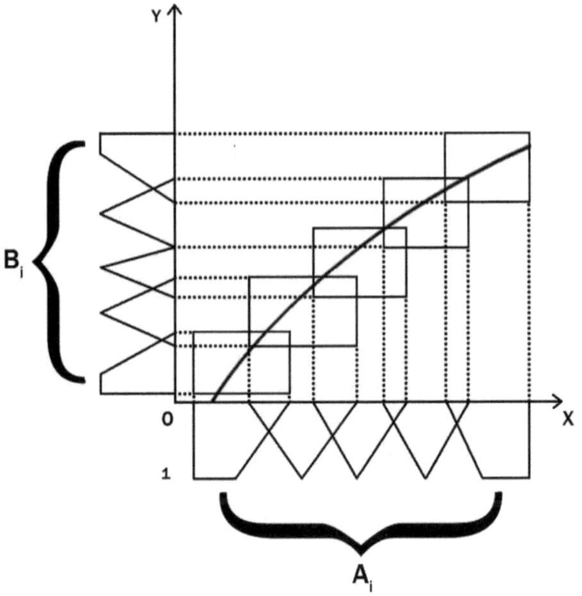

Fig. 6.9 Primordial structuring of a bank of fuzzy rules for a single input, single output system

From a cartesian product, it is possible to extract a subset, named *relation R*, whose elements (ordinated pairs having a certain property) belong completely to the set $X \times Y$. By generalization, we can see that in the fuzzy domain, a fuzzy relation (FR) is a subset of the cartesian product of two universes of discourse X and Y where the individual ordered pairs belong to FR at a certain level. If we think about X and Y as all the input and output variables respectively, FR represents the mapping $Y = f(X)$. In particular, covering the input–output space by means of all the possible cases by partially overlapping patches, each patch can be considered as a fuzzy rule *IF x IS A THEN y IS B* where A and B are fuzzy sets. The set of patches generates a bank of fuzzy rules which, when adequately combined among themselves by inference, create a primordial fuzzy system managing $Y = F(X)$ (Fig. 6.9) [8, 14]. Obviously, the generalization to multi-outputs and multi-outputs approaches is immediate [9, 13].

6.2.2 *From Fuzzy Sets to Fuzzy Systems: Mamdani's and Sugeno's Formulations*

Fuzzy systems are systems of rules of *IF ... THEN* type containing linguistic terms to express the membership of a variable to a certain fuzzy set [9]. Differently from expert systems, fuzzy inference systems (FISs) can have multiple rules active at the same time, and their inputs/outputs are numeric variables, not symbols or terms. Furthermore, it is possible, by means of relations, to approximate any function, including nonlinear ones. A simple fuzzy rule gets the form *IF x IS A THEN y IS B* in which A and B are the linguistic values defined by the fuzzy sets on X and Y, respectively. The *IF* part of the rule (indicated as p) is known as the *antecedent* or *premise*, while the *THEN* part (indicated as q) is said *consequent* or *conclusion*. Let us notice that A is represented by a real number ranging in $[0, 1]$, so the antecedent is an interpretation which returns a single number ranging on $[0, 1]$. On the other hand, B is represented as a fuzzy set, so the consequent is an assignment associating the whole fuzzy set B to the output variable. In a fuzzy rule, the element *IS* is used in two completely different ways, depending on whether it appears in the antecedent or in the consequent. In terms of planning, it is the same distinction that exists between uniformity $==$ and assignment $=$, so a rule can be rewritten as *IF* $x == A$ *THEN* $y = B$.

Generally, the input for a rule is the current value for the input variable and the output is the whole fuzzy set. The interpretation of a fuzzy rule involves distinct parts: first, the antecedent is assessed (input *fuzzification* and application of fuzzy operators); then, the result is applied to the consequent (*implication*) [9]. In the case of bivalent logic, the *IF* ... *THEN* rules do not present particular problems: if the premise is true, then the conclusion is also true. The question we ask is the following: relaxing the hypothesis to operate in a bivalent logic, if the antecedent is a fuzzy statement, how is the conclusion influenced? The answer is simple: if the antecedent is true with a certain degree, then the consequent is true with the same degree. In other words, in binary logic, $p \rightarrow q$ (p and q are both true or both false) [10]. In FL, indicating by μ a generic membership value, $\mu_{antecedent}(p) \rightarrow \mu_{consequent}(q)$ (partial antecedents imply partial consequents). Obviously, a rule can have multiple parts. For example, consider *IF* x_1 *IS* $A1$ *and* x_2 *IS* $A2$ *THEN* y *IS* B in which all the sections of the antecedent are computed simultaneously and associated to a single value using logic operators. Similarly, the consequent can also have multiple parts. In this case, all the consequents are equally influenced by the antecedent result: the consequent specifies a fuzzy set to assign to the output and the implication modifies that fuzzy set by the degree specified in the antecedent. In brief, the interpretation of the fuzzy rules is made up of five steps:

1. *Inputs fuzzification*: A membership degree is associated with all the statements in the antecedent;
2. *Application of the fuzzy operators to multiple antecedents*: They are applied to the fuzzy operators and the antecedent is considered active at the lowest membership degree;
3. *Application of the implication to the consequent of a rule*: A whole fuzzy set is given to the output. This is represented by an FMF that is chosen to indicate the "quality" of the consequent. If the antecedent is only partially true, then the fuzzy set is cut according to the chosen implication method. Formally, the implication produces a fuzzy set with membership function equal to

$$\mu_{A \rightarrow B}(\cdot, \cdot) = \tau(\mu_A(\cdot), \mu_B(\cdot)) \tag{6.12}$$

where τ is a T-norm operator. If τ is the min operator:

$$\mu_{A\rightarrow B}(\cdot,\cdot) = \min(\mu_A(\cdot), \mu_B(\cdot)) \tag{6.13}$$

4. *Aggregation*: The outputs of all the rules are aggregated in a single fuzzy set of outputs;
5. *Defuzzification*: Finally, the result is defuzzified, namely it is associated with a real value by a particular procedure such as the *center of gravity* computation which associates with each fuzzy set its barycenter according to Guldino-Pappo's formula as in

$$\textit{Defuzzified Value} = \textit{Center of Gravity} = \frac{\int_{\textit{Possible Values}} x \, \mathrm{d}x\mathrm{d}y}{\int_{\textit{Possible Values}} \mathrm{d}x\mathrm{d}y}. \tag{6.14}$$

In the discrete case, the above equation becomes:

$$\textit{Defuzzified Value} = \textit{Center of Gravity} = \frac{\sum_{\textit{Possible Values}} x \; \Delta x \Delta y}{\sum_{\textit{Possible Values}} \Delta x \Delta y}. \tag{6.15}$$

Alternatively, using the minimum of the maxima:

$$\textit{Defuzzified Value} = \min_A(\max(A)) \tag{6.16}$$

or, using the mean of the maxima:

$$\textit{Defuzzified Value} = \frac{1}{|\textit{Height}(A)|} \sum_{\textit{core}} \textit{Height}(A). \tag{6.17}$$

which computes the weighted mean of all the values constituting the fuzzy set using as weights the correspondent membership values. The choice of the method depends on the context. However, some criteria exist to measure the quality of the method. For instance, properties such as the continuity, the not ambiguity, the plausibility, and the computational simplicity (except for the center of gravity which is too expensive) can be used.

Fuzzy Inference

It represents the input/space–output/space mapping by means of FL. The main typologies of fuzzy inference are *Mamdani*'s and *Sugeno*'s models [9]. In Mamdani's model (Fig. 6.10), with multiple antecedents rules such as

$$\textit{IF } x_1 \textit{ IS } A1 \textit{ AND } x_2 \textit{ IS } A2 \textit{ THEN } y \textit{ IS } B \tag{6.18}$$

the application of the operators occurs by means of *AND* or *OR* connectors. In the first case, one has the form

$$\min(\mu_{A1}(x_1), \mu_{A2}(x_2)) \tag{6.19}$$

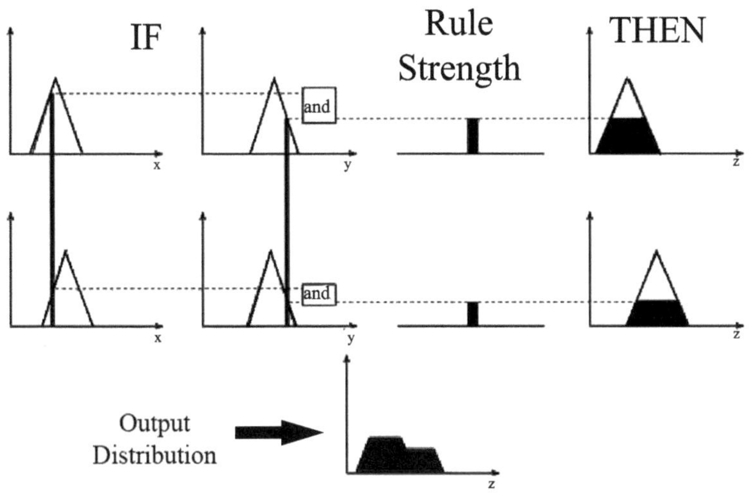

Fig. 6.10 Example of Mamdani's fuzzy inference (two rules, *AND* connector): after inputs fuzzification and min operator application, the operations of cutting-off and aggregation of the output fuzzy sets are well visible. Finally, the center of gravity of the obtained figure determines the numeric output value

or

$$\mu_{A1}(x_1) \times \mu_{A2}(x_2). \tag{6.20}$$

In the second case, one has

$$\max(\mu_{A1}(x_1), \mu_{A2}(x_2)) \tag{6.21}$$

or

$$\mu_{A1}(x_1) + \mu_{A2}(x_2) - \mu_{A1}(x_1) \times \mu_{A2}(x_2). \tag{6.22}$$

The implication is operated by the cutting off of the output fuzzy sets, while the aggregation is performed through the overlapping of the truncated output fuzzy sets. Finally, defuzzification translates the overlapping into a numeric value by its center of gravity. Differently from the above, Sugeno's model keeps the same steps as Mamdani, but uses some singletons as the output fuzzy sets, so a rule may get the form

$$IF\ x_1\ IS\ A1^j\ AND\ x_2\ IS\ A2^j\ THEN\ y = f(x_1, x_2) \tag{6.23}$$

where $y = f(x_1, x_2)$ is a polynomial function such as

$$\alpha_1 x_1 + \alpha_2 x_2 + c \tag{6.24}$$

and

$$\alpha_j = \tau(\mu_{A1^j}(\cdot), \mu_{A2^j}(\cdot)). \tag{6.25}$$

In this case, defuzzification is operated by means of a weighted mean of the singletons cutoffs

$$\textit{Defuzzified Value} = \frac{\sum_j \alpha_j f_j(\cdot,\cdot)}{\sum_j \alpha_j}. \tag{6.26}$$

The degree of the polynomial function is indicative of the order of the model. However, there are simplified models where fuzzification can occur also as follows:

$$\textit{Defuzzified Value} = \frac{\sum_j \alpha_j b_j}{\sum_j \alpha_j} \tag{6.27}$$

where b_j depends on the inference order.

Differently from Mamdani's procedure (which requires an expensive implementation of the defuzzifier and has consequents that are easy to interpret), Sugeno's approach (and its simplified modifications) does not require a defuzzification procedure, but has some difficulty in interpreting consequents. To reduce errors in the planning, excellent results have been obtained by means of fuzzy inference procedure directly extracted from the numerical data.

6.2.3 *Network Structuring of Fuzzy Systems: The Neuro-Fuzzy Approach*

Fuzzy inferences do not require learning mechanisms and, even if the accuracy is reduced, the corresponding FIS (that are less computationally expensive than other models), can be considered a good prototype of a quick output evaluation. A neural net is classified per typology and the learning algorithm is chosen according to the problem (optimization, classification, ...) [9]. In a FIS domain, the rules structure suggests the net architecture and typology. A fuzzy system is built up by a set of rules in the typical form:

$$\textit{IF } X_i = A_{i2} \textit{ AND } X_2 = A_{i4} \textit{ THEN } Y_1 = C_{i3} \textit{ AND } Y_2 = C_{i1} \tag{6.28}$$

where:

1. X_i are inputs (antecedents);
2. Y_i are outputs (consequents);
3. A_{ij} and C_{ij} are fuzzy set defined for the jth variable and the ith line.

The membership functions characterizing each fuzzy set can be chosen among the most common ones (gaussian, trapezoidal, triangular). The advantage of a neuro-fuzzy net, with respect to a neural one, is in the fact that the structure can be represented by "linguistic rules." The knots of a neuro-fuzzy network does not contain any weights, as it commonly happens in a neural network, but a decisional structure proper of FL. The network training occurs by means of back-propagation or genetic algorithms, with a trade-off between the achievement of excellent performance and a noteworthy increase in the complexity of the network design itself. Obviously, it is always possible to provide adaptations of fuzzy inferences by means of some tuning of the parameters characterizing FMF and to practice inference extraction directly from numeric data. For example, one may set the optimal number of clusters and practice their initialization by minimizing an appropriate cost function by means of an iterative optimization procedure: once the cluster centers are localized, each of them can be considered as a fuzzy rule. In particular, the extraction procedure for the bank of fuzzy rules, starting from the cluster centers, determines the number of rules and the FMFs so that by means of linear least-square estimation, the consequents of each rule are determined. The extraction procedure determines the number of rules and the membership functions of the antecedent to obtain then the consequent by linear least square estimation (Generate Fuzzy Inference System (GENFIS)—MatLab®). In this way, the conceived FIS, lends itself to be adaptively optimized by tuning the membership functions based on input/output pairs. In particular, the adaptive neuro-fuzzy inference system (ANFIS) routine [7], applied to a FIS tunes it through a learning algorithm based on input/output pairs. The net-type structure leads to the formulation of the gradient vector for the correction of parameters applying a number of optimization routine or reducing an measurement error (typically, the sum of the squared difference between actual and desired outputs).

6.3 Fuzzy Similarity and Neuro-Fuzzy System to Localize and Reconstruct Defects

Within industrial production environments, quality control has a strategic importance, This is particularly true in those sectors where public safety and manufacturing quality are imperatively required. In such a context, NDT/non-destructive evaluation (NDE) techniques have a preferential position because they present the undisputed advantage of leaving the specimen unaltered after the inspection. Moreover, NDT/NDE can deliver signals that are generally easy to analyze in order to look for cracks and/or flaws. In this section, an example of NFS is presented, to solve detection and classification problems. The example concerns the information processing involved in the assessment of carbon fiber reinforced plastics (CFRP) panels by means of a data set analysis of experimental measures collected by ultrasonic transducer (UT) (with one-dimensional signal treatment). In particular, starting from an experimental dataset, the example describes the design of an NFS where inputs are represented by features typical of given classes of defectiveness. The system provides as output the class of defects itself and its operation is based on a reduced number of rules getting some desiderable properties (easiness in system adaptation, upgrade by expert knowledge). Furthermore, the absence of conflicts among the rules is guaranteed.

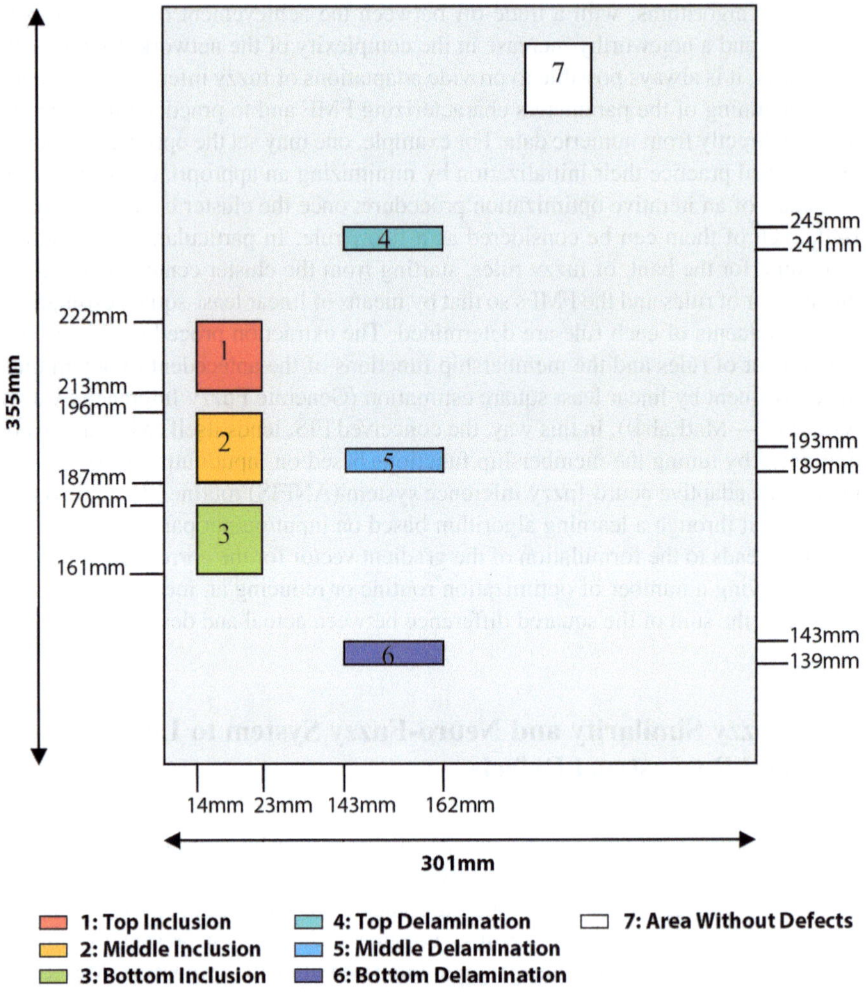

Fig. 6.11 CFRP panel (301 × 355 × 4 mm) exploited for UT database construction. The figure displays the area without defects, as well as *top*, *middle*, and *bottom* inclusion and delamination areas

6.3.1 An Example of Signals Classification Using a Neuro-Fuzzy Approach

6.3.1.1 The Experimental Database

In order to build an experimental database, different kinds of defects were obtained on sample artificial panels of CFRP. The different kinds of defects are: top, middle, and bottom delamination; top, middle, and bottom inclusion; porosity. Two sample specimens were considered. All the types of defects with the exception of porosity

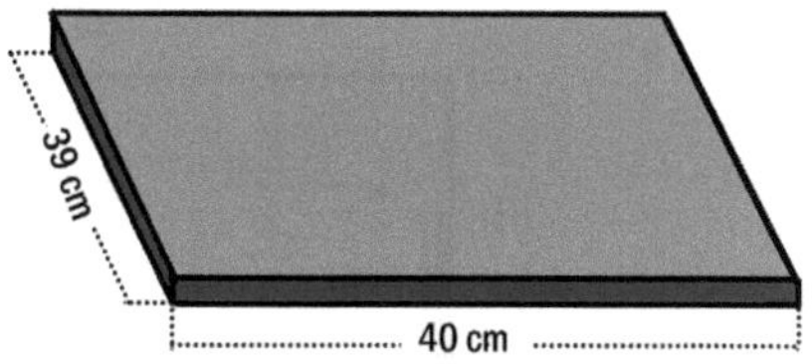

Fig. 6.12 CFRP panel (39 × 40 × 3) with porosity

were produced on different areas of the first specimen (Fig. 6.11), while porosities were obtained on another panel (Fig. 6.12). The so obtained defects were analyzed at the *Laboratory of Electrical Engineering and NDT/NDE* of the University of Reggio Calabria. The protocol for producing defects proved able to provide characteristics similar to flaws caused by normal wear. Delamination-like and inclusion flaws were obtained in a 301 × 355 × 4 mm CFRP panel. Within this specimen, we used defect-free areas in order to obtain benchmarks. Porosities were produced into the second CFRP panel, with dimensions 39 × 40 × 3 mm. A 5 MHz UT probe was used for the inspection with a 100 MHz sampling frequency; the speed of the ultrasonic wave on CFRP is 3000 m/s. Specimens were immersed into water. Usually, a UT benchmark signal, i.e., a UT signal depicting an undefected area, is characterized by a front-wall echo (also defined as top-echo), resulting from the interaction between the exciting ultrasonic wave and the top surface, and a bottom-echo, obtained since the reflection of the wave on the bottom of the specimen. If other peaks are within this range (Fig. 6.13), then discontinuities of the material are present, usually representing defects. According to the Nyquist's theorem, by fast Fourier transform (FFT) one should assure that the main frequency of the signal is within half the sampling frequency of the ultrasonic wave (Fig. 6.14). Since qualitative inspection is not enough for classifying defects, because different kinds of defects could produce similar UT signals (see Figs. 6.15 and 6.16), the solving of an inverse problem is required. Considering that attention is focused on what happens between the top and bottom surfaces of the specimen, we preprocessed the UT signals in order to extract useful information between front and bottom echoes. Uncertain cases, which are well known in industrial environments, require to be satisfactorily solved. In this case, the use of supporting classification algorithms with a low computational complexity is appreciated, in order to avoid the acquisition of sophisticated hardware for practicing computation on UT signals. For our applications, the experimental database was divided in two parts: the training section and the testing section. In Table 6.1 the database distribution is displayed in terms of number of signals for each category of defect (including the class of signals without defects).

6.3.1.2 The Basic Idea

The international scientific community has been so far consolidating two main research trends with reference to defect characterization.

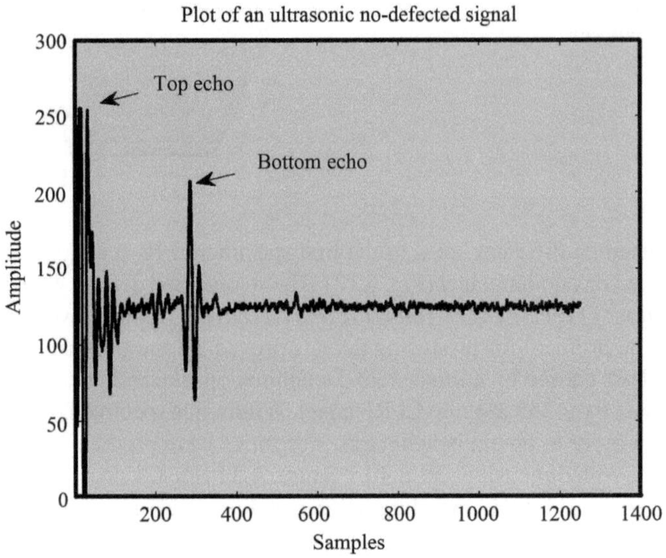

Fig. 6.13 Ultrasonic signal in CFRP specimen without defects. *Top* and *bottom* echoes zones are well visible

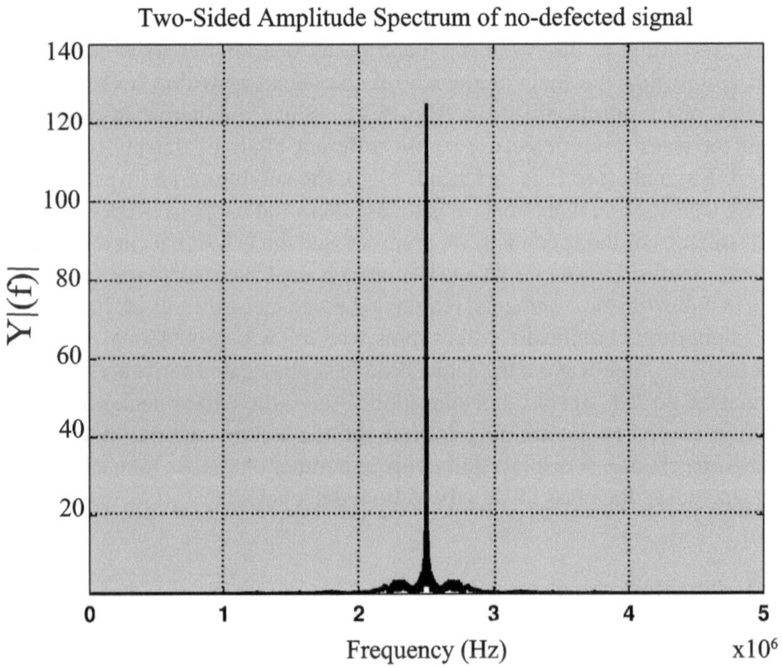

Fig. 6.14 FFT of a signal in which it is possible to note that the main frequency is approximately equal to half the sampling frequency as the Nyquist theorem requests

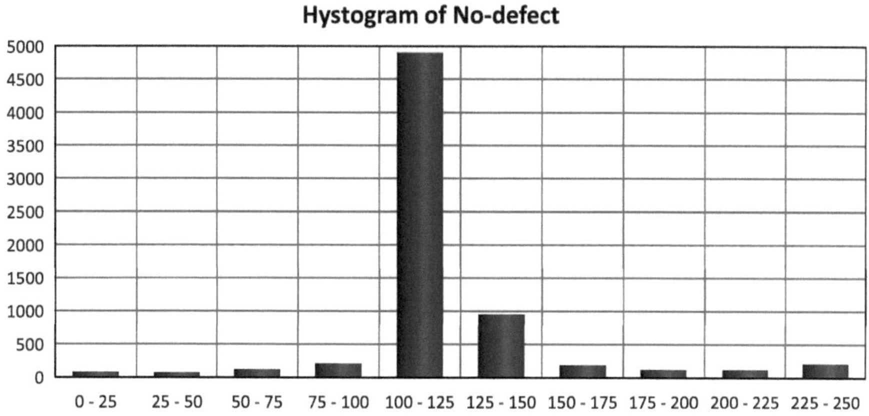

Fig. 6.15 Histogram for class of signals without defects

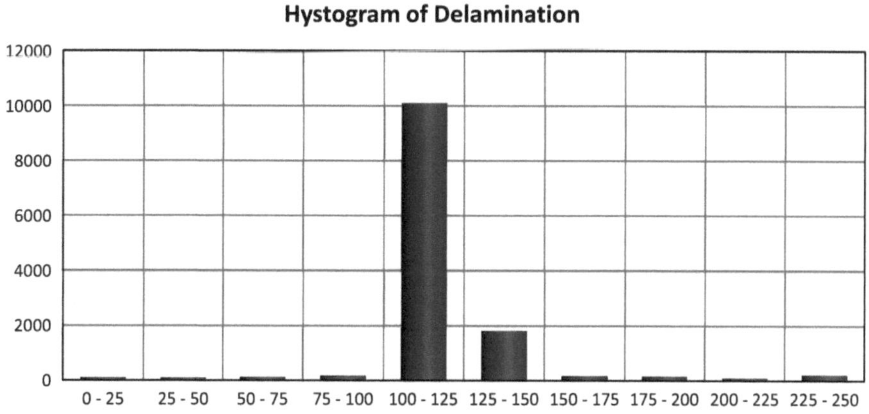

Fig. 6.16 Histogram for class of signals with delamination

Table 6.1 Distribution of signals in training and testing databases

	Training	Testing
Top delamination	42	21
Middle delamination	45	24
Bottom delamination	37	22
Top inclusion	35	28
Middle inclusion	38	25
Bottom inclusion	38	21
Porosity	40	25
No defects	45	30

The first approach, by means of optimization schemes, aims at searching the solution of the characterization problem by the iterative minimization of some cost function (whose actual formulation derives from the specific problem under study). This is practiced by manipulating the experimental data at disposal by means of particular analytical–numeric instruments derived from the characterization parameters thanks to interpolation procedures (that are nearly always multidimensional).

The second approach, that needs simulated or experimental databases, leads to the solution by means of specific learning procedures (that operate with or without supervision aid). In such a context, the appeal of the fuzzy approach is enhanced, thanks to the intrinsic ability to approximate multi-input/multi-output functional links (including linear ones) through *IF* ... *THEN* linguistic structures (conditional fuzzy linguistic structures) where both the antecedents and the consequents use fuzzy quantities (represented, implemented, and manipulated as reported in the previous sections) in place of numeric quantities that would be less suited for the computation of quantities where uncertainty is inherent.

In any case, whatever it is the procedure of resolution of the problem, the purpose is always to associate to a particular set of experimental measures the position and the typology of defect. FISs are mathematical devices possessing an inherent ability to tackle and exploit the uncertainty present in the data in order to achieve given goals. Hence, a problem of defect detection and classification can be formulated in terms of a research of the best mapping between the set of quantities characterizing the defects[1] and the classes of defects[2] (so intending set of signals affected by the same type of defect), as in

$$f(input_1,\ input_2,\ \ldots,\ input_n) = output \tag{6.29}$$

where f represents a fuzzy inference. An evident advantage of the approach is that the rules can be derived directly by visual inspection of the physical aspects of the problem under study. Furthermore, it is clear that the rules can be integrated or enhanced by means of expert knowledge. For this reason, visual extraction of the inference, in spite of a noteworthy conceptual interest, cannot find any comparison in the practice. From here on, one needs to dispose of an extraction procedure of fuzzy inferences directly from the numeric data. As far as our objectives are concerned, a convenient choice is to use the Matlab® GENFIS toolbox. This is justified by the fact that it generates models easy to deploy in "net"-type schemes, optimizable by standardized procedures (such as ANFIS) as described in the previous sections [7].

6.3.1.3 Choice of the Inputs and the Output

To localize a defect (in terms of epicenter) it is necessary to take into consideration the cartesian coordinates in the scanning plan (being the scanning practiced on the

[1] Set of inputs.
[2] That is, the output.

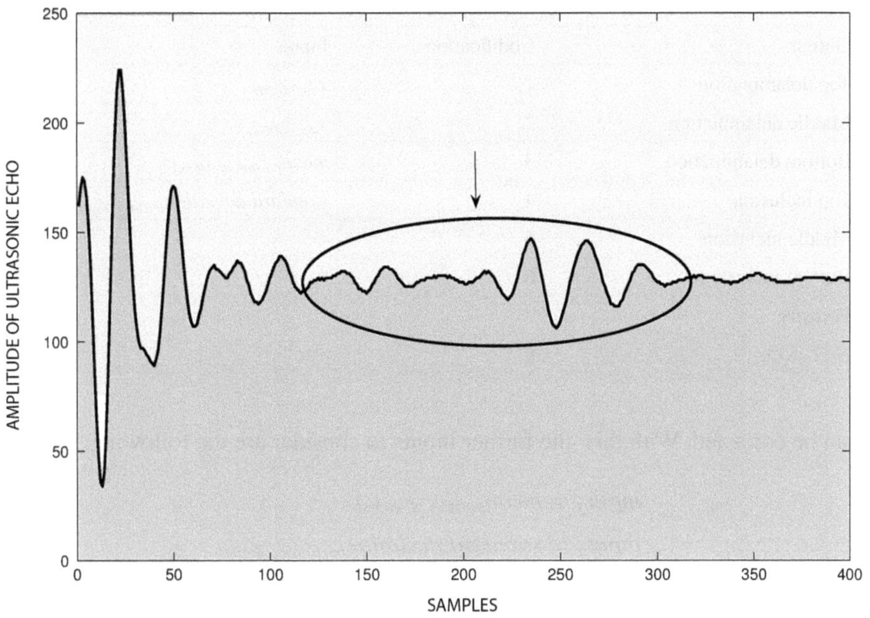

Fig. 6.17 Typical UT signal in CFRP panel where porosity takes place

same side where the signal is generated). This provides the first two inputs of the system:

$$input_1 = x_{coordinate} \qquad (6.30)$$
$$input_2 = y_{coordinate}$$

The choice of the input parameters indicating the typology of defect (delamination, inclusion, and porosity) and the hypocentric position of the same (top, middle, and bottom) is more complex. As is well known, it is the epicentric position of a defect[3] can be coarsely estimated by a simple visual inspection of a signal (see for example Fig. 6.17). Conversely, nothing can be said about its nature and depth. However, signals that are similar among them (that is affected by the same defects or by no defects) present similar features, so from each class of defect (or class of signals without any defects), it is possible to extract features (for example, of statistical type) characterizing each given class. In particular, signals affected by the same defect (or without defects) present overlapping ranges of mean values and standard deviation. In particular, from the portion of signals affected by the same defects (in which the presence of defects is likely) the mean values and the standard deviations typical for that particular defect (involving both the typology and the hypocenterness)

[3] The correct position will be guaranteed by the first two inputs of the procedure.

Table 6.2 Inputs/outputs parameters and their codifications

Output	Codification	Inputs
Top delamination	1	$x_{coordinate}$
Middle delamination	2	$y_{coordinate}$
Bottom delamination	3	$mean_{portion\ of\ signal}$
Top inclusion	4	$standard\ deviation_{portion\ of\ signal}$
Middle inclusion	5	
Bottom inclusion	6	
Porosity	7	
No defects	8	

can be extracted. With this, the further inputs to consider are the following:

$$input_3 = mean_{portion\ of\ signal} \tag{6.31}$$
$$input_4 = standard\ deviation_{portion\ of\ signal}$$

The output is represented by a simple codification of both the classes of signal with defects and the class of signals without defects. Finally, the required mapping can be formulated in terms of the following multiple input, single output (MISO) system

$$codification = f(x_{coordinate}, y_{coordinate}, \tag{6.32}$$
$$mean_{mathit portion\ of\ signal}, standard\ deviation_{portion\ of\ signal}) \tag{6.33}$$

Table 6.2 summarizes the aforementioned parameters used for inputs and outputs and the relative codifications.

6.3.1.4 Planning of the Fuzzy Inference Through MatLab® GENFIS System

Let us start the discussion on the planning of fuzzy inferences, by considering a well-known argument of fuzzy literature: with respect to a given number of input and input–output pairs, the mere visual analysis of the plots allows the building of *artisan* fuzzy behaviour rules. Conversely, in online applications, output parameters can be roughly estimated by artisan fuzzy systems. In both cases, one needs to extract fuzzy inferences by using numerical algorithms.

In our case, the tuning of the system requires eight FISs to be calculated: one for each class of defect and one for the class of signals without defects.[4] GENFIS system, starting from a training database, generates a Sugeno-type FIS structure exploited for ANFIS procedure training [7]. Fuzzy rules extraction, carried out according to the criteria explained before, leads to the formulation of linguistic structures in which the

[4] Top, middle, and bottom delamination; top, middle, and bottom inclusion; porosity; no defects.

number of multiple antecedents is equal to the inputs number. Hence, the GENFIS system generates a fuzzy rule in the typical form:

$$IF\ x_{coordinate}\ IS\ (MF_{x_{coordinate}})_t\ AND\ y_{coordinate}\ IS\ (MF_{y_{coordinate}})_t\ AND \tag{6.34}$$

$$mean_{portion\ of\ signal}\ IS\ (MF_{mean_{portion\ of\ signal}})_t\ AND \tag{6.35}$$

$$standard\ deviation_{portion\ of\ signal}\ IS\ (MF_{standard\ deviation_{portion\ of\ signal}})_t \tag{6.36}$$

$$THEN\ codification\ IS\ value_t \tag{6.37}$$

in which $(MF_{input})_t$ represents the t-th label of the membership function of the generic input and $value_t$ is the codification value of the output associated with the rule t. Membership function labeling is automatically done in growing numbers. On the other hand, the increase in labeling does not necessarily turn into an increasing sequence of the fuzzy sets of each variable. The analysis of the outputs allows to state that a good estimation of the output parameters can be done starting from a reduced number of rules. In principle, FISs were built, and they could be improved by proper tricks, but the reduced number of patterns would generate a number of rules too large to be calculated without an increase in the computational load (do not forget that, with each sample point, a fuzzy rule could be associated, leading to overfitting).[5] If this happens, the system learns from the memory, not from the examples. Finally, the generated FIS shows, in this way, a lower computational complexity. The obtained FISs (each one for each kind of defect and for signals without defects) are optimized by means of a sort of "tuning procedure" of the membership functions parameters by ANFIS procedure [7] which, acting by learning processes on the FIS net-type structure, obtains the gradient vector[6] and, by means of the application of particular optimization subroutines, the parameters are "tuned" by a reduction of error functional data.

6.3.1.5 The Obtained Results

The implementation of the combined GENFIS and ANFIS procedures for detecting and classifying defects confirms essentially the basic idea of the approach [7]. In other words, FISs exploitation, with a reduced number of rules (as marked in Table 6.3), on the one hand, guarantees the adaptability and, on the other hand, allows the updating by means of expert knowledge (inserting, for example, further rules). Such features have to be considered assured due to the low cardinality of the set of rules characterizing each FIS, which moves away risks of possible conflicts and contentious debates. Moreover, the reduced number of rules does not invalidate the quality of the obtained results (both in terms of detection and classification) with peaks of maximum performance for some defects (Table 6.3). However, an enhancement of

[5] The step of the scanning procedure must be set taking into account the diameter of the hole with respect to the length of the sensor.

[6] The gradient vector can be considered as a sort of measure of the goodness of the input/output modeling.

Table 6.3 GENFIS+ANFIS approach for detecting and classifying defects

	No. of rules	Detection (%)	Classification (%)
Porosity	10	99	97.5
Top delamination	9	100	98.4
Middle delamination	10	100	95.2
Bottom delamination	11	100	96.4
Top inclusion	9	99	95.9
Middle inclusion	11	99	97.1
Bottom inclusion	13	98	97.6
No defects	12	100	97.3

the results can be obtained through a more careful choice of the features to be used as inputs.

6.4 Conclusion

The main conclusion that can be drawn is that, by FIS, systems can be designed capable to practice detection and classification for ultrasonic applications. The advantages of this approach can be summarized as follows:

1. It is possible to explain the input system in terms of rules with a direct physical interpretation;
2. Black box-type structures can be improved augmenting them by means of knowledge expressed in terms of fuzzy rules;
3. The main limitation is that FISs are efficient but are in the presence of a reduced number of inputs that involves the use of techniques of data compression and/or inputs ranking procedure. However, these systems are very economical compared to other techniques in terms of computational complexity, making them particularly attractive for real-time applications. Finally, through the use of learning techniques, the models can be refined to increase their competitiveness.

References

1. Cacciola M, La Foresta F, Morabito FC, Versaci M (2007) Advanced use of soft computing and eddy current test to evaluate mechanical integrity of metallic plates. NDT E Int 40(5):357–362
2. Cacciola M, Calcagno S, Morabito FC, Versaci M (2008) Computational intelligence aspects for defect classification in aeronautic composites by using ultrasonic pulses. IEEE Trans Ultrason Ferroelectr Freq Control 55(4):870–878
3. Chiu SL (1994) Fuzzy model identification based on cluster estimation. J Intell Fuzzy Syst 2(3):267–278
4. Dubois D, Prade H (1988) Possibility theory. Springer, Berlin

5. Goguen JA (1969) The logic of inexact concepts. Synthese 19(3):325–373
6. Hanss M (2005) Applied fuzzy arithmetic. Springer, Berlin
7. Jang JSR (1993) ANFIS: adaptive-network-based fuzzy inference system. IEEE Trans Syst Man Cybern 23(3):665–685
8. Kóczy LT, Tikk D (2000) Fuzzy rendszerek. TypoTEX, Budapest
9. Kosko B (1996) Fuzzy engineering. Prentice-Hall, New York
10. Kuncheva LI (2000) Fuzzy classifier design. Springer, Berlin
11. Morabito FC, Versaci M (2001) A fuzzy neural approach to localizing holes in conducting plates. IEEE Trans Magn 37(5):3534–3537
12. Morabito FC, Calcagno S, Versaci M (2002) A soft computing approach to reconstruct defects on metallic plates. Int J Appl Electromagn Mech 14(1):489–494
13. Sanchez E (1979) Inverses of fuzzy relations. Application to possibility distributions and medical diagnosis. Fuzzy Sets Syst 2(1):75–86
14. Zimmermann HJ (1992) Fuzzy set theory and its applications, 2nd revised edn. Springer, Berlin

Chapter 7
Innovative Fuzzy Techniques for Characterizing Defects in Ultrasonic Nondestructive Evaluation

Mario Versaci, Salvatore Calcagno, Matteo Cacciola, Francesco Carlo Morabito, Isabella Palamara and Diego Pellicanò

Abstract Classification of defects in ultrasonic nondestructive testing (NDT)/nondestructive evaluation (NDE) has a role of primary importance in all those applications in which the knowledge of the typology of defect is crucial for the manufact destination. In such a context, the necessity to have efficient investigation instruments for a correct classification analysis emerges clearly. A defect, even when invisible to the naked eye, can be revealed as the cause that reduces the similarity of a measured signal with respect to a reference. Considering the fuzziness intrinsic in the signals, the reliance on fuzzy techniques to evaluate similarity appears desirable. Two fundamental achievement of research in such field, which both derive from the fuzzy thinking and which share common traits, are computing with words (CW) and the concept of fuzzy similarity (FS). In CW, a word is considered as a label of a fuzzy set of points clustered by similarity (*granule*) which lead to a particular formulation of a bank of fuzzy rules structured per classes. FS is an evaluation index of similarity among entities (for example signals) particularly useful for the formation of specific classes. Both approaches are based on computational linguistics (for example, descriptive formalism in natural language). This chapter is conceptually divided into two parts: the first one is dedicated to the development of detection and classification techniques of

M. Versaci (✉) · S. Calcagno · M. Cacciola · F. C. Morabito · I. Palamara · D. Pellicanò
DICEAM Department, Mediterranea University, Via Graziella Feo di Vito,
Reggio Calabria, Italy
e-mail: mario.versaci@unirc.it

S. Calcagno
e-mail: calcagno@unirc.it

M. Cacciola
e-mail: matteo.cacciola@unirc.it

F. C. Morbito
e-mail: morabito@unirc.it

I. Palamara
e-mail: isabella.palamara@unirc.it

D. Pellicanò
e-mail: diego.pellicano@unirc.it

P. Burrascano et al. (eds.), *Ultrasonic Nondestructive Evaluation Systems*,
DOI 10.1007/978-3-319-10566-6_7

defectiveness for ultrasonic NDE by means of CW, while the second one, with the same purpose, proposes an approach based on FS.

7.1 Introduction

In many industrial processes for the production of manufacts, detection and classification of defects represent crucial tasks for the choice of manufact destination. Therefore, they necessitate suitable investigation methodologies characterized by a high degree of efficiency and, also, by an acceptable computational load. In this context, two innovative investigative approaches for the characterization of defects in ultrasonic nondestructive testing (NDT)/nondestructive evaluation (NDE) are herein described. The first approach makes use of the concept of computing with words (CW) in which a word is considered as a label of a fuzzy set of points put together for similarity (concept of granule) which leads to a typical bank of fuzzy rules structured per classes. The second approach springs from the consideration that a defect (even when invisible at the naked eye) can be considered as the cause of a reduction in similarity of the corresponding signal, referred to a reference signal obtained from a target free of defects. Taking into account the inherent fuzziness present in measured signals, the use of fuzzy similarity (FS) techniques is clearly desirable.

7.2 Computing with Words and Fuzzy Entropy for Ultrasonic NDE

7.2.1 Computing with Words in Ultrasonic NDE

In CW, the objects of computations are words drawn from common language [16, 17]. Inspiration for CW comes from the remarkable ability of the brain to formulate complex reasonings without any references to well-defined parameters, without the need to explicitly practice important computations and with the ability to give correct interpretations to approximate estimations of some quantities [16, 17]. Certainly, it does not deal with precise measurements, but the main difference between measure and perception resides in the level of precision they are able to express: in the first case, we speak about *crisp* values while, in the second one, we deal with *fuzzy* numbers or, even better, with fuzzy *granules* which represent sets of objects (or elements) not distinguishable among them or anyway drawn together by criteria such as similarity or functionality [5, 10, 15]. The instance of a *granule*, g, is the word, w, which can be thought as the label of a *granule*. Often, a word can be *atomic* or *composite*. For example, the feature of being *deep*, related to a defect in a specimen, is an atomic *word*, while the expression *very deep* is a composite *word*. The *granules* have also some features defining the range, the sizes, or the state within they operate. The passage from a *granule* to another one is not sharp, but gradual

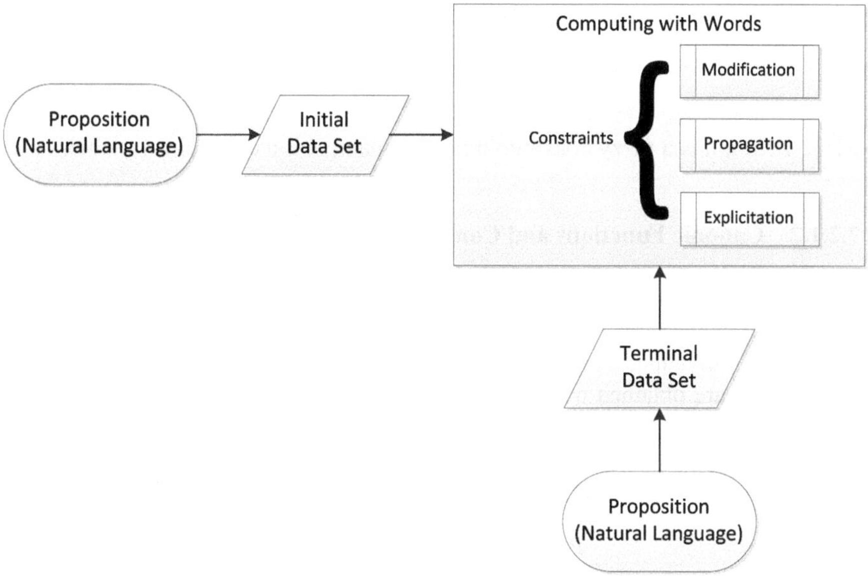

Fig. 7.1 Structure of CW: from IDS to TDS through the modification of the explicitation propagation and modification of the bank of fuzzy constraints

and represents a fuzzy constraint on a variable. For instance, in the proposition "the defect is deep" the adjective *deep* represents the label of the granule *deep*, while the fuzzy set *deep* represents a fuzzy constraint on the depth of the defect. Formally, CW is a system made up of an initial data set (IDS) and a terminal data set (TDS): starting from IDS, made up of a series of questions formulated in a natural language (NL), we obtain the answers (still expressed in a NL and constituting the TDS) and propositions in NL are made clear in canonical forms representing the constraints of the previous ones (Fig. 7.1). In particular, starting from IDS we go toward TDS by an explicit propagation and eventual modification of the bank of fuzzy constraints. Propagating the constraints by rules, the antecedents are transformed in consequents and, at the end, translated in NL by means of linguistic approximation procedures [16, 17]. But, when is it right to use CW techniques? When the inaccuracy of the signals does not justify the use of numerical values carrying out solutions in NL at a low-computational load.

7.2.1.1 Canonic Formulation of Fuzzy Constraints

To formalize fuzzy constraints, let us consider a *proposition* in a NL assimilable to a sort of fuzzy constraints network which, after aggregation, converge in an overall fuzzy constraint represented by an expression in canonic form (or canonic function)[1] as follows [16, 17]:

$$Z\ IS\ Relation \tag{7.1}$$

where *Relation* is a fuzzy constraining relation and Z is a constraint variable.

7.2.1.2 Canonic Functions and Constraining Relations

Essentially, two types of canonic functions exist: the *disjunctive* and the *conjunctive* ones. The first ones are obtained using exclusively terms embracing all the variables of a function in denied or direct form and put them in relation by operator *OR*; the second ones are obtained using exclusively terms embracing all the variables of a function in denied or direct form and put them in relation by operator *AND* [16, 17]. The purpose of a canonical form resides in putting in evidence the fuzzy constraints nested in *proposition* by the formalization:

$$proposition \rightarrow Z\ IS\ Relation \tag{7.2}$$

in which the arrow symbolizes the operation of explicitation.[2] In such a context, the explicitation of the *proposition* can be realized through an explanatory database (ED), which represents a set of relations to define the meaning of *proposition*, giving back the constrained variable Z and then acting on the ED to obtain the constraining relation *Relation*.

The following two examples show the steps of formation of *Relation* upon the correct formulation of the ED [16, 17]. If *Relation* is a generic fuzzy relation

$$Z\ IS\ manipulator(R) \rightarrow Z\ IS\ function(Relation) \tag{7.3}$$

where *manipulator* can be written as *not*, *very*, ... , and *function*(*Relation*) represents the way in which *manipulator* manipulates *Relation*. If $manipulator = not$, then $function(Relation) = not\text{-}Relation$ (complement). If $manipulator = very$, then $function(Relation) = {}^2Relation$ (square on the left). In terms of membership function:

$$membership\ function_{not\text{-}Relation}(y) = 1 - membership\ function_{Relation}(y) \tag{7.4}$$

$$membership\ function_{{}^2Relation}(y) = \left(membership\ function_{Relation}(y)\right)^2 \tag{7.5}$$

where y is an element of *Relation*. The purpose of such rules is to highlight the coincidence of the main laws ruling the propagation of the fuzzy constraints with those ruling the fuzzy inference [16, 17]. In the following, two explicative examples of ED structuring as to the ultrasonic NDE problems.

---

[1] It derives from the necessity to get a function from the table of truths.

[2] $\rightarrow$ represents the translation into CW of the transition from the antecedent to the consequent in a syllogism.

Example 7.1
If the starting proposition is:

$$proposition = \text{“}The\ Defect\ is\ not\ Very\ Deep\text{”} \tag{7.6}$$

it is possible to extract the ED through the formulation of the following construction [16, 17]:

$$\begin{aligned} ED = {} & Population_{Defects}[Typology\ of\ Defect\ Depth] \\ & + \text{Deep}[Depth; membership\ function(\cdot)] \end{aligned} \tag{7.7}$$

where:
1. $Population_{Defects}$ is a relation where *Typology of Defect* and *Depth* are the topics;
2. Deep is a relation with topics *Depth* and $\mu(\cdot)$;
3. The symbol + represents the logic operation of disjunction.

So, the constrained variable Z is the deepness which can be expressed in terms of ED with the following script [16, 17]:

$$Z = Depth(Defect) = {}_{Depth}Population[Typology\ of\ Defect = Defect] \tag{7.8}$$

explaining the mechanism acting on the ED which gives back the constrained variable Z. In particular, a *typology* is immediately associated to “Defect” and the result is projected on *Depth*, producing the deepness of the defect [3]. The constraining relation $Relation = Very\ Deep$ can be expressed in terms of squaring on the left through the following script [16, 17]:

$$Relation = ({}^{2}Deep) = Deep[Depth; membership\ function^{2}(\cdot)] \tag{7.9}$$

where, *very* is intended as square on the left. Similarly, *not* can be represented as a complement [16, 17]:

$$Relation = Deep[Depth; 1 - membership\ function(\cdot)] \tag{7.10}$$

Example 7.2
As further example let us consider the proposition:

$$proposition = \text{“}Defect\ in\ a\ Little\ Class\ \ is\ near\ \ a\ Big\ Class\text{”} \tag{7.11}$$

where the extrapolation of the ED leads to the following formulation:

$$ED = Population[Typology\ of\ Defect; Belonging\ Class]$$

$$+ \; little[Class; membership\,function(\cdot)]$$
$$+ \; near[Class_1; Class_2; membership\,function(\cdot)]. \qquad (7.12)$$

In this case:

$$Z = Belonging\ Class(Defect)$$
$$= {}_{Belonging\ Class}Population[Typology\ of\ Defects = Defect] \qquad (7.13)$$

and

$$Relation = Little[Belonging\ Class; membership\,function(\cdot)] \cap$$
$$_{Class_1}near[Class_2 = BigClass] \qquad (7.14)$$

where:
1. The first element is the fuzzy set of the little classes;
2. The second element is the fuzzy set of the little classes near the big one;
3. $\cap$ indicates the fuzzy intersection of the two fuzzy sets.

So, the methodology has been conceived to formulate constructions belonging to the form (7.1).[3] Obviously, such constraints can be considered as particular cases of the refined variety of shadings of linguistic constraints whatever ones NL possesses [3].

7.2.1.3 Generalization, Propagation, and Manipulation of Fuzzy Constraints

As the final purpose is represented by the writing of banks of fuzzy rules managed by inferential mechanisms, the script of the propositions in the IDS holds a role of primary importance [16, 17]. Generally, they are conjunctions or disjunctions of constraints as written in Eq. (7.1), so-called *basic constraint* but, sometimes, some generalizations formalized by [16, 17]:

$$Z\ IS\text{-}variable\ copula\ R \qquad (7.15)$$

are used in which *IS-variable copula* is a particular variable said *copula* which confirms the way how Z is constrained by the *Relation* which performs through the assignment of values to the discrete variable *variable copula* in the terms expressed by Table 7.1.

[3] In fact, the conditional statement *IF Z IS Relation then K IS H* can be easily restored in the form *K IS H IF Z IS Relation*.

Table 7.1 Interpretation of the values of the discrete variable *r*

Values	Interpretation	Values	Interpretation
eq	Equal	*us*	Usuality
con	Conjunctive	*ranfuset*	Random fuzzy set
dis	Disjunctive	*ranset*	Random set
probval	Probability value	*rouset*	Rough set
prob	Probabilistic	*fugr*	Fuzzy graph

For example:

1. If $variable\ copula = dis$, then the constraint *Z IS dis Relation* is reduced to *Z IS Relation*
2. If $variable\ copula = con$, then the constraint *Z IS con Relation* maintains the same membership value
3. If $variable\ copula = eq$, then the constraint is showed by the equal symbol
4. If $variable\ copula = prob$, then the constraint *Z IS prob Relation* is probabilistic one
5. If $variable\ copula = ranset$, then the constraint *Z IS ranset Relation* is a random one.

Once IDS proposition is expressed in terms of canonic forms, we pass to the structuring of fuzzy constraints propagation which, generally, occurs with the same rules of the fuzzy inference. However, in scientific literature, many rules of fuzzy constraints propagation have been proposed to manage the eventual modifications [16, 17] . In particular, if R_1 and R_2 are two fuzzy relations, we can structure disjunctive, conjunctive, projective, surjective, and derived rules. In particular, the disjunctive ones can be written as follows:

$$IF\ Z\ IS\ Relation_1\ OR\ Z\ IS\ Relation_2\ THEN\ Z\ IS\ (Relation_1 \cup Relation_2) \tag{7.16}$$

which, as for $variable\ copula = con$, it becomes:

$$IF\ \ Z\ IS\text{-}con\ Relation_1\ OR\ Z\ IS\ Relation_2\ THEN\ \ Z\ IS\ (Relation_1 \cap Relation_2) \tag{7.17}$$

or:

$$\begin{aligned} &IF\ Z_1\ IS\ Relation_1\ AND\ Z_2\ IS\ Relation_2\ THEN \\ &[(Z_1, Z_2)\ IS\ (Relation_1 \times M) \cup (N \times Relation_2)] \end{aligned} \tag{7.18}$$

in which $(Relation_1 \times V)$ and $(U \times Relation_2)$ are, respectively, cylindrical extensions of $Relation_1$ and $Relation_2$. The conjunctive constructions can get the forms:

$$IF\ \ Z\ IS\ Relation_1\ AND\ Z\ IS\ Relation_2\ THEN\ \ Z\ IS\ (Relation_1 \cap Relation_2) \tag{7.19}$$

$$IF\ Z_1\ IS\ Relation_1\ AND\ Z_2\ IS\ Relation_2\ THEN$$

$$(Z_1, Z_2)\ \mathit{IS}\ (\mathit{Relation}_1 \times \mathit{Relation}_2). \tag{7.20}$$

A possible projective rule can get the form:

$$\mathit{IF}\ (Z_1, Z_2)\ \mathit{IS}\ \mathit{Relation}_1\ \mathit{THEN}\ Z_2\ \mathit{IS}\ \sup\{\mathit{Relation}_1\} \tag{7.21}$$

while a surjective one can be written as follows:

$$\mathit{IF}\ Z\ \mathit{IS}\ \mathit{Relation}_1\ \mathit{THEN}\ (Z_1, Z_2)\ \mathit{IS}\ (\mathit{Relation}_1 \times M) \tag{7.22}$$

However, by linguistic modifiers such as *very*, *enough*, and *not*, the necessity to carry out a modification of the constraint structuring is expressed. For example, if we indicate by *Relation* a fuzzy relation and by *modifier* any modifiers, the script

$$Z\ \mathit{IS}\ (\mathit{modifier\ Relation}) \rightarrow Z\ \mathit{IS}\ \mathit{function}(\mathit{Relation}) \tag{7.23}$$

can be considered as a formalization of how the modifier manipulates the *Relation*. The writing of the derived rules is sort of complex. The scientific community deepened the research in this field as testified by important works where the structuring of the derived rules has been geared toward the proposal of structures like generalized *modus ponens* and *modus tollens* or even sillogisms of more refined making. The interested reader is invited to refer to more specific papers [16, 17].

7.2.1.4 TDS Construction

To make the TDS, we need to start from the generalized extension principle written as follows [16, 17]:

$$\begin{aligned} &\mathit{IF}\ \mathit{function}(Z_1, Z_2, \ldots, Z_n)\ \mathit{IS}\ \mathit{Relation} \\ &\mathit{THEN}\ \mathit{query}(Z_1, Z_2, \ldots, Z_n)\ \mathit{IS}\ \mathit{query}(\mathit{function}^{-1}(\mathit{Relation})) \end{aligned} \tag{7.24}$$

in which:

1. $Z_1, Z_2, \ldots, Z_n$ are variables;
2. $\mathit{function}(Z_1, Z_2, \ldots, Z_n)\ \mathit{IS}\ \mathit{Relation}$ is a constraint inferred by IDS;
3. $\mathit{query}(Z_1, Z_2, \ldots, Z_n)$ is a query constraining TDS.

It is demonstrable how the generalized extension principle leads to the formulation of the TDS by the writing of the following constraining maximization problem:

$$\mu_{\mathit{query}(Z_1, Z_2, \ldots, Z_n)}(\mathit{constraint}) = \sup_{(s_1, s_2, \ldots s m_n)} (\mathit{membership\ function}_{\mathit{Relation}}(f(s_1, s_2, \ldots, s_n))) \tag{7.25}$$

with the constraint $\mathit{constraint} = q(s_1, s_2, \ldots, s_n)$. If, for example, the problem under study is related to the search of the mean value of the deepness of a defect, pointing by hpd_j the population of the heights of defects, by $\mathit{membership\ function}_{\mathit{deep}}(\mathit{hpd}_i)$, $i =$

$1, 2, \ldots, n$ the relative values of membership and by $\frac{1}{N}\sum_i hpd_i$ the mean value of the depth, then the solution can be got through the resolution of the maximization problem:

$$membership\ function_{(constraint)} =$$

$$\sup_{hpd_1,\ldots,hpd_N}\left(membership\ function_{maximum}\left(\frac{1}{N}\sum_j membership\ function_{deep}(hpd_i)\right)\right) \tag{7.26}$$

with $constraint = \frac{1}{N}\sum_i hpd_i$. solvable, for example, through the use of genetic algorithms or other specific methodologies.

7.2.2 *Design of a Procedure for Classifying Defects by means of CW*

The purpose of this section is to devise a procedure for characterizing defectiveness in carbon fiber reinforced plastics (CFRP) specimens in terms of classification (and, implicitly, also of detection) through a bank of fuzzy rules and CW [4, 9, 16, 17]. In particular, starting from an experimental database of ultrasonic signals on samples in CFRP, a bank of fuzzy rules is carried out by extraction of statistical parameters and managed by Mamdani's inference. A bank of fuzzy rules in the CW domain is carried out to evaluate and compare the obtained results of classification.

7.2.2.1 The Experimental Database

As explained in Chap. 6, starting from a sample artificial CFRP panel (301 × 355 × 4 mm) immersed into water, top, middle, and bottom delamination/inclusion and defect-free areas were obtained. Furthermore, porosities were obtained on another panel (39 × 40 × 3 mm). Investigations were carried out at Laboratory of Electrical Engineering and NDT/NDE of University of Reggio Calabria. A 5 MHz ultrasonic transducer (UT) probe was used with 100 MHz sampling frequency. The speed of the ultrasonic wave on CFRP is 3000 m/s. Usually, an UT signal related to an undefected area is characterized by a top-echo (interaction between the exciting ultrasonic wave and the top surface) and a bottom-echo reflection of the wave on the bottom of the specimen). The presence of other peaks detects discontinuities of the material (defects)[4]. Starting from the idea that similar defects produce similar ultrasonic signals, it is not possible to classify defectiveness by means of qualitative inspection of the signals. In this way, we need imperatively to solve an inverse problem after preprocessing UT signals to extract information between front and

[4] The main frequency of the signal is the half of the sampling frequency of ultrasonic wave

Table 7.2 Distribution of signals in training and testing databases

	Training	Testing
Top delamination	42	21
Middle delamination	45	24
Bottom delamination	37	22
Top inclusion	35	28
Middle inclusion	38	25
Bottom inclusion	38	21
Porosity	40	25
No defects	45	30

bottom echoes particularly when doubtful cases take place. In addition, the use of supporting classification algorithms by means of a low computational load is appreciated especially if hardware device is required [1, 2, 4, 9]. For a benefit of a simple reading, Table 7.2 redisplays the training and testing sections of the experimental database (number of signals for each category of defect and class of signals without defects).

7.2.2.2 Some Preliminary Considerations

From the analysis of a generic signal without any defects, it comes out that between the top and the bottom echoes relevant anomalies are not present. On the contrary, in signals affected by defects, between the two echoes they are well visible alterations of the same signals. But, because similar defects produce similar signals, defectiveness are not distinguishable by a simple visual inspection of the signal [3, 4, 9]. Furthermore, such similarities affect also the features of the signals in the sense that similar defects produce signals with similar features [5, 10, 15]. For example, panels affected by particular defects produce signals with mean and standard deviations of such type of defect, which generate, in this way, a particular class. So, we have set the basis for a hypothetical classification procedure which, starting from the numerical values of mean and standard deviation of a signal (or a section of it), it marks the class of membership and, so, the defectiveness. Table 7.3 visualizes the inputs of the system (mean and standard deviation) and the eight classes of outputs (three for delamination, three for inclusion, one for porosity and one for the signals without defects) with relative codifications. Due to the fact that, in certain fields of applications, some defects are tolerated while others not much, let us consider only the defects thought as more dangerous (porosity, mean delamination, and mean inclusion) and the class without defects highlighted in Table 7.3 by the symbol (*). Finally, to underline the criticalities of eventual presence of borders, let us consider a further class of positions labelled by *Near* and *Far*, indicating the closeness or the distance of the defect from the edge. The classes of input (MN and SD) are subdivided into two and six subclasses respectively. Such a choice has been dictated by the fact that

Table 7.3 Inputs/Outputs classes codifications

Inputs	Codification	Outputs	Codification
Mean	MN1	Top delamination	TD
	MN2	Middle delamination (*)	MD
Standard deviation (SD)	SD1	Bottom delamination	BD
	SD2	Top inclusion	TI
	SD3	Middle inclusion (*)	MI
	SD4	Bottom inclusion	BI
	SD5	Porosity (*)	P
	SD6	No defects (*)	ND

Table 7.4 Inputs and outputs classes with codifications (defects more dangerous)

Inputs	Codification	Outputs	Codification
Mean	MN1	Porosity	P
	MN2	Middle delamination	MD
Standard deviation (SD)	SD1	Middle inclusion	MI
	SD2	No defects	ND
	SD3	Position	Near
	SD4		Far
	SD5		
	SD6		

the mean distribution is around the values of 100 and 140 of amplitude of ultrasonic echo (values on each they will be centered the gaussian function representative of the classes of inputs), while the standard deviation is distributed on an almost wide range of values. As it follows, the details of the procedure.

7.2.2.3 Fuzzification of the Variables

The process of granulation starts up by dividing the totality of the signals by making a distinction among inputs (mean and standard deviation) and outputs (classes of defectiveness and absence of defects) as visualized in Table 7.4 where, to each variable, a codification with label is associated. In particular, there are two codifications for the mean, six for the standard deviation, four for defectiveness (in which one of them is related to a class of signals without defects), and a position variable, labelled by *Near* and *Far*, indicating the closeness/distance of the defect from the edge. They are defined, then, the ranges of possible values for the inputs and the codifications for the outputs (Table 7.5). In short:

Table 7.5 Ranges of characterization of inputs and outputs classes (defects more dangerous)

Inputs	Range	Outputs	Codification
MN1	50–120	P	1
MN2	100–170	MD	2
SD1	5–5.99	MI	3
SD2	5.1–6.99	ND	4
SD3	5.2–5.49	Position Near	5
SD4	5.5–6.49	Position Far	6
SD5	6.5–7.19		
SD6	7.2–8.5		

Table 7.6 Classes and membership ranges

Inputs	Fuzzified class	Range
Mean	MN1	50–120
Mean	MN2	100–170
SD1	Extra small (ES)	5–5.99
SD2	Very small (VS)	5.1–6.99
SD3	Small (S)	5.2–5.49
SD4	Medium (M)	5.5–6.49
SD5	Large (L)	6.5–7.19
SD6	Very large (VL)	7.2–8.5

1. We choose only two membership functions for the mean because it is distributed around only two values: 100 and 140 of amplitude of ultrasonic echo and we center on them two gaussian functions representing the related membership functions
2. For standard deviation, to cover the whole distribution of values with a sufficient grade of diversification, it is necessary to make use of at least six classes[5]
3. Regarding the typology of defects, the classes are only three: porosity, mean delamination, and mean inclusion[6]
4. The last class is related to the signals without defects
5. Finally, a further variable indicates the closeness/distance of a defect from the edge.

The classification procedure requires, as it is well-known, the assignation to each variable of particular linguistic labels indicating their fuzziness. For example, a possible assignation of labels to the input variables is reported in Table 7.6 where the ranges of characterization of each single label are visualized.

[5] In this case, the knowledge of the expert is determining for a correct decision making.
[6] In order of decreasing dangerousness.

Table 7.7 Round up table of the parameters for the first sixteen signals. Starting from the scheme, a bank of fuzzy rules will be extracted

Signal	Mean	Standard deviation	Defect	Label	Position
$Signal_1$	MN1	ES	20	P	Near
$Signal_2$	MN1	M	60	MI	Far
$Signal_3$	MN1	S	40	MD	Near
$Signal_4$	MN1	L	100	ND	Far
$Signal_5$	MN2	VS	40	MD	Near
$Signal_6$	MN2	L	100	ND	Far
$Signal_7$	MN2	S	20	P	Near
$Signal_8$	MN2	VL	60	MI	Far
$Signal_9$	MN1	VS	40	MD	Near
$Signal_{10}$	MN1	VL	100	ND	Far
$Signal_{11}$	MN2	ES	40	MD	Near
$Signal_{12}$	MN2	L	100	ND	Far
$Signal_{13}$	MN1	ES	20	P	Near
$Signal_{14}$	MN1	M	60	MI	Far
$Signal_{15}$	MN2	S	20	P	Near
$Signal_{16}$	MN2	L	60	MI	Far
...	...	...	...	...	...
...	...	...	...	...	...

7.2.2.4 Bank of Fuzzy Rules Extraction

To correctly design the bank of fuzzy rules, it is enough to create a table with all the data carried out from the experimental database and, for each of them, to associate the class of membership. For the first sixteen signals of the database, such values are reported in Table 7.7 where it is possible to carry out the following rules:

1. (IF Mean is MN1 and SD is ES) THEN (Defect is P and Position is Near)
2. (IF Mean is MN1 and SD is M) THEN (Defect is MI and Position is Far)
3. (IF Mean is MN1 and SD is S) THEN (Defect is MD and Position is Near)
4. (IF Mean is MN1 and SD is L) THEN (Defect is ND and Position is Far)
5. (IF Mean is MN2 and SD is VS) THEN (Defect is MD and Position is Near)
6. (IF Mean is MN2 and SD is L) THEN (Defect is ND and Position is Far)
7. (IF Mean is MN2 and SD is S) THEN (Defect is P and Position is Near)
8. (IF Mean is MN2 and SD is L) THEN (Defect is MI and Position is Far)
9. (IF Mean is MN1 and SD is VS) THEN (Defect is MD and Position is Near)
10. (IF Mean is MN1 and SD is VL) THEN (Defect is ND and Position is Far)
11. (IF Mean is MN2 and SD is ES) THEN (Defect is MD and Position is Near)
12. (IF Mean is MN2 and SD is L) THEN (Defect is ND and Position is Far)

13. (IF Mean is MN1 and SD is ES) THEN (Defect is P and Position is Near)
14. (IF Mean is MN1 and SD is M) THEN (Defect is MI and Position is Far)
15. (IF Mean is MN2 and SD is S) THEN (Defect is P and Position is Near)
16. (IF Mean is MN2 and SD is L) THEN (Defect is MI and Position is Far)

which, after the elimination of the duplicative rules, can be reduced to the following list:

1. (IF Mean is MN1 and SD is ES) THEN (Defect is P and Position is Near)
2. (IF Mean is MN1 and SD is M) THEN (Defect is MI and Position is Far)
3. (IF Mean is MN1 and SD is S) THEN (Defect is MD and Position is Near)
4. (IF Mean is MN1 and SD is L) THEN (Defect is ND and Position is Far)
5. (IF Mean is MN2 and SD is VS) THEN (Defect is MD and Position is Near)
6. (IF Mean is MN2 and SD is L) THEN (Defect is ND and Position is Far)
7. (IF Mean is MN2 and SD is S) THEN (Defect is P and Position is Near)
8. (IF Mean is MN2 and SD is L) THEN (Defect is MI and Position is Far)
9. (IF Mean is MN1 and SD is VS) THEN (Defect is MD and Position is Near)
10. (IF Mean is MN1 and SD is VL) THEN (Defect is ND and Position is Far)
11. (IF Mean is MN2 and SD is ES) THEN (Defect is MD and Position is Near)
12. (IF Mean is MN2 and SD is L) THEN (Defect is MI and Position is Far)

where the rules 4 and 6 can be compacted in a single rule:

(IF Mean is MN1 or Mean is MN2 and SD is L) THEN (Defect is ND and Position is Far).

But, as regards to the mean variable, classes 1 and 2 cover completely the whole of the possible values, so the rule becomes:

(IF SD is L) THEN (Defect is ND and Position is Far).

In a very similar way, rules 5 and 9 can be compacted as follows:

(IF SD is VS) THEN (Defect is MD and Position is Near).

At the end, the bank of fuzzy rules gets the form:

1. (IF Mean is MN1 and SD is ES) THEN (Defect is P and Position is Near)
2. (IF Mean is MN1 and SD is M) THEN (Defect is MI and Position is Far)
3. (IF Mean is MN1 and SD is S) THEN (Defect is MD and Position is Near)
4. (IF SD is L) THEN (Defect is ND and Position is Far)
5. (IF SD is VS) THEN (Defect is MD and Position is Near)
6. (IF Mean is 2 and SD is S) THEN (Defect is P and Position is Near)
7. (IF Mean is 2 and SD is VL) THEN (Defect is MI and Position is Far)
8. (IF Mean is 1 and SD Deviation is VL) THEN (Defect is ND and Position is Far)
9. (IF Mean is 2 and SD is ES) THEN (Defect is MD and Position is Near)
10. (IF Mean is 2 and SD is L) THEN (Defect is MI and Position is Far)

Table 7.8 Detection and classification obtained through FIS and CW: comparison of the results

FIS vs CW	FIS detection (%)	FIS classification (%)	CW detection (%)	CW classification (%)
P	99	95	99	94.8
MD	100	96.5	97	96
MI	98	97.2	100	96.12
ND	100	96.9	100	95.74

Implementing by Mamdani's inference with Gaussian membership functions and defuzzification through the center of gravity, we obtain the values reported in Table 7.8.

7.2.2.5 ED Construction and Translation of the Bank of Fuzzy Rules into the CW Domain

The procedure of the ED construction is applied directly to the bank of fuzzy rules previously drawn. If we consider rule 1, the first antecedent is composed by the proposition

$$p_1 = X_1 \; is \; R_1 = MN \; is \; L \tag{7.27}$$

where $X_1 = MN$ and $R_1 = L$. *MN* is a *parameter* which defines a feature, so *MN* can be considered as $Characteristics_{MN}$. Moreover, not all the values of *MN* are taken into account, but only those falling on [50–170] (their domain). Then, ED for proposition *MN is L* can be expressed as:

$$\begin{aligned} ED = {} & Characteristics_{MN}[Parameter; Domain] \\ & + L[6.5\text{–}7.19; \mu(\cdot)] = Characteristics_{MN}[MN; 50\text{–}170] \\ & + L[120\text{–}170; \mu(\cdot)] \end{aligned} \tag{7.28}$$

in which $Characteristics_{MN}$ is a relation which topics are *Parameter* and *Domain*. *L* is a relation with topics *Domain* and $\mu(\cdot)$. Finally, the symbol "+" represents the disjunction operator. The constrained variable X_1 is the domain of the variable *MN* in terms of ED, and it can be expressed in the form:

$$X_1 = Domain(MN) = {}_{Domain}Caracteristics_{MN}[Parameter = MN] \tag{7.29}$$

Such expression specifies the procedure acting on ED and giving X_1 . In particular, *Parameter* is immediately associated to *MN* and the result is projected on *Domain* producing *Domain of the Main*. The relation R_1 corresponds to:

$$R_1 = L[6.5\text{–}7.19; \mu(\cdot)]. \tag{7.30}$$

Finally, proposition p is definitely expressed as follows:

$$_{Domain}Caracteristics_{MN}[Parameter = MN]\ is\ L[6.5\text{–}7.19; \mu(\cdot)]. \tag{7.31}$$

Similarly, for the second antecedent, the ED counts as:

$$\begin{aligned} ED &= Characteristics_{SD}[Parameter; Domain] + ES[5\text{–}5.99; \mu(\cdot)] \\ &= Characteristics_{SD}[SD; 5\text{–}8.5] + ES[5\text{–}5.99; \mu(\cdot)] \end{aligned} \tag{7.32}$$

In this case the variable is constrained to the domain of the possible values (peak-peak of the phase) which, in terms of ED, can be expressed like this:

$$X_2 = SD = Domain(SD) = {}_{Domain}Characteristics_{SD}[Parameter = SD] \tag{7.33}$$

relation R_2 corresponds to

$$R_2 = ES[5\text{–}5.99; \mu(\cdot)] \tag{7.34}$$

and proposition p_2 can be expressed as:

$$_{Domain}Characteristics_{SD}[Parameter = SD]\ is\ ES[5\text{–}5.99; \mu(\cdot)]. \tag{7.35}$$

First rule consequent can be written as

$$q = X_0\ is\ S = Class\ is\ Defect\ P \tag{7.36}$$

and its ED counts as:

$$\begin{aligned} ED &= Characteristics_{Defect}[Parameter; Possible\ Values] + P[1; \mu(\cdot)] \\ &= Characteristics_{Defect}[Class; 1\text{–}6] + P[1; \mu(\cdot)]. \end{aligned} \tag{7.37}$$

constrained variable is the whole of the possible values of the class of defects which, in terms of ED, can be expressed as:

$$\begin{aligned} X_0 &= Class = PossibleValues(Class) \\ &= {}_{PossibleValues}Characteristics_{Defect}[Parameter = Class]. \end{aligned} \tag{7.38}$$

Relation S corresponds to the expression

$$S = P[1; \mu(\cdot)] \tag{7.39}$$

while the consequent can be expressed in the form:

$$_{PossibleValues}Characteristics_{Defect}[Parameter = Class]\ is\ P[1; \mu(\cdot)]. \tag{7.40}$$

We go on likewise for the second consequent (position of defect) obtaining:

$$_{PossibleValues}Characteristics\_Position[Parameter = Class]\ is\ Near[5; \mu(\cdot)]. \tag{7.41}$$

Finally, the bank of fuzzy rules, written in terms of CW, becomes:

Fuzzy Rule 1.

$$
\begin{aligned}
&IF\ {}_{Domain}Characteristics_{MN}[Parameter = MN]\ is\ MN1[50\text{–}120; \mu(\cdot)]\ and\\
&\quad {}_{Domain}Characteristics_{SD}[Parameter = SD]\ is\ ES[5\text{–}5.99; \mu(\cdot)]\\
&THEN\ {}_{PossibleValues}Characteristics_{Defect}[Parameter = Class]\ is\ P[1; \mu(\cdot)]\ and\\
&\quad {}_{PossibleValues}Characteristics\_Position[Parameter = Class]\ is\ Near[5; \mu(\cdot)].
\end{aligned} \tag{7.42}
$$

Fuzzy Rule 2.

$$
\begin{aligned}
&IF\ {}_{Domain}Characteristics_{MN}[Parameter = MN]\ is\ MN1[Domain; \mu]\ and\\
&\quad {}_{Domain}Characteristics_{SD}[Parameter = SD]\ is\ M[5.5\text{–}6.49; \mu(\cdot)]\\
&THEN\ {}_{PossibleValues}Characteristics_{Defect}[Parameter = Class]\ is\ MI[3; \mu(\cdot)]\ and\\
&\quad {}_{PossibleValues}Characteristics_{Position}[Parameter = Class]\ is\ Far[6; \mu(\cdot)].
\end{aligned} \tag{7.43}
$$

Fuzzy Rule 3.

$$
\begin{aligned}
&IF\ {}_{Domain}Characteristics_{MN}[Parameter = MNt]\ is\ MN1[50\text{–}120; \mu(\cdot)]\ and\\
&\quad {}_{Domain}Characteristics_{SD}[Parameter = SD]\ is\ S[5.2\text{–}5.49; \mu(\cdot)]\\
&THEN\ {}_{PossibleValues}Characteristics_{Defect}[Parameter = Class]\ is\ MD[2; \mu(\cdot)]\ and\\
&\quad {}_{PossibleValues}Characteristics_{Position}[Parameter = Class]\ is\ Near[5; \mu(\cdot)].
\end{aligned} \tag{7.44}
$$

Fuzzy Rule 4.

$$
\begin{aligned}
&IF\ {}_{Domain}Characteristics_{SD}[Parameter = SD]\ is\ L[6.5\text{–}7.19; \mu(\cdot)]\\
&\qquad THEN\ ND[4; \mu(\cdot)]\ and\\
&{}_{PossibleValues}Characteristics_{Positit
ion}[Parameter = Class]\ is\ Far[6; \mu(\cdot)].
\end{aligned} \tag{7.45}
$$

Fuzzy Rule 5.

$$
\begin{aligned}
&IF\ {}_{Domain}Characteristics_{SD}[Parameter = SD]\ is\ VS[5.1\text{–}6.99; \mu(\cdot)]\\
&THEN\ {}_{PossibleValues}Characteristics_{Defect}[Parameter = Class]\ is\ MD[2; \mu(\cdot)]\ and\\
&{}_{PossibleValues}Characteristics_{Posititition}[Parameter = Class]\ is\ Near[5; \mu(\cdot)].
\end{aligned} \tag{7.46}
$$

Fuzzy Rule 6.

$$IF\ {}_{Domain}Characteristics_{MN}[Parameter = MN]\ is\ MN2[50\text{–}120; \mu(\cdot)]\ and$$

$$_{Domain}Characteristics_{SD}[Parameter = SD] \text{ is } S[5.2\text{–}5.49; \mu(\cdot)]$$
$$THEN\ _{PossibleValues}Characteristics_{Defect}[Parameter = Class] \text{ is } P[1; \mu(\cdot)] \text{ and}$$
$$_{PossibleValues}Characteristics_{Positition}[Parameter = Class] \text{ is } Near[5; \mu(\cdot)]. \quad (7.47)$$

Fuzzy Rule 7.

$$IF\ _{Domain}Characteristics_{MN}[Parameter = MN] \text{ is } MN2[100\text{–}170; \mu(\cdot)] \text{ and}$$
$$_{Domain}Characteristics_{SD}[Parameter = SD] \text{ is } VL[7.2\text{–}8.5]$$
$$THEN\ _{PossibleValues}Characteristics_{Defect}[Parameter = Class] \text{ is } MI[3; \mu(\cdot)] \text{ and}$$
$$_{PossibleValues}Characteristics_{Positition}[Parameter = Class]. \quad (7.48)$$

Fuzzy Rule 8.

$$IF\ _{Domain}Characterisics_{MN}[Parameter = MN] \text{ is } MN1[50\text{–}120; \mu(\cdot)] \text{ and}$$
$$_{Domain}Characteristics_{SD}[Parameter = SD] \text{ is } VL[7.2\text{–}8.5; \mu(\cdot)]$$
$$THEN\ _{PossibleValues}Characteristics_{Defect}[Parameter = Class] \text{ is } ND[4; \mu(\cdot)] \text{ and}$$
$$_{PossibleValues}Characteristics_{Positition}[Parameter = Class] \text{ is } Far[6; \mu(\cdot)]. \quad (7.49)$$

Fuzzy Rule 9.

$$IF\ _{Domain}Characteristics_{MN}[Parameter = MN] \text{ is } MN2[100\text{–}170; \mu(\cdot)] \text{ and}$$
$$_{Domain}Characteristics_{SD}[Parameter = SD] \text{ is } ES[5\text{–}5.99; \mu(\cdot)]$$
$$THEN\ _{PossibleValues}Characteristics_{Defect}[Parameter = Class] \text{ is } MD[2; \mu(\cdot)] \text{ and}$$
$$_{PossibleValues}Characteristics_{Positition}[Parameter = Class] \text{ is } Near[5; \mu(\cdot)]. \quad (7.50)$$

Fuzzy Rule 10.

$$IF\ _{Domain}Characteristics_{MN}[Parameter = MN] \text{ is } MN2[100\text{–}170; \mu(\cdot)] \text{ and}$$
$$_{Domain}Characteristics_{SD}[Parameter = SD] \text{ is } L[6.5\text{–}7.19; \mu(\cdot)]$$
$$THEN\ _{PossibleValues}Characteristics_{Defect}[Parameter = Class] \text{ is } MI[3; \mu(\cdot)] \text{ and}$$
$$_{PossibleValues}Characteristics_{Positition}[Parameter = Class] \text{ is } Far[6; \mu(\cdot)]. \quad (7.51)$$

Once the bank of fuzzy rules is extracted (about 46 rules)[7], it is right to extract the TDS through Eq. (7.25). As the problem under study regards the research of

[7] $|set\ of\ bank\ of\ fuzzy\ rules| < 10\,\%|inputs|$

the typology of a defect, indicating with d_j the population of the typologies of the defects ($j = P, MD, MI, ND$) and with $\mu_{Defect}(d_j)$ the related values of membership, the solution is obtained by the resolution of the following maximization problem [16, 17]:[8]

$$\mu(z) = \sup_{(d_P, d_{MD}, d_{MI}, d_{ND})} \left(\mu_{maximum} \left(\frac{1}{N} \sum_j \mu_{Defect}(d_j) \right) \right) \tag{7.52}$$

with $z = \frac{1}{N} \sum_j d_j$. The results are summarized in Table 7.8 in which it comes out with the comparability of the percentages got by Mamdani's inference and CW. The misclassified patterns can be, however, considered afferent to a further class of doubtful cases, for which the system still points out the presence of an error, but without classifying it.

The obtained light decreasing of percentages by this latter approach is mostly due to the slightness of the number of intervals reserved to the inputs variables, with a consequent repercussion on the reduced number of rules which makes almost rough evaluation. Obviously, a major refining of the partition of the inputs/outputs space will enhance the performances, reaching the optimal condition if they will use algorithms of automatic extraction of inferences directly from the numerical data at disposal [16, 17].

7.2.3 How to Classify Materials Defects by Minimal Fuzzy Entropy Decisional Model

An approach based on fuzzy entropy (FE) concepts could be exploited for solving classification problems. Particularly, let μ_{jk} represent the fuzzy membership of a jth defect belonging to a kth class (with $k = 0, \ldots, N$, with N being the possible categories of analysed defects). Then, fuzzy partition creates N layers, where each one denote how much a single defect belongs to a particular class (membership level). Within this framework, the fuziness of a classification procedure can be evaluated by a suitable Shannon index indicated as Shannon entropy (SE) [11]. So, fuzzy entropy of a defect, H, i.e., its statistic information content, can be calculated according to the following equation:

$$H = -\sum_{k=1}^{N} \mu_{jk} \ln(\mu_{jk}) \tag{7.53}$$

where $\ln(\mu_{jk}) = 0$ when $\mu_{jk} = 0$. Fuzzification using Shannon entropy fuzzification (SEF) is carried out by a minimal fuzzy entropy decisional model (MFEDM). Starting

[8] Through genetic algorithms.

from a set of elements $X = \{x_1, \ldots, x_n\}$, let us denote with A a k-elements fuzzy set ($k < n$) defined on a subset of input space. Within this space, let $\mu_A(x_i)$ be the membership value of x_i to A. If $C_1, C_2, \ldots, C_m$ are the m classes in which the n elements are partitioned, then $S_{C_j}(x_n)$ denotes the set of elements belonging to the class j. Moreover, if we consider:

$$D_j = \frac{\sum_{x \in S_{C_j}} \mu_A(x)}{\sum_{x \in X} \mu_A(x)} \tag{7.54}$$

SEF element of jth class in an interval can be defined as:

$$SEF_{Class}(A) = -D_j \log_2 D_j. \tag{7.55}$$

Finally, the nonprobabilistic entropy can be computed as:

$$SEF(A) = \sum_{j=1}^{m} SEF_{C_j}(A). \tag{7.56}$$

With these statements, we can obtain a FIS from an L-dimensional set of training patterns $p \in X$, i.e., $p = p_1, \ldots, p_L$ by a subtractive clustering. Here, each one of the L features (inputs) belongs to the jth class C_j, $1 \leq j \leq m$, and the FIS has an user-defined number of a fuzzy membership function n_{fmf} for each input. Considering the rth input ($1 \leq r \leq l$) of each training pattern p_t, we have number of intervals equal to $(n_{fmf} + 1)$. Thus, the boundaries of fuzzy intervals are:

1. $[\min(p_t, r),\ cL]$, for the left-most interval
2. $[c_s - 1,\ c_s + 1]$, for each sth inner interval
3. $[c_{nfmf} - 1,\ \text{map}_t(r)]$, for the right-most interval.

For each sth interval, $SEF(A_s)$ is calculated according to the equations above introduced. The SEF of the considered rth input S (SEF_r) is the sum of all $SEF(A_s)$, $s = 1, \ldots, (n_{fmf} + 1)$. Denoting as SEF_{tot} the SEF's total amount, the procedure is stopped and the relative fuzzy inference system is returned when SEF_{tot} has a minimum.

7.3 Measures of Fuzzy Similarities to Characterize Defects in Ultrasonic NDE

For many reasons, NDT/NDE ultrasonic inspection techniques deliver signals affected by errors, so generating a high-intrinsic grade of inaccuracy and imprecision. This necessarily reflects on the quality of identification procedures through the informative content of the signals themselves [3, 4, 9]. Hence, it is imperative to practice the analysis with procedures capable of taking into account the vagueness present in the signals delivered by NDT/NDE measurements. In many application fields, there are defects that hinder the quality of a manufact more than others. This creates the need to practice both the operations of detection and classification. However, experience teaches that "similar" effects often result in signals that are "similar" to each other. This makes it hard to classify defects a mere visual inspection of the corresponding signals. The international scientific community is particularly active in this field, as testified by the publication of works reporting good results obtained through a variety of methodological approaches (tomography, neural networks, wavelets, ...), but without putting any emphasis on the problem of the similarity. The idea here developed starts from the assumption that a defect (often not visible at the naked eye) is revealable through the evaluation of the similarity of the signal containing it, with a signal without defects translating, in this way, the detection of a defect into an equivalent problem of evaluation of distances. Given the presence of vagueness in ultrasonic signals, a possible fuzzy approach appears natural. In the scientific literature, moreover, the fuzzy approach in signal processing is widely spread, as testified by the vast amount of publications developing procedures of analysis and synthesis of fuzzy and neuro-fuzzy systems. But FS, even well-known in fields like the biomedical engineering, in NDT/NDE as for the characterization of defects still appears as an open problem. Moreover, each signalling of defect presents typical intervals of values of similarity so, through such comparison, it is possible to deduct the typology of the defect (classification). If the revelation is not possible, the approach detects the presence of defect without classifying the typology. The advantages are obvious because if from one hand the localization and the classification occur with a low-computational charge, on the other hand the doubtful cases occurs in terms of presence/absence of the defect. So, there emerge two possible operational protocols: the first one for defect localization and the second one for its classification [2, 4, 9].

Protocol for Defect Localization To localize a defect, we start from the assumption that similar signals among them create a class labelled with A_j (with j typology of defect), while signals without defects create a class labelled with B. The localization of defect C occurs through the evaluation of the similarities between the generic pair x-A_j and the similarity between x and B [5, 10, 15].

Protocol for Defect Classification As for the classification, each signalling of defect present intervals of possible values of similarity. For comparison it is deduced the typology of the defect. If the surveying is not possible, the approach records the presence of defect without classifying the typology (similarities dissimilar from those typical of the given class) creating a new class of defects (doubtful cases) [5, 10, 15].

7.3.1 The Fuzzy Similarities: Similarities as Distance

To understand how FSs work, it is necessary to start from a universe of the discourse X and to define on it the fuzzy power set $F(X)$. If A and B are fuzzy sets belonging to $F(X)$, it is possible to define their similarity through the formulation of a function

S defined in this way [4, 5, 9, 10, 15]:

$$S : F(X) \times F(X) \mapsto [0, \infty] \quad (7.57)$$

enjoying the following properties:

1. $S(A, B) = S(B, A)$
2. If $A \subset B \subset C$ then $S(A, B) \geq S(A, C)$ and $S(B, C) \geq S(A, C)$, where A, B, and C are fuzzy sets of $F(X)$
3. $S(D, D) = \max_{A,B \in F(X)} S(A, B), \quad \forall D \in F(X)$
4. $S(C, C^c) = 0, \quad \forall C \in P(X)$ where $P(X)$ represents the whole of all the crisp sets.

The first condition confirms the property of symmetry of the fuzzy similarity (FS): if a fuzzy set A is similar to the fuzzy set B with a certain value, also B is similar to A with the same value. The second condition, as a matter of fact, suggests us that the measure of similarity is monotonic [11]. Moreover, the third condition imposes the reflectivity. In other terms, a generic fuzzy set is similar to itself with the maximum grade. Finally, the last condition establishes that a generic fuzzy set is totally dissimilar from its complementary. To underline the nature of the approach it is right to normalize the function of similarity.[9] In particular, we need to reformulate the approach as follows:

$$S : F(X) \times F(X) \mapsto [0, 1] \quad (7.58)$$

and so the four properties become:

1. $S(A, B) = S(B, A)$
2. If $A \subset B \subset C$ then $S(A, B) \geq S(A, C)$ and $S(B, C) \geq S(A, C)$ where A, B, and C are fuzzy sets of $F(X)$
3. $S(A, A) = 1$
4. $S(A, A^c) = 0$.

The possible approaches of analytic formalization of FS measures are two. The first approach faces the problem through the formulation of single samples as points in a space, so the FS comes out from the computation of a distance. The second approach evaluates the similarity through sets of logical predicates as set of features [5, 10, 11, 15].

7.3.1.1 FS as Distances Among Points in a Given Space

By this approach the FS is concretely the measure of a distance among points in a given space and, to be defined, it should be fixed, in that given space, the modality of computation of the distance. In the scientific literature, there are many ways in

[9] In fuzzy logic all is referable to a problem of measure quantifiable in a value belonging to the interval [0, 1].

which FS are computed, all exploiting the membership function concept [4, 5, 9, 10, 15]. For example, if we point by A and B two fuzzy sets (in our case, a class of defect and one without defects), by $\mu_A(x_i)$ and $\mu_B(x_i)$ the membership values to A and B, and by $\|\mu_A(x_i) - \mu_B(x_i)\|$ a measure of fuzzy distance, the four possible functions of similarity are as follows [5, 8, 12, 13]:

$$Similarity_1(A, B) = \frac{1}{n}\sum_{i=1}^{n}\frac{\min(\mu_A(x_i) - \mu_B(x_i))}{\max(\mu_A(x_i) - \mu_B(x_i))} \tag{7.59}$$

$$Similarity_2(A, B) = \frac{\sum_{i=1}^{n} 1 - \|\mu_A(x_i) - \mu_B(x_i)\|}{n} \tag{7.60}$$

$$Similarity_3(A, B) = 1 - \sum_{i=1}^{n}\frac{\|\mu_A(x_i) - \mu_B(x_i)\|}{\mu_A(x_i) + \mu_B(x_i)} \tag{7.61}$$

$$Similarity_4(A, B) = \frac{1}{1 + \|\mu_A(x_i) - \mu_B(x_i)\|} \tag{7.62}$$

where:

1. n is the number of elements contained in A and B
2. $\|\mu_A(x_i) - \mu_B(x_i)\| = \sum_{i=1}^{n} |\mu_A(x_i) - \mu_B(x_i)|$ if Hamming distance is considered
3. $\|\mu_A(x_i) - \mu_B(x_i)\| = \sqrt{\sum_{i=1}^{n} |\mu_A(x_i) - \mu_B(x_i)^2|}$ if Euclidean distance is considered [6–8, 12, 13].[10]

Certainly, $Similarity_k$ ($K = 1, 2, 3, 4$) satisfied the general properties of similarity and, in particular, if $\mu_A(x_i) = \mu_B(x_i) = 1$ the two classes are exactly similar. Of course, it is stated the problem of the choice of the membership function μ which, however, is to be found in the application under study, dimensioning its parameters and taking into account features directly extracted by the databases at disposal. However, the application of similarity suggests that the signals under examination are affected, in a certain measure, by fuzziness. So, it is imperative evaluating the fuzziness of a signal through a numerical index which, globally, provides an indication of how much fuzziness is contained in the signal itself [14]. In the scientific literature, the fuzziness of a signal is exaustively quantifiable through the joint evaluation of the linear index (LI) and the FE [4, 9, 12]:

$$LI = \frac{2}{n}\sum_{n=1}^{n}\min(\mu_A(x_i), (1 - \mu_A(x_i))) \tag{7.63}$$

$$FE = \frac{1}{n}\sum_{i=1}^{n}(M_i - N_i). \tag{7.64}$$

[10] Obviously, other measures of fuzzy distance can be taken into consideration. The choice is often dictated by the particular application under examination.

with

$$M_i = -\mu_A(x_i)\log(\mu_A(x_i)) \tag{7.65}$$

$$N_i = (1-\mu_A(x_i))\log(1-\mu_A(x_i)) \tag{7.66}$$

If *LI* and *LE* get low values, the fuzziness contained in the signals is reduced, while the presence of fuzziness is highlighted in the increase of their values.[11]

7.3.1.2 Fuzzy Similarities by Exploiting Set of Features

This approach describes the classes using set of features by means of logic of the predicates [14]. Specifically, starting from the three generic signals, x, y, and z, the similarity between A and B gets the form [5, 15]:

$$Similarity(A,B) = F(A \cap B, A - B, B - A) \tag{7.67}$$

and F is chosen so that

$$Similarity(x,y) > Similarity(x,z) \tag{7.68}$$

all the times that:

1. $A \cap C \subseteq A \cap B$
2. $A - B \subseteq A - C$
3. $B - A \subseteq A$.

The procedure usually starts from the evaluation of a generalized index (GI) defined by [12]:

$$GI(A,B,\epsilon,\theta) = \frac{f(A \cap B)}{f(A \cap B) + \epsilon \times f(A - B) + \theta \times f(B - A)} \tag{7.69}$$

in which:

1. f represents an monotonically increasing function
2. A and B are predicates of the measurement
3. ϵ and θ are positive numerical parameters of importance, determining the importance of A and B profiles
4. $A \cap B$, the features present both in A and in B
5. $A - B$, the features of A which are absent in B
6. $B - A$, the features of B which are absent in A.

For example, if A resulted more important than B, *GI* would get the form:

$$Similarity_5(A,B) = \frac{\sum_{i=1}^{n} \min(\mu_A(x_i), \mu_B(x_i))}{\sum_{i=1}^{n} \min(\mu_A(x_i), \mu_B(x_i)) + \Sigma + \Upsilon} \tag{7.70}$$

---

[11] Obviously, other formulations of indexes are possible, but depending on the application under study.

where:

1. $\Sigma = \epsilon \times \min(\mu_a(x_i), 1 - \mu_B(x_i))$
2. $\Upsilon = \theta \times \min(1 - \mu_A(x_i), \mu_B))$
3. $Similarity_5(A, B) = 1$ when $\mu_A(x_i) = \mu_B(x_i)$.

Equation 7.70, obviously, represents the FS expressed by means of a set of features.

7.3.2 Detection and Classification Problems in Terms of Similarity Measures

This section illustrates a possible procedure to detect and classify defects by exploiting the FS concept. In particular, the following steps summarize the underlying idea of the approach. In the procedure, FS, solves the problem of the classification of defects in a rather exhaustive way, thanks to its reduced computational complexity [2, 4, 9].

Step 1. Preprocessing signals

The first step is dedicated to the evaluation of the fuzziness of a signal and the link between it and the eventual presence of a defect. In particular, after the computation of particular indexes (for example, the LI and FE, as in Eqs. 7.63 and 7.64), the fuzziness of the signal is evaluated, in the sense that the low values of such indexes are interpreted as a symptom of reduced fuzziness and, thus as an unlikely presence of defects. Conversely, high values of LI and FE are a symptom of high fuzziness with a marked possibility of presence of a defect.

Step 2. Feature' extraction

This step is very delicate because from the right choice of the features extracted from the signals afferent to a same class of defects (or no defects), it is possible to build the membership functions. For example, it is possible to extract a set of features indicating the informative content of a set of similar signals by particular statistical evaluations. In fact, because similar defects determine similar signals, the signals indicating a same typology of defect (or absence of defect) get similar statistic parameter values. If, for example, we analyze the typical histogram of a UT signal with a given typology of defect, it gets the typical form of a Gaussian so, the extraction of only two statistic parameters is enough: the mean and the standard deviation.

Step 3. Fuzzy Membership Functions (FMFs) Construction

If we point with $m_{i,j}$ and $\sigma_{i,j}$ mean and standard deviation of the ith signal belonging to jth class, the mean value of the means and the maximum value of the standard deviation determine the mean and the standard deviation of the Gaussian membership function indicating the generic class A_j with j the jth defect and the class B of signals without any defect. In, particular:

$$m_{A_j} = \frac{\sum_1^N m_{i,j}}{N_j} \tag{7.71}$$

where N_j is the number of signals in the generic class A_j and

$$\sigma_{A_j} = \max\{\sigma_{i,j}\} \tag{7.72}$$

$$\mu_{A_j}(x) = \mathrm{e}^{\frac{-(x-m_{A_j})^2}{2\sigma_{A_j}^2}}. \tag{7.73}$$

Similarly, computing m_B and σ_B for class B of signals without defects, we obtain:

$$m_B = \frac{\sum_1^N m_{i,B}}{N_B} \tag{7.74}$$

$$\sigma_B = \max\{\sigma_{i,B}\} \tag{7.75}$$

$$\mu_B(x) = \mathrm{e}^{\frac{-(x-m_B)^2}{2\sigma_B^2}}. \tag{7.76}$$

So, the obtained membership functions will be exploited for the computation of the fuzzy similarities [4, 5, 9, 10, 15].

Step 4. FMFs comparison between FSs for the classification of defects

After the FMFs construction, for each class of signals affected by a particular defect (A_j) and for the class of signals without any defects (B), we go on computing the FS values (Eqs. 7.59, 7.60, 7.61, 7.62, and 7.70) caring, as for $Similarity_5$, to fix previously the values of the parameters of importance (ϵ and θ). If it is not possible to highlight the defect, the procedure reveals the presence of the defect (if the values of FS start to be distant from the unit) without classifying the typology if such values are far from the typical ones of a certain class A_j. For this reason, the presence of a defect in an unknown signal is underlined by the comparison of the same signal with a known benchmark (for example the signal without defects) while, for each class of defect, each detection presents intervals of possible values of FS. In Fig. 7.2 the four steps of the procedure are visualized.

7.3.3 An Example of Signal Treatment by Means of FS Approach

7.3.3.1 The Detection and Classification Procedure

Starting from the experimental database (see Sect. 7.2.2.1), LI and FE indexes have been evaluated for each single class of signals with defects A_j and for the class

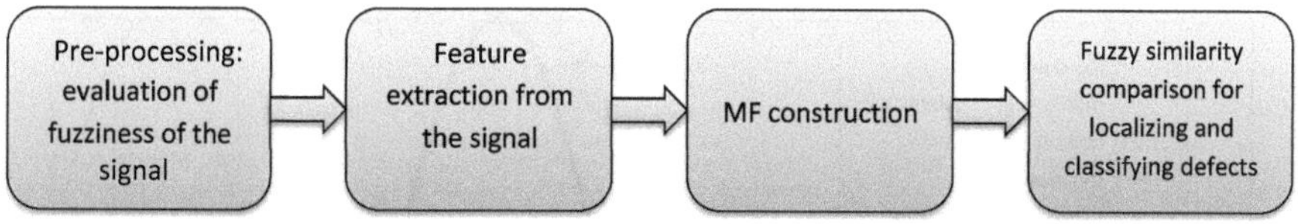

Fig. 7.2 Step of detection and classification procedure exploiting FS concepts

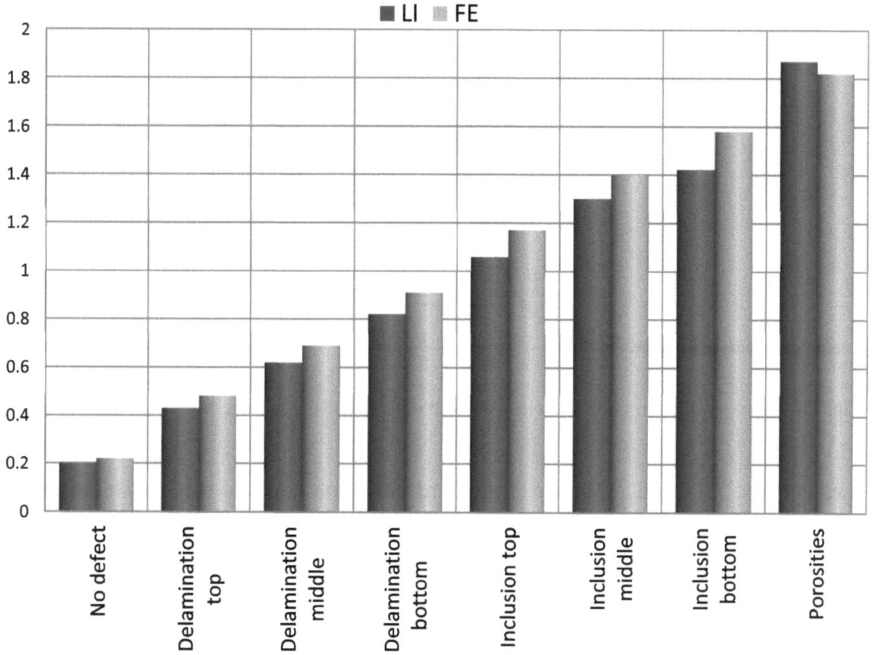

Fig. 7.3 Computation of LI and FE for each category A_j and for category B

of signals without defects B. In particular, in Fig. 7.3 it comes out that the class of signals affected by porosity possesses very-high values of the above-mentioned indexes, so the incidence of the defectosity is low[12]. On the contrary, the class without defects presents lower values of indexes of fuzziness. The remaining classes present an increasing trend of LI and FE values, passing from delamination (top, middle, and bottom) to inclusion (top, middle, and bottom) respectively [4, 9].

After, to design the fuzzy Gaussian membership functions for the computation of the FSs, they have been evaluated, for each class A_j and class B, the mean and standard deviation of each signal (Fig. 7.4). The Gaussian membership function identifying each class (Eqs. 7.73 and 7.76) is characterized by a mean equal to the

[12] The porosity, however, even with low values of FE and LI, has not to be left in many applications (especially in aeronautics); it is the defect of major dangerousness

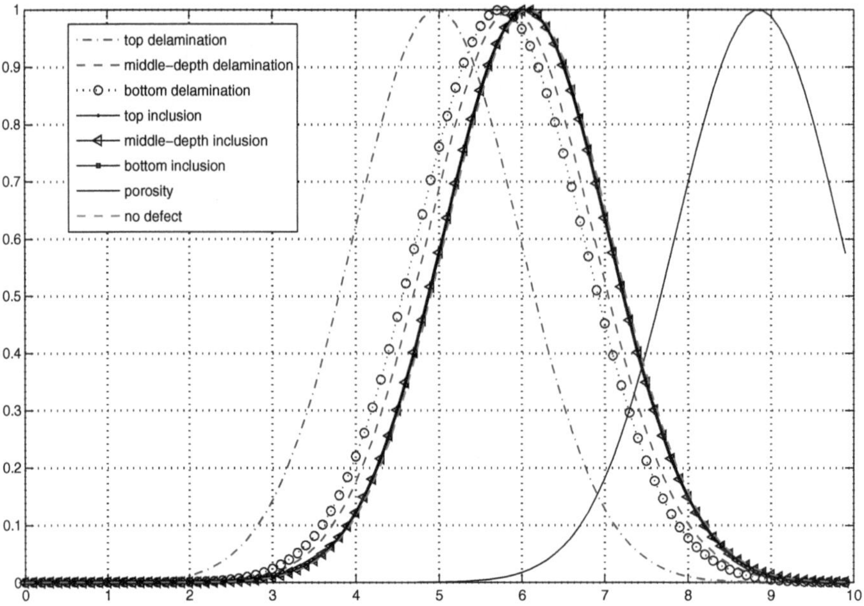

Fig. 7.4 Gaussian FMF related to a class without defects and a generic class A_j

means (Eqs. 7.71 and 7.74) and a standard deviation equal to the maximum of the single standard deviations (Eqs. 7.72 and 7.75). The graphic details are visualized in Fig. 7.4.

By an example, we report the classification by means of comparison of FSs values (Fig. 7.5) of a signal with unknown defectiveness (dark-grey bars), by means of comparison with FS belonging to the class of signals affected by top-delamination (black-bars). Light-grey bars represent FS of benchmark signals[13].

Finally, Figs. 7.6, 7.7, 7.8, 7.9, and 7.10 show the classification procedure. In particular, figures show different values of fuzzy similarities for different classes of defect. Dark-bars give information of bench-marking signals (no-defect) while grey-bars represent similarity values for different class of defect. The comparison among bar-values is useful to detect and classify defects. The procedure has taken out the presence of defect in the 100 % of the cases under examination, while more than the 90 % of the defects pointed out have been correctly classified. The remaining misclassified cases can be considered concerning a class characterized by signals with doubtful defects the system already takes out and classifies [5, 15].

[13] For *Similarity*$_5$ computation: $\epsilon = \theta = 0.5$

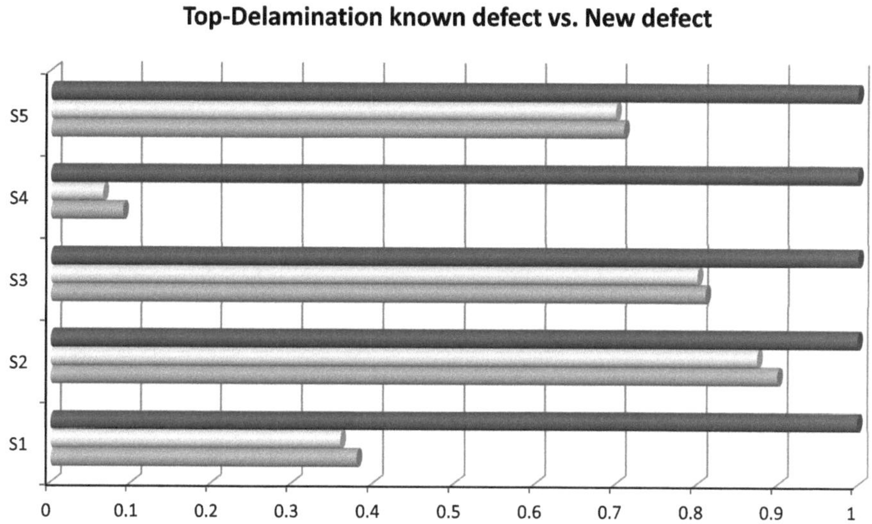

Fig. 7.5 Example of classification of an unknown signal by means of FS comparison

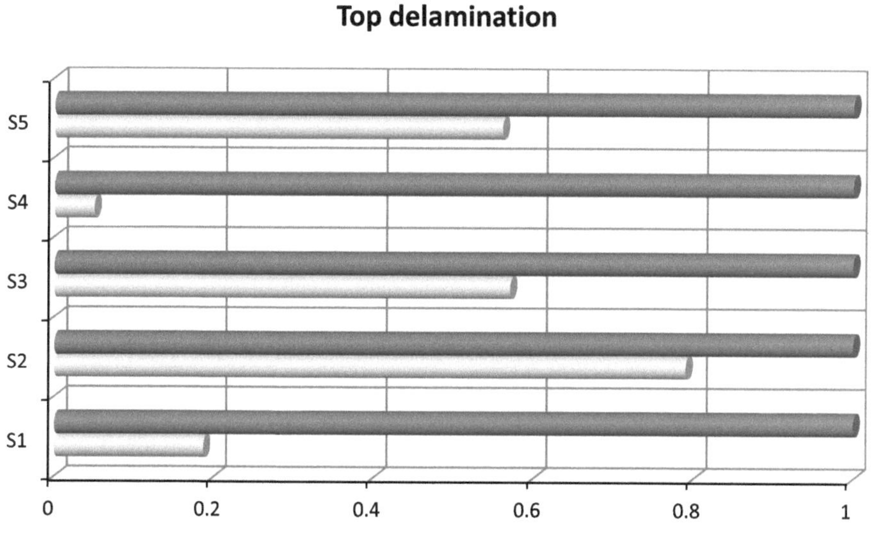

Fig. 7.6 FS values related to top delamination

7.4 Conclusions

Industrial manufacturing processes require powerful investigation techniques for quality control, but at the same time need them to be characterized by a low degree of computational complexity, so that they can be applied in real-time. In this context, two techniques for the characterization of defects in ultrasonic NDT/NDE have been

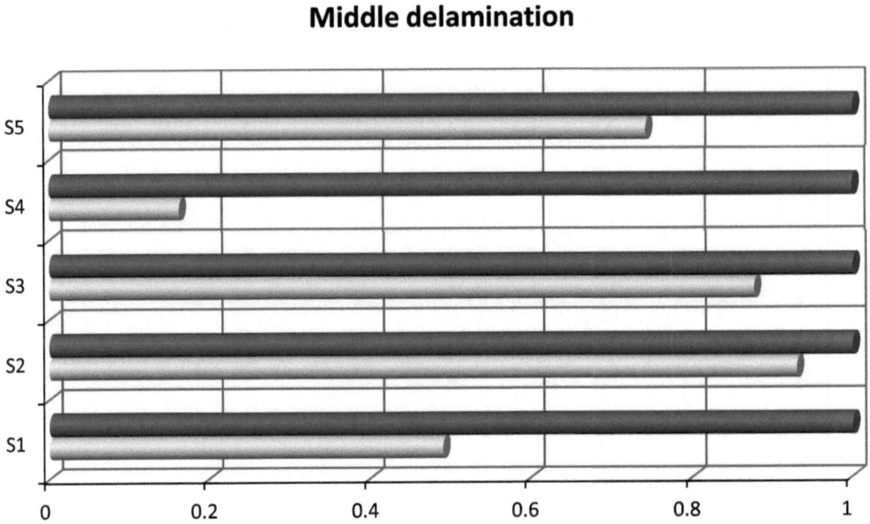

Fig. 7.7 FS values related to middle delamination

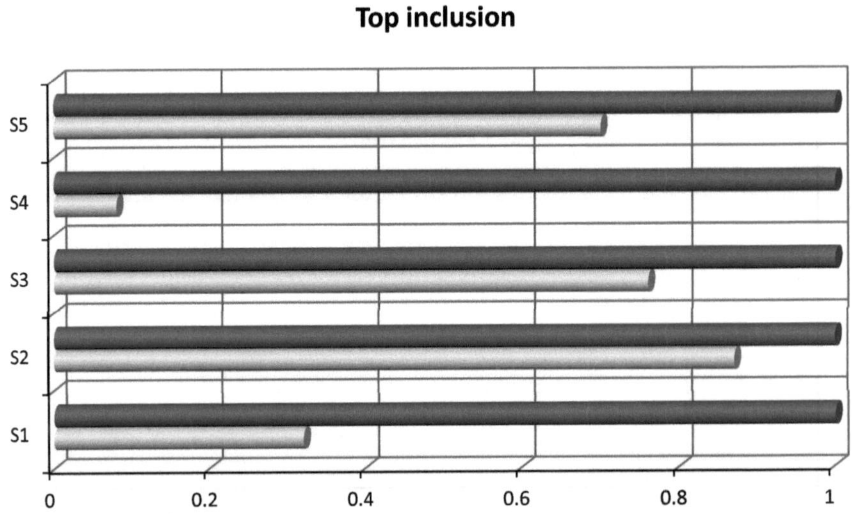

Fig. 7.8 FS values related to top inclusion

described: the CW and FS approaches for detecting and classifying defects in CFRPs. The first approach, which considers a class of defects such as a fuzzy granule in some space, leads to the formulation of banks of fuzzy rules managed by fuzzy inferences with advances linguistic structure with results that in terms of the classification are comparable to those obtainable by much more sophisticated fuzzy inference techniques. Obviously, more refined formulations of fuzzy granules can still be

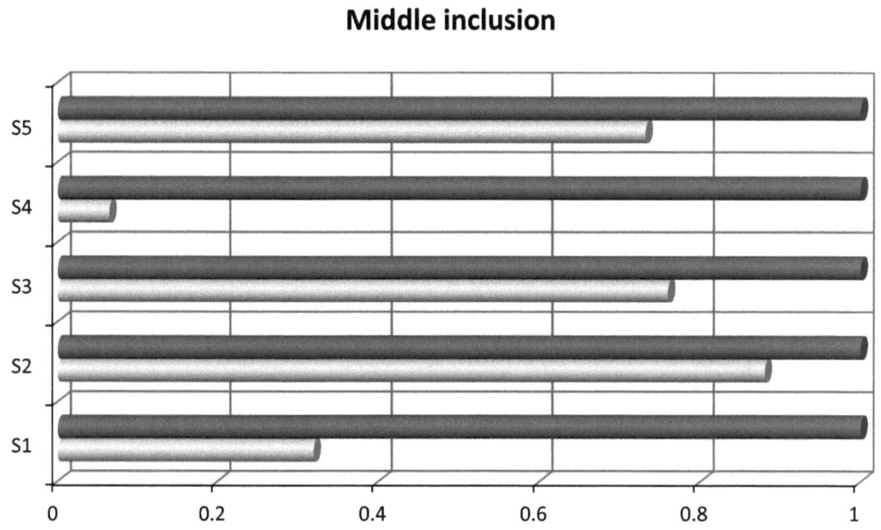

Fig. 7.9 FS values related to middle inclusion

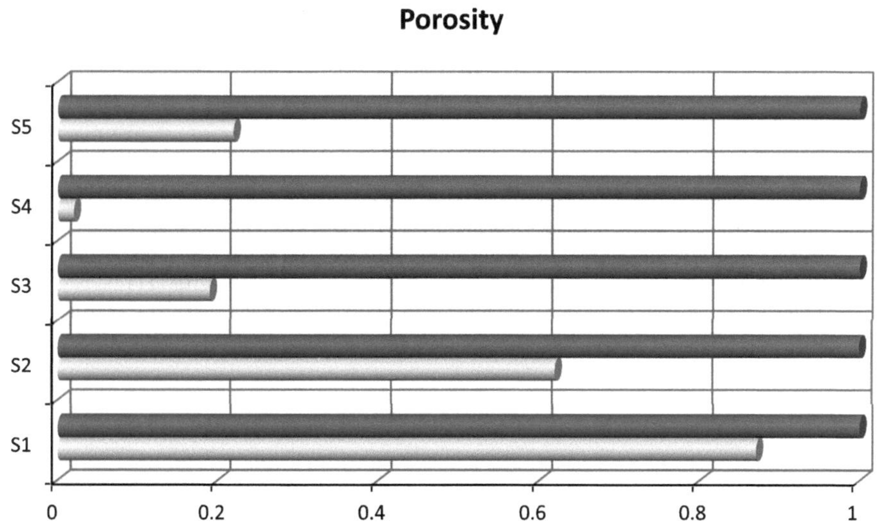

Fig. 7.10 FS values related to porosity

desirable for classifying uncertain cases. The second approach, structured on fuzzy similarities, bases the classification process on the fuzzy computation of the reduction of similarity among signals. The reported results can be considered particularly encouraging. On the one hand, they lead to correct classification ratios entirely comparable with those obtainable by much more sophisticated heuristic techniques. Secondly, they enable the creation of a further class of defects comprising uncertain cases.

References

1. Cacciola M, Calcagno S, Morabito FC, Versaci M (2007) Swarm optimization for imaging of corrosion by impedance measurements in eddy current test. IEEE Trans Magn 43(4):1853–1856
2. Cacciola M, Calcagno S, Morabito FC, Versaci M (2008) Computational intelligence aspects for defect classification in aeronautic composites by using ultrasonic pulses. IEEE Trans Ultrason Ferroelectr Freq Control 55(4):870–878
3. Cacciola M, Calcagno S, LaganÃ F, Morabito FC, Pellicanò D, Palamara I, Versaci M (2012) A new approach to evaluate defects in metallic plates based on computing with words and fuzzy entropy. Int J Meas Technol Instrum Eng 2(2):20–28
4. Cacciola M, Calcagno S, Morabito FC, Palamara I, Pellicanò D, Versaci M, Acciani G, Dimucci A (2013) Fuzzy similarities for the optimization of the operational parameters related to ultrasonic phased array probes for industrial application. In: 18th International Conference on Digital Signal Processing (DSP), 2013, IEEE, pp 1–6
5. Chaira T, Ray AK (2009) Fuzzy image processing and applications with MATLAB. CRC Press/Taylor Francis, Boca Raton
6. Chiu SL (1994) Fuzzy model identification based on cluster estimation. J Intell Fuzzy Syst 2(3):267–278
7. Kosko B (1996) Fuzzy engineering. Prentice-Hall, New York
8. Kuncheva LI (2000) Fuzzy classifier design. Springer, Berlin
9. Pellicanò D, Palamara I, Cacciola M, Calcagno S, Versaci M, Morabito FC (2013) FS measures for detection and classification of defects in CFRP. IEEE Trans Ultrason Ferroelectr Freq Control 60(9):1917–1927
10. Sander W (1989) On measures of fuzziness. Fuzzy Sets Syst 29(1):49–55
11. Tolias YA, Panas SM, Tsoukalas LH (2001) Generalized fuzzy indices for similarity matching. Fuzzy Sets Syst 120(2):255–270
12. Tversky A (1977) Features of similarity. Psychol Rev 84(4):327
13. Wang WJ (1997) New similarity measures on fuzzy sets and on elements. Fuzzy Sets Syst 85(3):305–309
14. Williams J, Steele N (2002) Difference, distance and similarity as a basis for fuzzy decision support based on prototypical decision classes. Fuzzy Sets Syst 131(1):35–46
15. Zadeh LA (1971) Similarity relations and fuzzy orderings. Inf Sci 3(2):177–200
16. Zadeh LA (1996) Computing with words. IEEE Trans Fuzzy Syst 4(2):103–111
17. Zadeh LA (1996) Fuzzy logic = computing with words. IEEE Trans Fuzzy Syst 4(2):103–111

Part IV
Real World Applications of Ultrasonic Non-destructive Evaluation Systems: Industrial Application Issues

In this part, we discuss how the techniques described in the previous chapters can be applied to different non-destructive evaluation (NDE) systems of industrial relevance. In particular, Chap. 8 shows the potentiality of the ultrasonic investigation approach to study the masonry structures, where the through-transmission (TrT) is applied in order to detect voids inside stone walls. Chapter 9 describes how the quality control through ultrasonic inspection of forgings of large dimensions for energy industry can achieve significant increase in the signal-to-noise ratio (SNR), and therefore, an improvement in the inspection capability, with the application of the pulse-compression (PuC) technique. Chapter 10 describes a protocol based on 'soft computing' techniques to check the integrity of carbon fiber reinforced plastics (CFRP) structures. As vagueness almost always affects the signals , the characterisation of defects from a geometric point of view is performed by formulating a fuzzy subsethood operator (FSO) based approach and in this way we translate the problem in the computation of a sort of 'fuzzy distance' in order to obtain a robust classification of data. In Chap. 11 we present a survey of industrial applications of non-contact ultrasonic techniques. Objects that are moving on an assembly line, hot materials and materials that cannot risk contamination, require the use of non-contact inspection techniques. The lack of contact poses specific requirements, due to the acoustic impedance mismatch in the air layer and the consequent energy losses. The chapter describes some examples of setups, transducer arrangements and signal processing strategies capable of addressing such requirements. In Chap. 12, the frequency modulated continuous wave (FMCW) is presented as a possible alternative implementation of the PuC technique to perform time-of-flight (ToF) measurements. We show that the FMCW allows to relax both the hardware and software requirements with respect to the standard PuC procedure based on the application of the matched filter.

Chapter 8
Ultrasonics for the Diagnosis of a Trachite Stone Wall

Sara Carcangiu, Augusto Montisci and Renato Forcinetti

Abstract The aim of this chapter is to show the potentiality of a novel approach to ultrasonic investigation of masonry structures. To obtain useful, rapid, and relatively low-cost informations, the ultrasonic "Direct Transmission Technique" is applied on the structure under exam. This technique uses a beam of ultrasonic waves emitted by a transducer and received by another one, placed on the opposite surface of the structure. The waves pass through the (sample under test) and are reflected and refracted by incidental discontinuities encountered along the path. The characteristics of the ultrasonic vibration are then investigated in order to obtain interesting information on the conditions of the materials. In particular, the attention has been focused on the correlation between the average density of the ray paths and the characteristics of acquired signals. This feature has been extracted from signals obtained by direct measurements on a real stone masonry with a known internal cavity. Then measurements have been related to the known geometry to verify the actual correlation with the characteristics of the traversed paths.

8.1 Introduction

Worldwide, stone masonry buildings are an integral part of the historical building heritage. To achieve the conservation and restoration of such structures, it is necessary to evaluate their functionality and, if possible, their load capacity. However, practicing evaluation on these buildings can be difficult because of the extremely limited a-priori knowledge or experience on their original design. Furthermore, the assessment of their actual conditions should not affect the status and functionality of the building, and should be performed on limited budgets. To obtain valid assessment results, reliable input parameters are required, and effective monitoring

S. Carcangiu (✉) · A. Montisci · R. Forcinetti
Dipartimento di Ingegneria Elettrica ed (DIEE), University of Cagliari, Cagliari, Italy
e-mail: s.carcangiu@diee.unica.it

A. Montisci
e-mail: amontisci@diee.unica.it

R. Forcinetti
e-mail: renato.forcinetti@diee.unica.it

P. Burrascano et al. (eds.), *Ultrasonic Nondestructive Evaluation Systems*,
DOI 10.1007/978-3-319-10566-6_8

methods are needed to test the input parameters themselves. Due to all these reasons, the inspection and control of structural conditions of the buildings is more and more an interesting topic, both for their correct management and rehabilitation, and for the challenging implications that it involves.

The data collection for the evaluation of the quality of the structure, the integrity verification of structural components, and identification of mechanical and physical properties of a material, is based on tests that can be divided into two classes: nondestructive and destructive. Selection of the method and the procedure to be followed, to test the structures, depends on the ability to obtain dependable results without being too invasive [4].

Although destructive tests are very reliable, unfortunately they tend to be expensive and time consuming. Furthermore, one can end up operating in bad spots, or causing some deterioration of the structure itself [7]. For this reason, among all the possible methods of assessment, particular emphasis is nowadays given to the development of nondestructive testing (NDT) techniques, including procedures for data evaluation and decision-making support. These techniques have the advantage of delivering useful information quickly, and, most important, without compromising the condition of structures and infrastructures [8].

Beyond providing direct diagnosis, NDT can drive technicians in the selection of samples to be collected for further laboratory tests. The set of destructive and nondestructive techniques, in conjunction with the obtained information, also allows to reduce the number of tests to be performed and, consequently, the diagnosis costs. In the framework of NDT techniques, the ultrasonic method is based on the well known principle that wave propagation depends basically on the medium through which it travels. Therefore, it is possible to put in relation the changes in measurable parameters due to the passage of a wave through a material with the changes in physical properties of the material itself [1, 3].

One of the simplest way to get relevant information about the wave propagating through a certain material is to measure the wave propagation velocity. Indeed, this is a traditional application of ultrasonic techniques. Moreover, there are other important features, such as attenuation, scattering, and frequency content associated to the elastic wave power, which allow to get further relevant information about the material. This is due to the dependence of wave propagation on the properties of the medium through which waves travel. In fact, parameters such as density, viscosity, and homogeneity can affect material absorption or attenuation properties depending on the characteristics of the material itself. Furthermore, wave propagation may be affected by the presence of discontinuities or defects, resulting in wave amplitude variations, changes in the direction of propagation, reflections, or changes in the frequency content of scattered signals [2, 5].

In the following, this principle is highlighted and a new procedure for the nondestructive diagnosis of masonry structures is described. The proposed procedure permits to detect and analyze the presence of material defects, such as fractures, cavities, or inhomogeneities which might reduce the structural stability.

Fig. 8.1 Trachyte stone wall

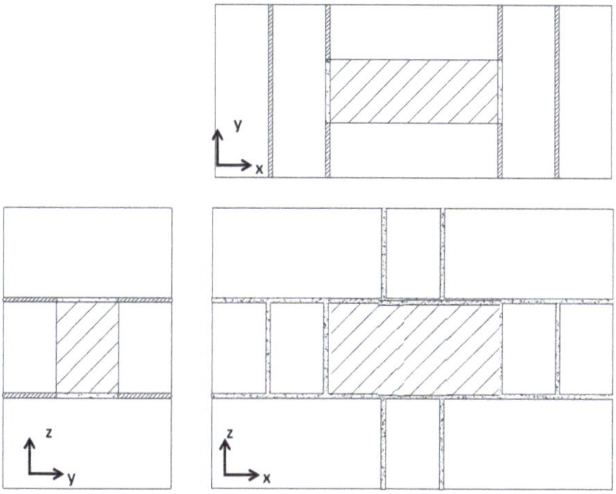

Fig. 8.2 Sections that intercept the internal cavity of the trachyte stone wall

8.2 Experimental Application

A series of trials have been performed on a trachyte stone wall of sizes $90 \times 38 \times 62\ \text{cm}^3$, shown in Fig. 8.1, where the blocks, which have sizes $28 \times 38 \times 12\ \text{cm}^3$, are joined by mortar. The scheme of assembly is shown in Fig. 8.2. An internal block has intentionally been omitted in order to obtain an internal cavity. Our aim is to find a sign of such a cavity by means of external ultrasounds tests.

The test consists in applying the direct transmission technique (DTT) [9] at a number of test points of the wall, and in comparing the values of some features of the acquired signals. The expectation is that the action of passing through the cavity can deeply modify the signal, allowing the diagnosis to be based on such variations. First of all we need to establish what are the most suitable test points of the wall. The distance between the points in the grid depends on the size of the cavity that we expect to detect. An almost regular grid has been assumed for the test points, such

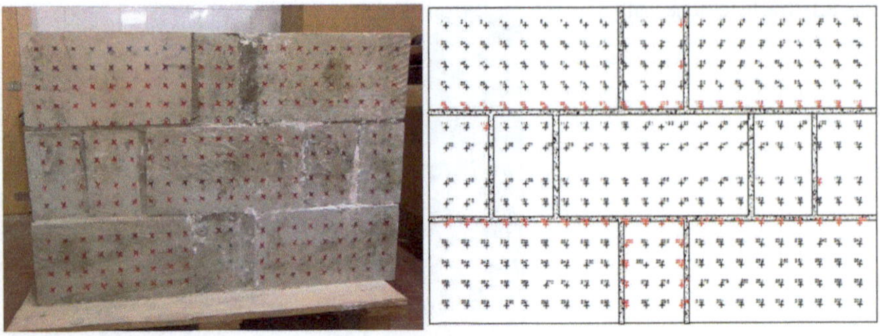

Fig. 8.3 Grid of test points

Fig. 8.4 The pulse generator PUNDIT LAB

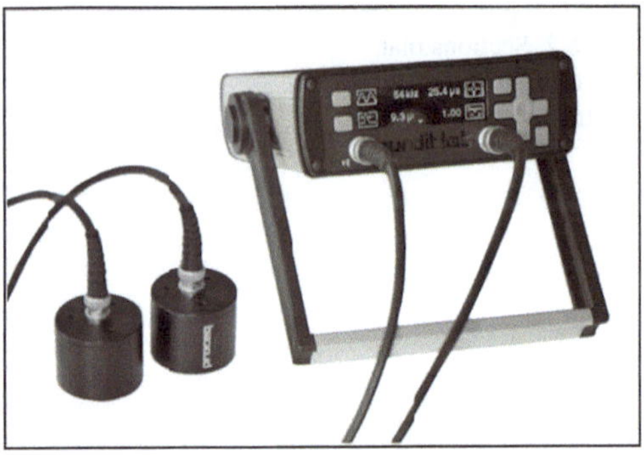

as shown in Fig. 8.3. The missing points are due to the fact that some parts of the wall surface are too irregular to suitably apply the ultrasonic transducers.

To guarantee the repeatability of the measures, the stimuli have been generated by means of an electronic signal generator, which can be driven by means of a personal computer. This allowed us to properly choose the form and the intensity of the pilot voltage signal.

The experimental setup comprises:

- A voltage pulse generator
- A set of transducers, operating both transmitter and receiver, with BNC connection to link them to the pulse generator
- An oscilloscope with provision of interfacing to a personal computer for visualizing the acquired signals and saving them for later processing
- A personal computer equipped with the proper software

The pulse generator used in this experiment is the PUNDIT LAB (Proceq [6]—Fig. 8.4)

Table 8.1 reports the main setup parameters.

By means of the software *PunditLink* it is possible to set the trigger time, the online acquisition of the data, the remote control of the equipment, including the

Table 8.1 Set up parameters

Parameter	Value
Frequency	54 kHz
Calibration time	25 μs
Pulse duration	9.3 μs
Temperature	± 1 °C
Voltage	500 V
Receiving gain	100×

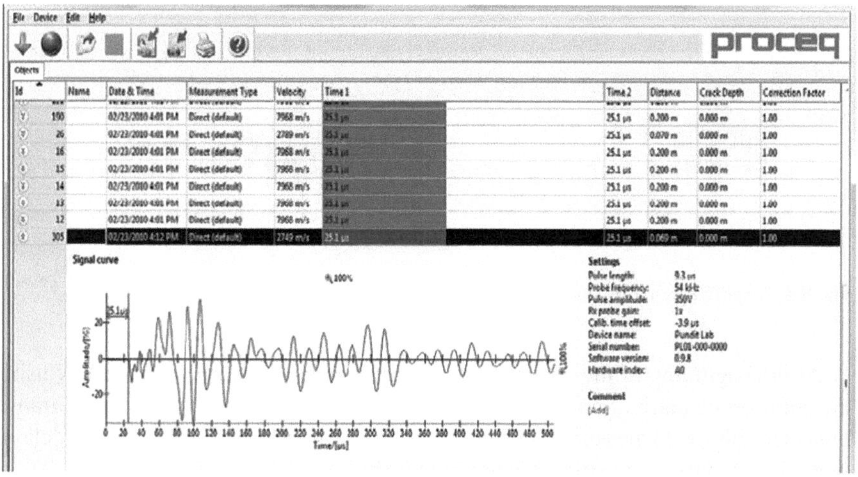

Fig. 8.5 GUI of PunditLink

Fig. 8.6 Tx/Rx transducers

direct storage of the data on the personal computer. In Fig. 8.5 the interface of the software is shown.

Together with the outfit of the pulse generator, a number of Tx/Rx transducers are provided (Fig. 8.6), each optimized for the use within a specific range of frequencies.

Fig. 8.7 Acquisition scheme

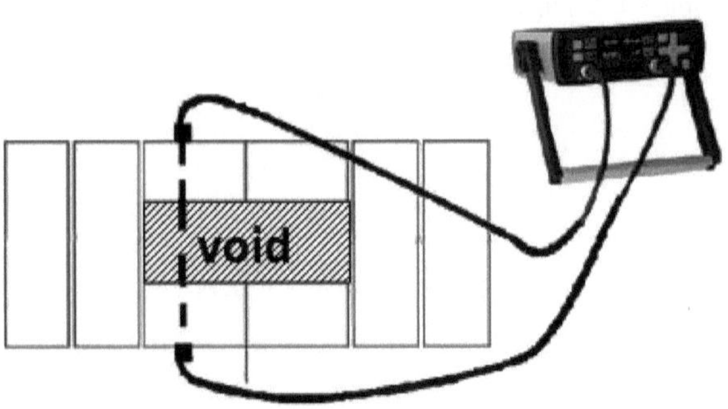

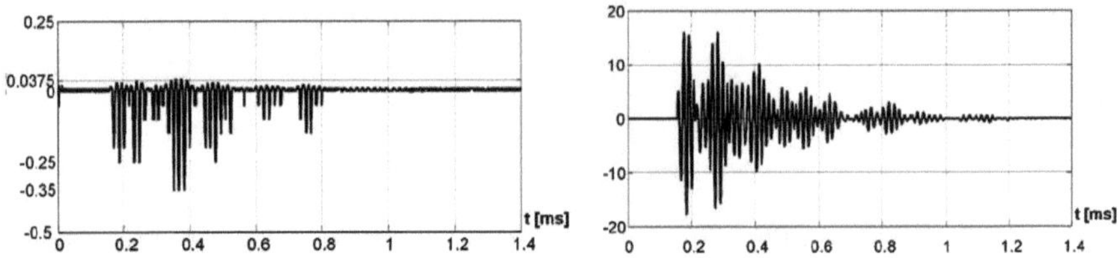

Fig. 8.8 Acquired signals in correspondence of mortar and trachyte paths respectively

At the beginning of the experiment, the equipment has to be calibrated by using the appropriate calibration bar. This procedure serves to eliminate from the transit time (see Table 8.1) measured on the wall, the systematic error due to internal delays. In the Fig. 8.7 the acquisition scheme is shown.

Several elements can affect the received signal, even if the transmitted signal is always the same. For example, the surface where the transducers are applied is one of the most influential aspect, affecting the coupling between the transducer and the wall. To improve the sound transfer, a proper gel has to be used to create an interface between the surfaces, maximizing the coupling and adapting the acoustic impedance. Nevertheless, those areas which are more strongly irregular, can be rather troublesome. Furthermore, the internal inhomogeneity of the means influences the transmitted signal, giving rise to completely different signals corresponding to paths that theoretically should have very-similar propagation parameters. For example, Fig. 8.8 shows two diagrams which corresponds to a path entirely constituted by mortar and another one made of trachyte.

Another influencing factor is the distance from the boundaries of the wall, due to the reflections that interfere with the main beam of signal. In Fig. 8.9 the two diagrams refers to two paths of trachyte, the former one next to the boundary and the latter one far from it. Finally, the measure is affected by the quantity of gel placed on the surface and by the hand of the operator.

Due to the irregularities of the surface, some test points which could derive from a regular grid happen to be not suitable for measurements. In the following Fig. 8.10, the test points used in the experiment are shown. As it can be seen, the difficulty in getting a good contact between the transducer and wall occur only in correspondence

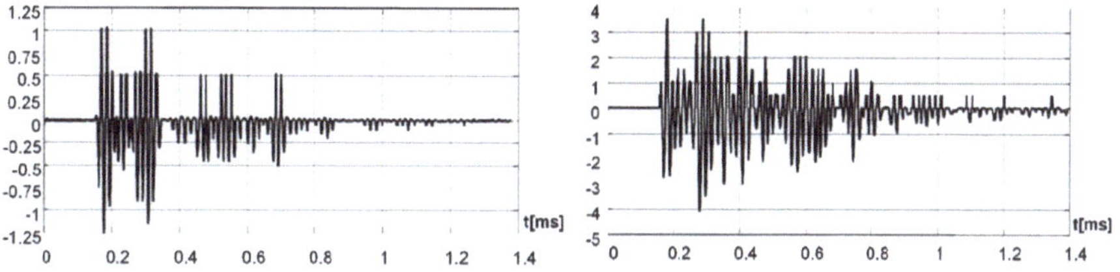

Fig. 8.9 Acquired signals in correspondence of paths near to and far from the boundary respectively

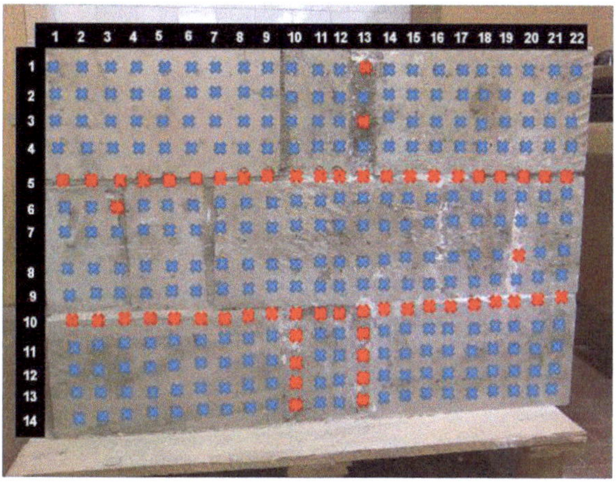

Fig. 8.10 Test points (*blue*) and suppressed points (*red*)

of the mortar, due to the fact that the blocks are not perfectly aligned. On the other hand, having such irregularity on the prototype used for the experiment makes it more realistic.

Overally we have 252 available tests points, each corresponding to an acquired signal. As shown in Figs. 8.8 and 8.9 the acquired signal is very sensitive to the presence of any kind of inhomogeneity of the means, but the problem is to distinguish the behavior of the object to be detected (in this case the cavity) with respect to any other cause of signal alteration. For this reason, we need to find some features of the signal that are both robust with respect to the variations within the same class of inhomogeneity, and sensitive with respect to the difference from other classes. In particular, the attention has been focused on the correlation between the average density of the ray paths and the characteristics of acquired signals.

8.3 Results

The frequency content of the acquired signal can provide suitable features. More specifically, the amplitudes of the most significant frequency components can be considered robust, while the phases or the low-energy components show a too high

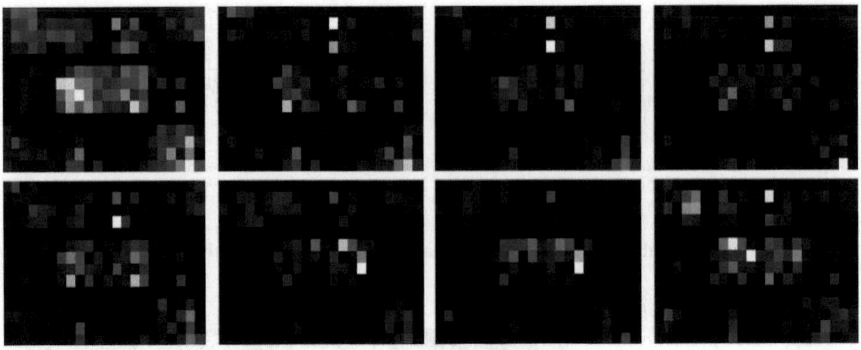

Fig. 8.11 Maps of amplitudes of the first eight frequency components

Fig. 8.12 Resulting map of eight components

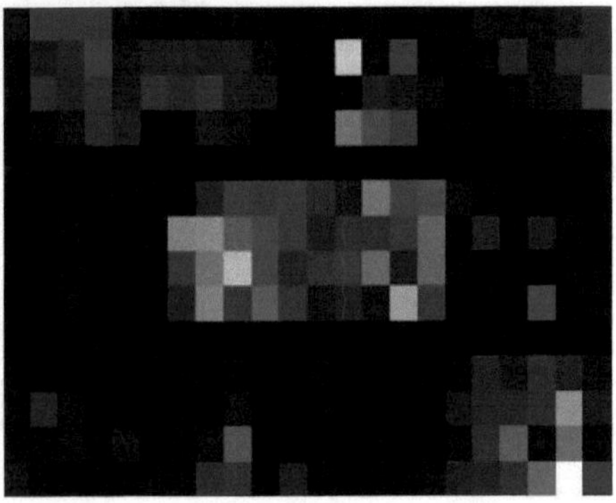

variability within the same class. We can construct a map of the wall for each frequency component, reporting the value of the amplitude described as a grey level. In Fig. 8.11 the maps corresponding to the first eight components are shown. Due to the high variability of the total energy of the signal, which is affected by several factors that are not related to the presence of the cavity, we preferred to normalize each signal with respect to the total energy. In this way, the difference between two signals acquired in different points resides in the shape of the diagram, rather than in its total energy. As we can see, no single feature alone can put in evidence the cavity, because individual signal features are sensitive to other artifacts, such as irregularities of the surface, presence/absence of plaster, and flaws of the mortar internally to the wall.

Nevertheless such artifacts have a different influence on the various components, therefore by adding together the eight components we can see that the internal cavity has a certain evidence (Fig. 8.12).

We can improve the readability of the result by applying a proper threshold to the values of the map, in order to enhance the clear pixels. Figure 8.13 represents the same map of Fig. 8.12 after the application of a threshold, corresponding to 20 % of the total range of variation.

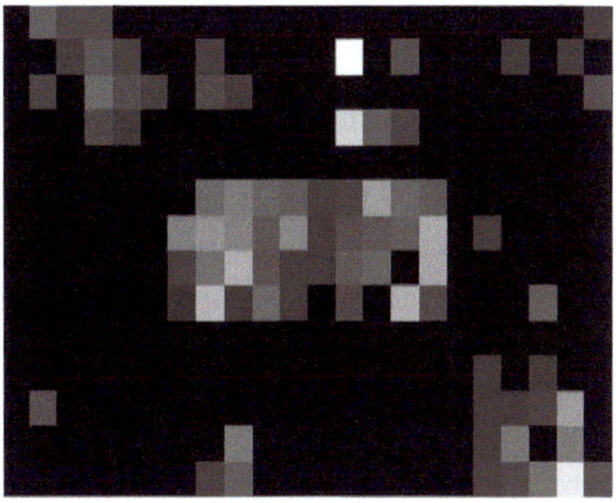

Fig. 8.13 Map with a threshold of 20 %

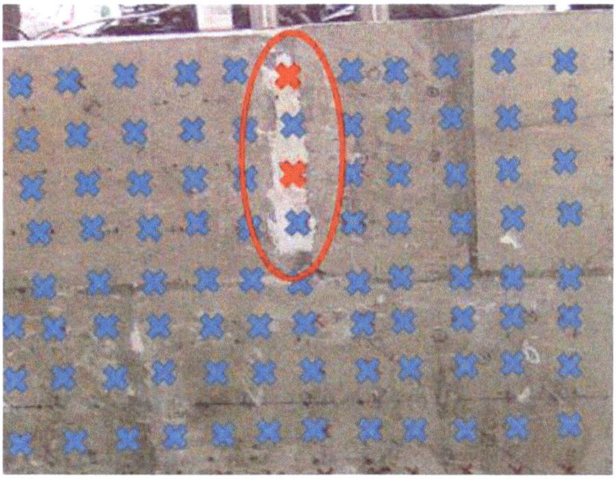

Fig. 8.14 Unreliable test points

In this last map, some false positive still remain, deserving some comments. First of all, we need to distinguish the artifacts depending on the irregularities of the surface from the ones due to internal inhomogeneities. Indeed, the former kind can be properly taken into account, by simply labeling the test point as potentially unreliable during the acquisition phase. This is the case, for example, shown in Fig. 8.14 where two points correspond to an area where the irregularity of the surface has been smoothed with plaster, whose propagation properties are totally different with respect to both mortar and trachyte. These two points constitute a false positive just over the cavity in the map in Fig. 8.13.

The other areas of the map where the grey level is comparable to that of the cavity, cannot be justified by defects on the surface of the wall, and actually they cannot be considered false positive. Indeed they correspond to paths that pass through layers of mortar, where plausibly some pockets of air remained trapped during the application. As a proof of this, we can observe that all the paths through the trachyte have dark pixels in the map.

8.4 Conclusion

The proposed technique results suitable for detecting voids inside stone walls. The quality of the manufacturing affects the performance of the test, but it appears to be sufficiently robust with respect to negligible flaws. A strong point of the method, for example with respect to tomography, is its very low computational cost and that the result of the diagnosis does not depend on solvability conditions, such as matrix singularities, or equations solvability, but at the same time it is able to resort to features slightly affected by the variations of signals belonging to the same class.

References

1. Berke M (1996) Nondestructive material testing with ultrasonics: introduction to the basic principles. Krautkamer & Company, Hurth
2. Kautz HE (2002) Acousto-ultrasonics to assess material and structural properties. Technical Report, National Aeronautics and Space Administration, NASA Glenn Research Center
3. Krautkrämer J, Krautkrämer H (1990) Ultrasonic testing of materials. Springer, Berlin
4. Masi A, Vona M (2007) Prove distruttive e non distruttive su materiali ed elementi strutturali di edifici esistenti in cemento armato. In: Atti del XII Convegno Nazionale dell'AIPnD
5. Perov DV, Rinkevich AB, Nemytova OV (2007) Interaction of pulse ultrasonic signals with reflectors of different types. Russ J Nondestructive Test 43(6):369–377
6. Pundit Lab (2013) Pundit Lab/Pundit Lab+ ultrasonic instrument–operating instructions. Proceq SA. http://www.abmbv.nl/files/proceq_pundit_lab_operating_instructions_en.pdf. Accessed June 2014
7. Sansalone M, Streett WB (1997) Impact-echo: non-destructive evaluation of concrete and masonry. Bullbrier Press, Ithaca
8. Taffe A, Mayerhofer C (2003) Guidelines for NDT methods in civil engineering. In: Proceedings of the 5th international symposium on NDT in civil engineering, pp 454–462
9. UNI ISO (2004) Natural stone test methods—determination of sound speed propagation. Published standard EN 14579:2004, UNI ISO

Chapter 9
Industrial Applications: Ultrasonic Inspection of Large Forgings

Luca Senni, Luigi Battaglini, Pietro Burrascano, Stefano Laureti and Marco Ricci

Abstract The quality control through ultrasonic inspection of forgings of large dimensions for energy industry still presents major limitations due to power losses occurring during the penetration of ultrasounds in the sample under examination. Here is reviewed the application of the pulse-compression (PuC) technique to this problem as a strategy to achieve significant increase in the signal-to-noise ratio (SNR) and therefore a corresponding improvement of the inspection capability. The performance of this technique, which exploits the linear chirp as excitation signal, and custom AVG curves for defect sizing, is analyzed both theoretically and experimentally and also the results of a thorough comparison between pulse-echo (PuE) and PuC are reported.

9.1 Introduction

In the field of nondestructive evaluation (NDE) for the steel industry, the ultrasonic inspection is the most widely adopted technical analysis. This is due to the ultrasound propagation conditions that characterize most metallic materials and which make the technique extremely powerful. In particular, ultrasonic NDE methods are

L. Senni (✉) · L. Battaglini · P. Burrascano · M. Ricci
Department of Engineering, Polo Scientifico Didattico di Terni,
University of Perugia, Terni, Italy
e-mail: luca.senni@unipg.it

L. Battaglini
e-mail: luigi.battaglini@gmail.com

P. Burrascano
e-mail: pietro.burrascano@unipg.it

M. Ricci
e-mail: marco.ricci@unipg.it

S. Laureti
School of Engineering, University of Warwick, Coventry, UK
e-mail: s.laureti@warwick.ac.uk

Dipartimento di Ingegneria, Polo Scientifico Didattico di Terni,
University of Perugia, Terni, Italy

P. Burrascano et al. (eds.), *Ultrasonic Nondestructive Evaluation Systems*,
DOI 10.1007/978-3-319-10566-6_9

Fig. 9.1 Forging process. **a** The process in action: the press acts on the metal when still hot. **b** The forged part (rotor) for energy industry with diameter larger than two meters

widely used for the verification of the quality of forged steel products dedicated to the energy industry (Figs. 9.1 and 9.5): after being forged, the parts are inspected to detect the presence of irregularities or defects that may be inside. Currently, this analysis is performed by following procedures based on pulse-echo (PuE) measurement technique.

To certify the quality of products, techniques capable of detecting millimeter defects at depth of meters are needed: each piece is accepted or rejected by the customer on the basis of the results of the ultrasound inspection. Moreover, as the forging realization process includes several heat treatment cycles, the possibility of being able to validate the samples in the early stages represents a significant improvement in the production process, especially when it comes on large-forged structures (due to the high cost of each thermal treatment). In the presence of large objects made of material characterized by a high attenuation of the ultrasonic waves, the traditional instruments based on the standard PuE method fail to ensure the requested inspection ability to assess the integrity of the sample.

In the inspection of such systems, it is therefore of utmost importance increasing the signal-to-noise ratio (SNR) and the resolution to allow the detection of sufficient small defects. To achieve this goal, ultrasonic pulse-compression (PuC) techniques for NDE of forgings with critical structural features [9, 11, 14, 17] can be applied. The basic idea is to exploit signals characterized by a wide band and high energy in order to improve both resolution and penetration compared to those obtainable with PuE [4, 8].

In developing the application described here, both hardware and software custom instruments were developed (see Example 9.1 for details), the procedure was then tested on real samples produced by "Società delle Fucine s.r.l.," an italian company in the ThyssenKrupp group. The data were compared with those obtained using commercial PuE instrumentation.

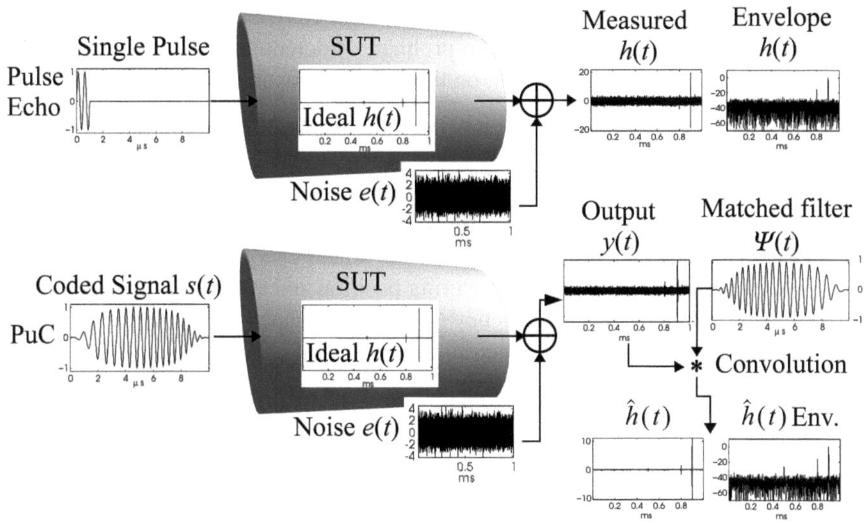

Fig. 9.2 Comparison of the PuE and PuC measurement procedures: in the PuE case (*top*), a short sinusoidal burst or a boxcar pulse is emitted (as explained in the paragraph); in the case of PuC technique, a coded signal $s(t)$ excites the SUT: the output signal $y(t)$ is collected. By convolving $y(t)$ with the time-reversed input signal Rev$[s](t)$, i.e., the matched filter, an estimate $\hat{h}(t) \approx h(t)$ is attained. The function $\hat{\delta}(t)$ approximates the Dirac delta

9.2 Pulse Echo vs Pulse Compression on High Dissipating Materials

As explained through the book, see for instance Introduction and Chap. 4, ultrasonic inspection actually consists of measuring the impulse response $h(t)$ of the sample under test (SUT) with respect to a mechanical wave excitation. In standard PuE method, the impulse response $h(t)$ of the system under inspection is estimated by exciting the sample with a short pulse $\tilde{\delta}(t)$ and then recording the system response $\tilde{h}(t) = [\tilde{\delta} * h](t)$; on the contrary in a PuC measurement scheme an estimate $\hat{h}(t)$ of $h(t)$ is retrieved by (Fig. 9.2): (i) exciting the system with a coded signal $s(t)$; (ii) measuring the output of the coded excitation $d(t) = [s * h](t)$; (iii) applying to the output the so-called matched filter $\Psi(t)$ [17]. At the end of the procedure we have:

$$\hat{h}(t) = [y * \Psi](t) = \underbrace{[\mathrm{Rev}[s] * s](t)}_{\hat{\delta}(t)} * h(t) = [\hat{\delta} * h](t) \approx \hat{h}(t). \qquad (9.1)$$

Furthermore, while most PuE systems, due to the short duration of the excitation pulse, rely on a single transducer that acts as both transmitting transducer (TX) and receiving transducer (RX), PuC-based schemes usually employ two distinct transducers for the transmission and the reception tasks (Fig. 9.7). The increased complexity of the resulting procedure is justified by the benefits it can provide in terms of resolution and SNR enhancement. To gain insight into the advantages of the

PuC let us first introduce some parameters that characterize the ultrasonic inspection of forged products, and introduce the problem of defect sizing. There are several parameters to be taken into account, on the side of ultrasonic inspection:

(a) The defect resolution, ρ: indicates the dimensions of the minimum detectable defect (MDD);
(b) The range resolution (RR): indicates the minimum distance in range at which two detectable defects can be distinguished;
(c) The SNR: indicates the ratio between the signal energy and the noise energy, determines the possibility of measuring position and amplitude of an echo;
(d) Ultrasonic signal central frequency (f_0) and the related signal bandwidth B: f_0 is related to ρ value and to penetration depth, while B is related to RR.

In the perspective of the specimen under control, the main parameters are:

(a) Defect size;
(b) Defect depth—sample dimensions—thickness of material between the probe surface and the defect itself; and
(c) Attenuation α: reduction in amplitude as a function of distance traveled in the sample by the ultrasonic signal, usually due to the transfer of energy to the wave propagation medium. Its counterpart in mechanical systems is the absorption or dissipation of mechanical energy into thermal energy due to friction. For a fixed depth of inspection and for a fixed noise energy, the SNR instead depends on the energy E_s of the emitted signal, i.e. on both its effective duration T and its amplitude.

A comparison between PuE and PuC can be interesting. But how to do that? We will show in the following that a comparison can be made of coarse in terms of SNR, which is expected to increase with PuC, but it is also more effective to compare the results in terms of MDD, that is a parameter specifically introduced in the forgings inspection and which directly quantify the defect detection capability. SNR and MDD are strictly related, as it will be explained in the following, but the latter take also into account the information about the excitation frequency and the geometrical and physical attenuation of the ultrasonic beam inside the SUT.

9.2.1 Coded Waveform for Applications on Large Forgings

Lets now motivate the choice of the excitation signal for this specific application, which fell on a windowed *linear chirp* (LChirp) signal. More strategies, based on both frequency modulation (FM) and pseudo noise (PN) waveforms were tested and compared: LChirp, *nonlinear chirp* (NLChirp), Maximum length sequence (MLS) and Golay complementary sequence (GCS) [5, 6, 12, 15, 16]. An exhaustive comparison of the two approaches (Chirp and PN sequences) is given in Chap. 4 and the reader is referred to Sects. 4.4 and 4.5 especially. Here we limit to recall some properties of the LChirp signal focusing on the most relevant ones for the procedure implemented. Nevertheless, it is worth to make some considerations about this choice. In forging inspection, as it will be clarified below, it is fundamental to have very low sidelobes level in order to be able to detect very small defects in large structures. In particular, since the defects signature is evaluated by comparison with the back-wall echo signal, and usually defects in cylindrical samples are searched from the closest surface to the axis, the far sidelobe level (FSL) should be minimized first of all. Furthermore, in the presence of very close defects, in order to fairly assess their dimensions also the near sidelobe level (NSL) should be maintained low. As discussed in Sect. 4.6, in the perspective of the sidelobes minimization, the optimal waveform is not uniquely identified but depends on the transducers bandwidth and also on the noise level. Quite generally, the probes used in ultrasonic inspection of forgings are narrow-band, in order to make use of the Abstand Verstärkung Grosse (AVG) curves, as described in the following section. In this case windowed LChirp tailored on the transducer bandwidth achieves the best performances in terms of sidelobes level. There is also another more subtle reason to prefer LChirp instead of PN sequences. Actually, large forgings can be assimilated to reverberating structures with losses: the ultrasound wave travels inside the SUT and it is trapped into the medium due to the huge impedance discontinuity between steel and air so that several round trips can be collected by the RX, especially in low-attenuating forgings or at low frequencies. It is therefore not trivial to establish a priori the duration of the $h(t)$ to be measured. This fact makes the use of acyclic pulse compression (APC) not feasible for the constraints that this approach entails. A similar limitation, even if with weaker implications, affects also the use of Acyclic GCS since, if the time interval between the execution of the two complementary sequences GCS_A and GCS_B is smaller than the mean trapping time of the ultrasounds inside the structure, part of the output signal relative to the first sequence superimposes to the response of the second sequence leading to an unwanted extra-noise term. For these reasons,LChirp had been showed to be the optimal choice for forgings inspection [13].

Let us now recall the main characteristics of a LChirp signal. LChirp is one of the most diffused waveform adopted in PuC applications and it is described by the expression: $s(t) = w(t)\sin(\Phi(t))$, where $\Phi(t) = 2\pi(f_{\text{start}}t + \frac{B}{2T}t^2) + \phi_0$ is the quadratic phase term and $w(t)$ is a windowing function that modulates the amplitude of $s(t)$ and that is not vanishing only in the interval $t \in [0, T]$. Due to the characteristic of LChirp, the amplitude modulation acts also a frequency band-pass filter. As said in the previous chapters, the matched filter impulse response, that is the reference waveform used to correlate the output data $d(t)$, in the simplest cases equals the time-reversed replica of the signal itself $\Psi(t) = \text{Rev}[s(t)]$, but here we consider the chance to vary the windowing function: $\Psi(t) = \tilde{w}(t)\sin(\Phi(-t))$. The convolution function of $s(t)$ and $\Psi(t)$ gives the approximation of the unit pulse, $\hat{\delta}(t)$. The simplest window is the rectangular one, but it is well known that if $w(t), \tilde{w}(t) = \theta(t) - \theta(t-T)$, where $\theta(t)$ is the Heaviside function, $\hat{\delta}(t)$ exhibits high sidelobes (as better explained through the book). The sidelobe reduction in the matched filter is carried out by choosing an appropriate window. There is a plethora of functions reported, such as Hanning, Kaiser, Blackman, Hamming, etc. [7]. More systematic window design optimization techniques are reported in literature [1]. In order to select the optimal windows pair, it must be taken into account that in ultrasound applications, bandpass filtering of the excitation signal occurs inherently from the transducer.

This has two main consequences: (i) for a fixed power the gain in SNR is lowered due to the attenuation of some spectral components from the transducer, (ii) on the other hand the effect of the transducer is equivalent to an additional weighting that, although suboptimal, is beneficial to sidelobe reduction. Ideally the optimal excitation signal should therefore be the one that exactly matches the overall spectral characteristic, including the filtering effect of the transducer [10].

9.3 Defect Sizing and AVG Curves

As previously said, the fundamental parameters to be estimated in forging inspection is the MDD, i.e., the size of the smallest defect than can be detected in the worst geometrical position. In the most usual case, large forgings show rotational symmetry, therefore, the geometrical position at the maximum distance from all possible locations of the probe is on the longitudinal axis of rotation. The amplitude of the echo signal, compared to the amplitude of additive noise (i.e., the SNR) can be shown to be fundamental parameter to which the MDD is related. According to a reliable and reproducible procedure, it is necessary to made measurement so to estimate the entity of the minimum discontinuity revealable by the ultrasonic inspection. There are several procedures to obtain these results, depending on whether the defect dimension is larger or smaller than the ultrasonic beam. In general, to have a reliable estimate of defect size, a tuning procedure must be performed: the received echo is compared with the echoes provided by a series of artificial reflectors, typically consisting of circular discs oriented normally to the beam axis (flat bottom holes), moreover, in the sizing operations, the geometric attenuation of the emission beam must be considered. Taking into account the results obtained for both the reflection from the artificial discontinuities and the geometrical attenuation, we obtain the echo amplitude attenuation trend as function of distance, namely the distance amplitude correction (DAC), and then the sizing of the discontinuity. The DAC curves-based calibration procedure can be adopted in repetitive cases, when both the geometry and type of material are not modified.

In fact, the availability of samples of all sizes appears to be rather expensive, especially when applied on extremely large specimens as in the case discussed here (see Figs. 9.1b and 9.5): the range of samples needed to build the DAC curve should necessarily be extremely high. For large objects the possibility to an ad hoc construction of suitable reference blocks, is not practicable.

A different technique is thus adopted, based on the numeric calculation of the ultrasonic pressure field collected by the RX in presence of a defect as a function of:

(a) The inspection frequency f_0;
(b) The equivalent radius of the defect assumed to be "flat bottom hole";
(c) The dimensions of the probes;

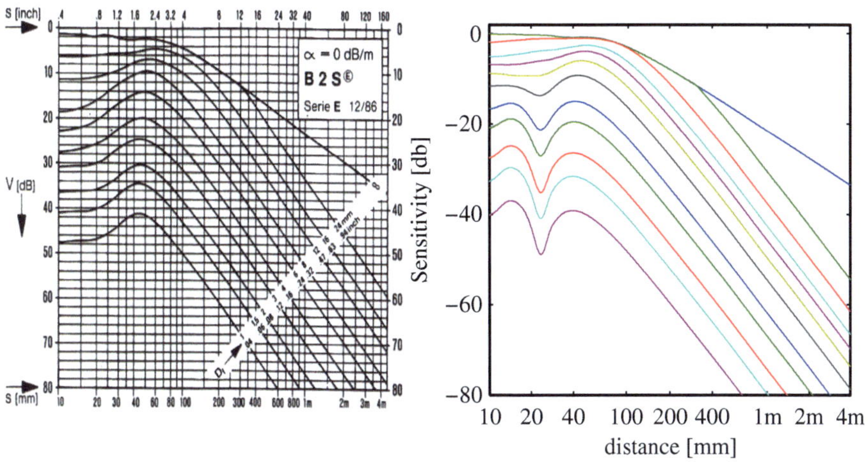

Fig. 9.3 Comparison between factory curves and simulated curves: single probe Kraut–Kramer B2S single element: ø24 mm, $f_0 = 2$ MHz. In the far-field, the curves overlap, in the near-field the numerical model also take into account interference for a defect placed along the symmetry axis of the probe

(d) The distance of the defect from the inspection surface.

In forging inspections this method is summarized in the well-known AVG-distance gain size (DGS) curves diagram that is computed by assuming the PuE configuration then with a TX coinciding with the RX. Practically, this technique involves comparison—typically manual—between the detected echo amplitudes, the amplitude of the backwall echo, and reference curves (Figs. 9.3 and 9.6). Moreover, since AVGs are calculated in the far-field by considering only the geometrical attenuation due to the beam divergence, during the sizing procedure, the physical attenuation α of the material has to be taken into account. It is worth to stress that α is derived by comparing the amplitude of the first back-wall echo with the amplitude of the second back-wall echo that is detected after the beam traveled two times inside the sample. This procedure assures that one can measure the effect of the physical attenuation by knowing the path went across by the ultrasonic beam between two reflections from the same surface. There is no other simple method to measure α apart from this one and indeed, accordingly with the AVG procedure, when it is not possible to detect at least two back-wall echoes, an overestimation of the minimum detectable defect (MDD) is achieved. Of course, if the MDD value is however smaller than the limit one, the impossibility to measure α does not constitute an issue but otherwise, and this is the most probable situation in large forgings inspection, it hampers the possibility to use the procedure. It is therefore of utmost importance to change the measurement setup or by lowering the inspection frequency—but this implies a higher geometrical attenuation and a worse resolution—or by enhancing the SNR to allow the detection of the second back-wall echo.

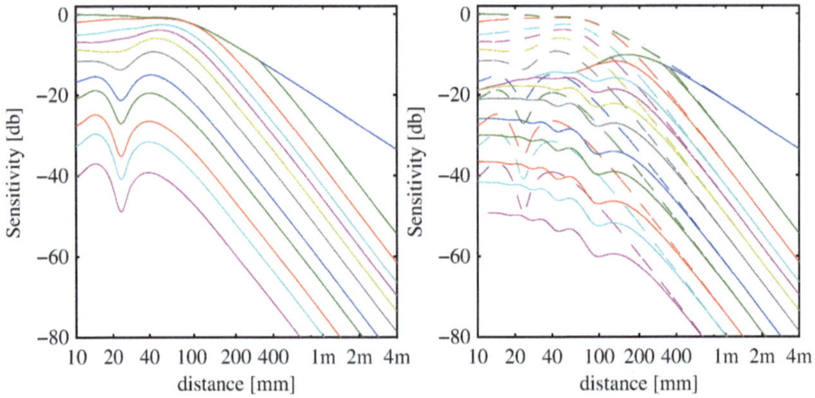

Fig. 9.4 Comparison between AVG-curves (two probes vs single probe). Single probe Kraut-Kramer B2S single element: ø24 mm, f_0 = 2 MHz, a pair of probes with ø24 mm, f_0 = 2 MHz, placed side by side. In the far-field the curves in the two configurations almost coincide, as expected. In the near field the two-probes configurations suffer for the small aperture angle of the ultrasonic beam

PuC can thus represent a fundamental tool in these cases but it is needed to calculate the AVG diagram for the new configuration. Indeed, whereas the PuE method relies on a single transducer that acts both as transmitter and receiver, PuC-based schemes usually exploit two distinct transducers for the transmission and reception tasks (see Fig. 9.7 in the following). In order to obtain a good evaluation of the discontinuity dimensions, the AVG curves are adapted for two probes and arbitrary geometry and are here showed in some particular cases (Fig. 9.4).

A numerical tool can be implemented that calculates the AVG curves by:

- Discretizing TX, RX probes and defect surfaces identifying an unitary element;
- Calculating the pressure field emitted by each element considered as a source and as a scatterer, using spherical wave approximation;
- Calculating the overall pressure field at the RX surface.

In it:

- The full interference model among TX-defects-RX has been considered avoiding the further simplification of piston;
- The phase interference between several elements contributions is taken into account;
- The pressure field at the receiver is calculated as the sum of the energies of all the RX elements.

With the help of the automatic AVG calculator tool, a virtual instrument (described in Example 9.4) performing PuC measurements was implemented. Flaw detectability potential, i.e., MDD, are strictly related to the SNR and SNR is in turn related to the frequency f_0 and to α.

Fig. 9.5 A tested rotor forged steel specimen with 3.2 m diameter: desired MDD is 1 mm on the axis

9.4 Examples and Observations

For the verification of methods and instrumentation set out above, measurement campaigns on large forgings (kindly provided by Società delle Fucine s.r.l., an italian member of the ThyssenKrupp group) have been performed. The production of such parts is practiced only in few mills in the world and it requires fairly sophisticated technologies and equipment. These forgings are almost entirely devoted to the energy industry and for heavy duty applications, therefore the presence of defects in such products may be considered particularly serious, and customer requirements are very stringent. A series of forgings was therefore chosen on which measurements were performed with the PuE (with traditional instruments, single probe, and narrowband) and PuC (with the virtual instrument developed, two probes, and broadband) techniques. The results are summarized in Figs. 9.8 and 9.9.

The first objective was to verify if PuC can increase the performances compared with PuE commercial devices. It has been shown that, in cases in which PuE-based commercial devices (Olympus EPOCH 4, GE USN60, Gilardoni) have difficulty in detecting the second echo from the bottom, a PuC-based procedure allows identifying up to 6–7 echoes of the bottom. It is important to emphasize that this result was obtained by using a reduced excitation power. In fact the signal input to the system does not exceed in amplitude the 20 V peak-to-peak and it is characterized by a value of the average voltage of 10 V. As a further advantage deriving from the application of PuC is that it allows better resolution on the dynamic impulse response. Indeed, using the same number of quantization steps of the analog to digital converter (ADC) for both PuE and PuC devices, in the second case the acquired signal has a much greater amplitude. This is because the PuC is less affected by quantization noise compared to a direct measurement of the impulse response [2, 3, 12]. Implementing the techniques of PuC and rebuilding the impulse response $h(t)$ high values of SNR are obtained that allow to detect defects of approximately 1–2 mm.

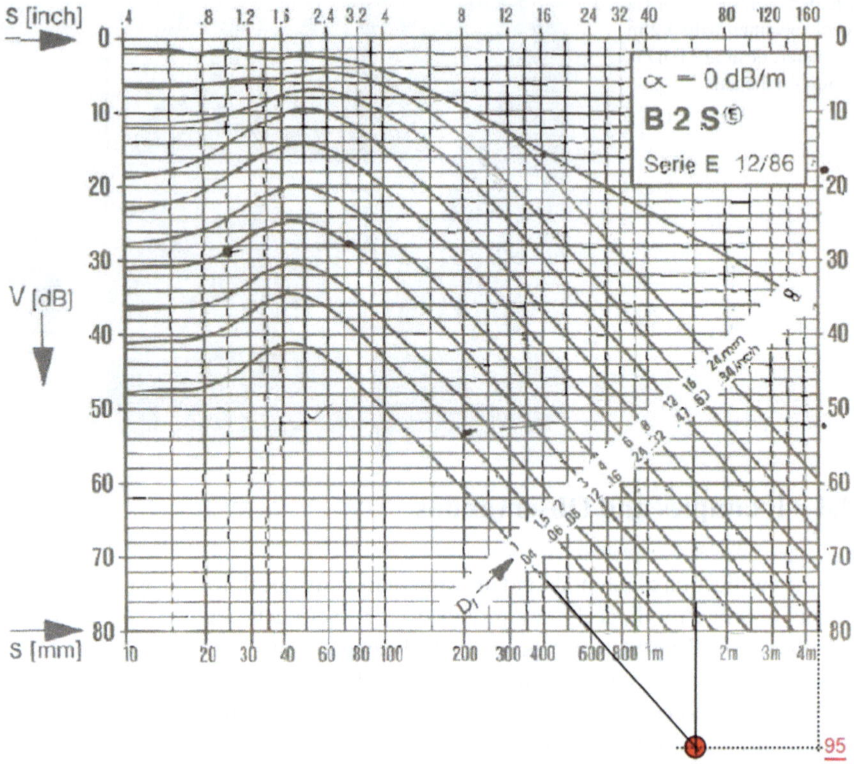

Fig. 9.6 Application of the AVG curves in the case of a large forging (3.2 m of diameter) with PuE measurement: the *red point* indicates the geometric attenuation for a 1 mm defect at the axis (1.6 m) of about 95 dB

To better explain all this, some results and an example are reported: a rotor-forged steel specimen of 3.2 m diameter (Fig. 9.5). The target is to find discontinuities with dimensions lower than 1 mm on the axis, depth 1.6 m with PuE/PuC method. The inspections were done at different steps of hardening resulting in a different attenuation of the ultrasound energy. Different windowing functions were applied.

PuE A single 2 MHz Probe from KrautKramer (B2S) was used for PuE measurement performed by a standard Olympus Epoch 4 flaw detector using a 400 V impulse amplitude. AVG curves (Fig 9.6) show a geometric attenuation for a 1 mm defect at the axis of about 95 db, the attenuation of the medium is approximately 2 db/m (on the axis is 3 db), losses due to coupling and transducer efficiency are approximately 30 db. Total attenuation results in 128 db. By exciting the probe with a 500 V pulse, a 1 mm defect returns an echo of about 200 V. In conclusion, PuE common devices could not have the necessary sensitivity and SNR.

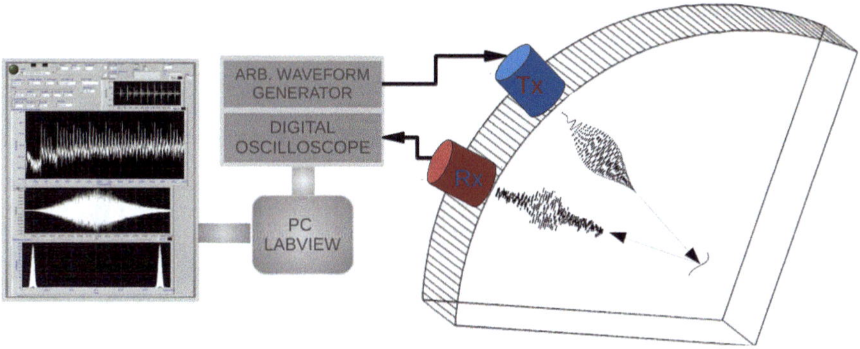

Fig. 9.7 Experimental setup scheme: two probes in pitch-catch configuration side by side along a circumference, excitation signal provided by an AWG (National Instruments PXI 100 MS/s), output signal amplified by a low noise amplifier, and then ADC 100 MS/s, all process is managed by a virtual instrument realized by LabView (screenshot)

PuC Two 2 MHz Probe from KrautKramer (B2S) were used for the PuC forgings inspection, the two probes are in pitch-catch configuration (side-by-side along a circumference) (Fig. 9.7); the excitation signal is provided by an arbitrary waveform generator (AWG) National Instruments PXI 100 MS/s; the Output signal is amplified by a low noise amplifier (LNA) and ADC 100 MS/s. All the process is managed by a virtual instrument realized with Labview. The MDD is calculated from: (i) geometric attenuation (AVG curves); (ii) physical attenuation between two bottom echoes; (iii) SNR between 1° echo and noise at the axis (SNR @ axis). Early stage: attenuation 4 dB/m; total sensitivity $28 + 6 + 44 = 78$ dB; MDD approximately 2 mm; Middle stage: attenuation 2.3 dB/m; total sensitivity $28 + 3.5 + 51 = 82.5$ dB; MDD approximately 1.5 mm.

In conclusion for that forging only the PuC technique can achieve the desired MDD target, and in general for a large diameter forged-steel specimen, the PuC technique has a gain in SNR between 5 and 20 dB, depending on attenuation values. This means that with PuC-based method, it is possible to reveal defects between 10 and 30 %, smaller than with standard PuE measurements (Figs. 9.8 and 9.9).

Example 9.1*(The Virtual Instrument)*
Since standard nondestructive testing (NDT) ultrasonic hardware does not exhibit the flexibility needed to fully implement the PuC procedure with respect to the signal generation, acquisition, and processing, a virtual instrument (VI) has been developed by means of the LabView software that allows: (i) the generation of the proper coded signal by means of an AWG module NI PXI-5412

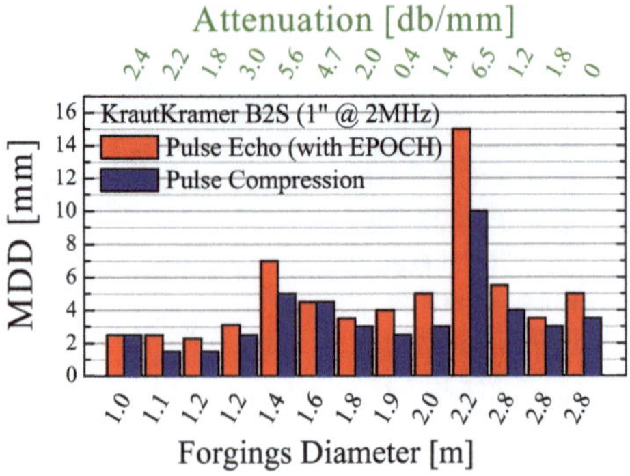

Fig. 9.8 First comparison: MDD for various forgings at 2 MHz. With KrautKramer B2S (narrowband probes: unambiguous AVGcurves), PuC achieves significant better results than standard PuE, especially for large forgings or high attenuation (early stages of hardening)

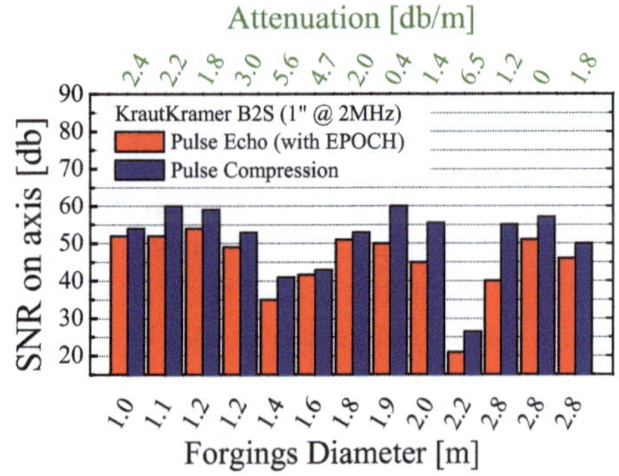

Fig. 9.9 Second comparison: SNR for various forgings at 2 MHz. For all forgings inspected, the pulse compression-based method achieve higher SNR values than standard PuE. In far field, the AVG for PuE and for PuC almost coincide

operating up to 100 MS/s; (ii) the synchronization between the excitation signal and the digitalisation of the receiving probe signal carried out by employing a 12-bit digital oscilloscope module NI PXI-5105 with maximum sample rate of 60 MS/s and with 60 MHz bandwidth; (iii) the real-time processing of the measured data through the convolution with the matched filter and the application of selectable processing algorithms; (iv) the simultaneous visualization of the acquired data $y(n)$, the reconstructed impulse response $h[n]$, its envelope $\mathrm{Env}[h](n)$ in dB scale, and its amplitude spectrum. All the parameters of the chirp signal—weighting function, carrier frequency, bandwidth, and duration—are set on the front panel and automatically updated. In Fig. 9.10, the front panel and block diagram of the VI are reported, respectively.

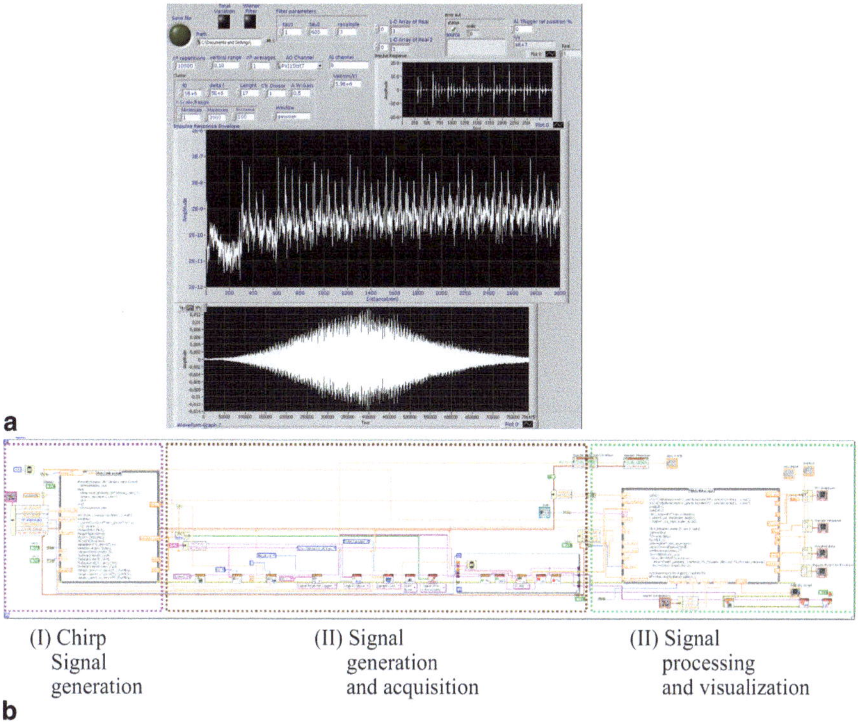

Fig. 9.10 **a** The front panel of the virtual instrument. At the *top*, there are all the controls regarding the generation and acquisition tasks: clock divisor, generation and acquisition rate, ultrasound velocity of the medium, vertical range, $\alpha(t)$ window pop-up menu, range of visualization (in m @ the ultrasound velocity), and the parameters of the chirp signal. In the same zone of the interface, one can also find the controls related to the files and the hardware management, as well as some indicators that allow to check the correct operation of the system. The *bottom* side of the panel is subdivided in four panes showing respectively (from *top* to *bottom*): the estimated impulse response $\hat{h}(n)$, its full-wave envelope $\mathrm{Env}[\hat{h}](n)$, the acquired data $d(t)$, and the amplitude spectrum of $\hat{h}(n)$. **b** A block diagram of the virtual instrument where the main steps of the procedure have been evidenced: (I) the numerical generation of the selected chirp signal $s(n)$; (II) the synchronous generation and acquisition of $s(n)$ and of the system output signal $y(n)$; (III) the data processing and the subsequent results visualization

9.5 Conclusion

An application of PuC techniques to the ultrasonic inspection of large dimensions forged-steel products with high dissipation has been presented and compared with standard PuE techniques. The results show that the chirp-based PuC technique allows to achieve an enhancement of the inspection capability even adopting low voltage excitation signals. By developing a suitable linear amplifier, able to provide higher voltage levels, the performance can be further increased contextually assuring high values of SNR and of range resolution. Moreover, high voltage of the excitation signals make it possible to maintain the gain of SNR at the level here adopted even when shorter coded signals are employed. In this scenario, a PuC-PuE device can be successfully implemented for the inspection of the forged parts.

Acknowledgements The authors would like to give special thanks to M. Calderini, S. Neri, R. Borgna, and Società delle Fucine s.r.l., an italian member of the ThyssenKrupp group, for making available the specimens with suitable characteristics and for the helpful discussions about the forgings inspection.

References

1. Adams JW (1991) A new optimal window. IEEE Trans Signal Process 39(8):1753–1769
2. Challis RE, Ivchenko VG (2011) Sub-threshold sampling in a correlation-based ultrasonic spectrometer. Meas Sci Technol 22(2):1–12
3. Derode A, Tourin A, Fink M (1999) Ultrasonic pulse compression with one-bit time reversal through multiple scattering. J Appl Phys 85(9):6343–6352
4. Gan TH, Hutchins DA, Billson DR, Schindel DW (2001) The use of broadband acoustic transducers and pulse-compression techniques for air-coupled ultrasonic imaging. Ultrasonics 39(3):181–194
5. Golay MJE (1961) Complementary series. IRE Trans Inf Theory 7(2):82–87
6. Golomb SW, Welch LR, Goldstein RM, Hales AW (1982) Shift register sequences, vol 78. Aegean Park Press, Laguna Hills
7. Harris FJ (1978) On the use of windows for harmonic analysis with the discrete Fourier transform. Proc IEEE 66(1):51–83
8. Misaridis T, Jensen JA (2005) Use of modulated excitation signals in medical ultrasound. Part I: basic concepts and expected benefits. IEEE Trans Ultrason Ferroelectr Freq Control 52(2):177–191
9. O'Donnell M (1992) Coded excitation system for improving the penetration of real-time phased-array imaging systems. IEEE Trans Ultrason Ferroelectr Freq Control 39(3):341–351
10. Pollakowski M, Ermert H (1994) Chirp signal matching and signal power optimization in pulse-echo mode ultrasonic nondestructive testing. IEEE Trans Ultrason Ferroelectr Freq Control 41(5):655–659
11. Rao N (1994) Investigation of a pulse compression technique for medical ultrasound: a simulation study. Med Biol Eng Comput 32(2):181–188
12. Ricci M, Senni L, Burrascano P (2012) Exploiting pseudorandom sequences to enhance noise immunity for air-coupled ultrasonic nondestructive testing. IEEE Trans Instrum Meas 61(11):2905–2915
13. Ricci M, Senni L, Burrascano P, Borgna R, Neri S, Calderini M (2012) Pulse-compression ultrasonic technique for the inspection of forged steel with high attenuation. Insight Non-Destruct Test Cond Monit 54(2):91–95
14. Rihaczek AW (1969) Principles of high-resolution radar. McGraw-Hill, New York
15. Sarwate DV, Pursley MB (1980) Crosscorrelation properties of pseudorandom and related sequences. Proc IEEE 68(5):593–619
16. Schroeder MR (1979) Integrated-impulse method measuring sound decay without using impulses. J Acoust Soc Am 66(2):497–500
17. Turin GL (1976) An introduction to digital matched filters. Proc IEEE 64(7):1092–1112

Chapter 10
Fuzzy Geometrical Techniques for Characterizing Defects in Ultrasonic Non-destructive Evaluation

Mario Versaci, Salvatore Calcagno, Matteo Cacciola, Francesco Carlo Morabito, Isabella Palamara and Diego Pellicanò

Abstract Many types of defectiveness can appear during the manufacturing of carbon fiber reinforced plastics (CFRP), putting at risk both safety and the quality of products. Therefore, a protocol to check the integrity of CFRPs is an important industrial requirement. It should involve non-destructive testing (NDT)/non-destructive evaluation (NDE), in order to be the least invasive as possible. When exploiting ultrasonic testing, there is not a one-to-one correspondence between the type of defect and the trend of the resulting signal. Thus, visual inspection of ultrasonic signals can be a really hard task, needing a considerable experience or a suitable computing support. In the latter case, it rises the problem of ill-posedness, precisely because of the complex correspondence between defects and signal trends. The scientific literature presents a number of studies aiming to approach this problem, focusing on heuristic techniques, but characterized by high-computational complexity. Conversely, for real-time applications, fast procedures are needed, with a low computational complexity. Experience in soft computing, even in frameworks different than NDT/NDE, can be valuable for implementing such low-time-execution algorithms. This is particularly true with respect to the handling of data affected by uncertainty and/or imprecision caused by sampling and noising of signals. Due to

M. Versaci (✉) · S. Calcagno · M. Cacciola · F. C. Morabito · I. Palamara · D. Pellicanò
DICEAM Department, Mediterranea University,
Via Graziella Feo di Vito, Reggio Calabria, Italy
e-mail: mario.versaci@unirc.it

S. Calcagno
e-mail: calcagno@unirc.it

M. Cacciola
e-mail: matteo.cacciola@unirc.it

F. C. Morbito
e-mail: morabito@unirc.it

I. Palamara
e-mail: isabella.palamara@unirc.it

D. Pellicanò
e-mail: diego.pellicano@unirc.it

P. Burrascano et al. (eds.), *Ultrasonic Nondestructive Evaluation Systems*,
DOI 10.1007/978-3-319-10566-6_10

its nature, it is convenient to approach the classification problem as a fuzzy matter, where ultrasonic signals resulting from the same kind of defect (i.e., same class of defectiveness) have similar statistic values. That is because classification problem can be seen as a fuzzy geometrical problem, where each class is taken into account as a specific family of fuzzy sets (fuzzy hyper-rectangles) inside a fuzzy unit hyper-cube. Thus, an ultrasonic signal depicting an unknown defect can be mapped as a point into the unit hyper-cube and it be classified there by means of its distance from the hyper-rectangles.

10.1 Introduction

Currently, the scientific community is particularly active in ultrasonic non-destructive testing (NDT)/non-destructive evaluation (NDE) domain for the characterization of defects in carbon fiber reinforced plastics (CFRP) in terms of detection and classification. Since the signals are nearly always affected by vagueness, it is necessary to formulate the problem using fuzzy approaches that can take into account the vague information content. Moreover, in many applications, it is useful to formulate the classification problem in "geometric terms" plunging ultrasonic signals in certain fuzzy geometrical spaces and translating them into points: in this way, the classification problem can be translated into a sort of distance evaluation among points. In this section, the authors examine the problem from a geometric point of view by formulating an approach based on fuzzy subsethood operator (FSO) to translate the characterization of defects in a sort of fuzzy distance and to assess how much a class of defects is a subclass of another one.

10.2 A Fuzzy Geometrical Point of View for Classifying Defects in Ultrasonic NDE

Characterization of defects in ultrasonic NDE by means of fuzzy inferences is based on the drawing up of particular sets of fuzzy rules which manage the interaction among fuzzy membership functions (FMFs) [1, 4, 9]. These ones, specifying the single fuzzy sets, are representative of the fuzziness of the inputs/outputs parameters of the same system of characterization [1]. In the hypothesis of system design of defect characterization in terms of *closeness/distance* among unknown defects and known defects, the problem can be translated into the conception of a geometrical approach evaluating the closeness/distance of a signal with unknown defect from classes of signals with known defects. So, if a signal (or the features extracted from it) can be represented by a point in the space, and a class of signals as sets of points enveloped by a solid figure, the problem is equivalent to a mere choice of minimum distance. In a generic n-dimensional space the distances are evaluated in tensorial form by the well-known relation

$$ds^2 = m_{ij}dx^i dx^j, \quad i = 1, \dots, n \quad \text{and} \quad j = 1, \dots, n; \tag{10.1}$$

(with m_{ij} metrical tensor), which, after integration, provides:

$$s = \int \mathrm{d}s = \int \sqrt{m_{ij}\mathrm{d}x^i\mathrm{d}x^j} \, . \tag{10.2}$$

However, to formalize the distances by Eq. (10.2) means, on one hand, to neglect the fuzziness of each signal and, on the other hand, to exploit procedures with high computational complexity. To overcome this issue, it looks natural to formalize a new geometrical approach for chacarterization of defects capable of taking into account the above-mentioned needs. In particular, in terms of classification, a protocol should be formalized so that:

1. The n-dimensional space is comprehensive of the fuzzy nature of the procedure;
2. From signals affected by the same typology of defect, it is possible to extract a number of n-tuples of features characterizing the defect;
3. Each feature, once properly fuzzified, is represented as a point in the n-dimensional space so that a class of signals (with the same type of defect) generates a cluster of points envelopable by a well-defined solid;
4. From a signal with unknown defect, an n-dimensional set of features is extracted and, after fuzzification, it represents a point in the space;
5. It is possible to formulate suitable metrics able to estimate how much volume a point cluster and a solid occupy in the n-dimensional space;
6. It is possible to define a correct distance acting among points and solids in the space.

In the following, some ideas are presented for carrying out classification defects in a real case under study by means of the above protocol.

10.2.1 Defects and Classes of Defects as Points and Hyper-Rectangles in an Unit Hyper-Cube

10.2.1.1 Fuzzy n-dimensional Space as Container of Fuzzified Signals Features and Classes of Signals

Usually, taking out features from signals means extracting from each of them an n-dimensional set of values which, by means of fuzziness, is translated into a new set of values belonging to $[0, 1]$. In this way, the new fuzzified set is viewed as a point inside a unit hyper-cube $I^n = [0, 1]^n$ in which each side is an unitary interval. If n is the number of extracted features, 2^n corners of I^n represent crisp subsets (fuzzy power set $F(2^X)$), while fuzzy subsets are located inside I^n [2, 4, 6]. With this, $F(2^X)$ is a fuzzy hyper-cube I^n and a fuzzy set is a point inside I^n in which the corners represent crisp sets (because the coordinates are sequences of only two values: zero and/or one), while any other point represents a fuzzy set (because the coordinates are sequences of numbers belonging to $[0, 1]$). For example, if the universe of the discourse X consist of only two elements $\{feature_1, feature_2\}$, the fuzzy power set can be written as[1]:

$$F(2^X) = \{\emptyset, \{feature_1, feature_2\}, feature_1, feature_2\} \tag{10.3}$$

and corners $(0,0)$, $(0,1)$, $(1,0)$, and $(1,1)$ represent crisp sets where the unit is indicative of the total presence of features, while zero gives indication of the whole absence.

10.2.1.2 Signals as Points Inside I^n

For a generic signal, a fuzzified set of n-features is representable as a point inside I^n whose coordinates are their fuzzified values. In other words, if $X = (feature_1, feature_2, \ldots, feature_n)$ then the fuzzy set

$$fuzzified(X) = (fuzzified(feature_j)) \qquad j = 1, 2, \ldots, n \tag{10.4}$$

is a point inside I^n where the coordinates are $fuzzified(feature_j)$.

10.2.1.3 Classes of Signals as Hyper-rectangles Inside I^n

Thinking about a class as a family of signals affected by the same typology of defects (or without defects), each signal can be mapped in I^n by means of the set of its fuzzified features. If the choice of the features is correct, then signals affected by the same type of defects generate nearby features (in a fuzzy sense). So, signals belonging to the same class map to points in I^n significantly close to each other. One can thus think of enveloping the totality of such points by a solid representing the borderline of the volume occupied by them. If for each extracted feature, $feature_j$, the ranges $[\max(fuzzified(feature_j)), \min(fuzzified(feature_j))]$ are considered, one can see that in I^n, they generate j hyper-rectangles (number of considered classes). In such a context, the classification problem is translated into the equivalent computation of distances between a point (correspondent to a signal affected by an unknown defect) and hyper-rectangles (classes of signals of known defects and classes of signals without defects). With this, it is clear that a necessity exists to understand how much volume a fuzzified set of features occupy inside I^n and to determine how to compute the distances among points (signals) and hyper-rectangles (classes of defects/absence of defects) in I^n.

10.2.1.4 How Much Volume Does a Fuzzified Set of Features Occupy Inside I^n?

Having to satisfy the need to evaluate how "big" a fuzzified set of features $\{fuzzified(features_j)\}$ is inside I^n,[2] it is necessary to provide the measure

---

[1] Cantor's theorem: if $|X| = n$, then $|F(2^X)| = 2^n$.

[2] $|\{fuzzified(features_j)\}| = n$.

$M(\{fuzzified(features_j)\})$ of how much volume it occupies in I^n. By means of the vector norm[3], $M(\{fuzzified(features_j)\})$ is computable as[4]:

$$M(\{(fuzzified(features_j)\}) = \left(\sum_{j=1}^{n} fuzzified(features_j))^p\right)^{\frac{1}{p}}. \quad (10.5)$$

For $p = 2$ (euclidean norm):

$$M(\{(fuzzified(features_j)\}) = \sqrt{\sum_{j=1}^{n} fuzzified(features_j))^2}. \quad (10.6)$$

For $p = 1$ ("taxi-cab" norm):

$$M(\{fuzzified(features_j)\}) = \sum_{j=1}^{n} \left|fuzzified(feature_j)\right|. \quad (10.7)$$

Finally, for $p = \infty$:

$$M(\{fuzzified(features_j)\}) = \max_{j=1,2,\ldots,n} \left|fuzzified(feature_j)\right| \quad (10.8)$$

Similarly, to evaluate the distance between two signals, H and K, it is sufficient to refer to Eq. (10.5), considering $(fuzzified(feature_j))_H - (fuzzified(feature_j))_K)$ instead of $fuzzified(feature_j)$.

10.2.1.5 Quantification of the Distances Among Fuzzified Sets of Features and Classes of Signals in I^n

The geometrical formulation of fuzzy sets in I^n, together with the FSO, can play a crucial role with respect to detection and classification problems in ultrasonic NDE. The starting point is the idea that fuzzy subsethood in I^n, identified by $FSO(A, B)$, and representing how much a fuzzy set A is a subset of another fuzzy set B (with A and B two fuzzy sets in I^n), can also represent how much a fuzzy set A belongs to the class represented by B. In our study, if $A = \{fuzzified(features_j)\}$ represents, in I^n, a signal with unknown defect and B represents a class of signals affected by a typology of defect (hyper-rectangle $[\max(fuzzified(feature_j)), \min(fuzzified(feature_j))]$),

---

[3] A fuzzy set A inside I^n is a point which can be seen as the extreme of a vector pointing to $\{fuzzified(features_j)\}$ from the vertex $(0, \ldots, 0)$.

[4] A norm is a function $\|\cdot\|$ defined on a real or complex space X ranging on $[0, +\infty)$ so that: (i) $\|x\| \geq 0$; (ii) $\|\alpha x\| = |\alpha| \cdot \|x\|$; and (iii) $\|x + y\| \leq \|x\| + \|y\| \; \forall x, y \in X$ and $\alpha \in \mathbb{R}$. Moreover, $\|x\| = 0$ iff $x = 0$.

FSO(A, B) can be formulated in many ways. Matematically, Sanchez proposed its computation to be based on cardinality and intersection operations [7]:

$$FSO_1(A,B) = \frac{|A \cap B|}{|A|}\,. \tag{10.9}$$

Exploiting inclusion and cardinality operations, Goguen [3] gave the measure

$$FSO_2(A,B) = \overline{|A| - |B|}\,. \tag{10.10}$$

Kosko's formulation computes FSO as follows [5, 8]:

$$FSO_3(A,B) = \frac{1 - \sum_{j=1}^{n} \max[0,\ (\mathit{fuzzified}(\mathit{feature}_j))_A - \mathit{fuzzified}(\mathit{feature}_j)_B]}{M(A)}\,. \tag{10.11}$$

where $M(A)$ can be formulated, for example, by means of the Euclidean norm. Obviously, for different purposes, other distances can be taken into account to solve the problems at hand. Of course,

$$FSO(A,B) \leq 1 \tag{10.12}$$

and

$$FSO(B,A) = 1 - FSO(A,B)\,. \tag{10.13}$$

Hence, due to the fact that $S(A, B)$ computes how much A is contained in B, $FSO(B, A)$ indicates how much B is contained in A.

10.3 Classification Procedures in Ultrasonic NDE by Means of Fuzzy Geometrical Approach

The basic idea of the proposed procedure springs from the consideration that a generic class of defects k is characterized by ranges of certain features (for examples, statistical parameters: mean, standard deviation, skewness, and kurtosis) falling into particular ranges of values. So, for each section of signals, those features are computed (labeled by $F_1^k, F_2^k, \ldots, F_n^k$, respectively). Then, for the four parameters, a multi-dimensional matrix F^k can be defined as follows:

$$F^k = \begin{pmatrix} F_1^1(j) & F_2^1(j) & \cdots & F_n^1(j) \\ F_1^2(j) & F_2^2(j) & \cdots & F_n^2(j) \\ F_1^3(j) & F_2^3(j) & \cdots & F_n^3(j) \\ F_1^4(j) & F_2^4(j) & \cdots & F_n^4(j) \end{pmatrix} \tag{10.14}$$

where m is the cardinality of the jth section of the signal. Each element of F^k, namely $F^k(x, y)$, is a set of values which represents the whole of the possible values of features x for class k.[5] Later on, two new matrixes $\min(F^k)$ and $\max(F^k)$ can be computed as follows:

$$\min(F^k) = \begin{pmatrix} \min\left(F_1^1(j)\right) & \min(F_2^1(j)) & \cdots & \min(F_n^1(j)) \\ \min(F_1^2(j)) & \min(F_2^2(j)) & \cdots & \min(F_n^2(j)) \\ \min(F_1^3(j)) & \min(F_2^3(j)) & \cdots & \min(F_n^3(j)) \\ \min(F_1^4(j)) & \min(F_2^4(j)) & \cdots & \min(F_n^4(j)) \end{pmatrix} \tag{10.15}$$

$$\max(F^k) = \begin{pmatrix} \max(F_1^1(j)) & \max(F_2^1(j)) & \cdots & \max(F_n^1(j)) \\ \max(F_1^2(j)) & \max(F_2^2(j)) & \cdots & \max(F_n^2(j)) \\ \max(F_1^3(j)) & \max(F_2^3(j)) & \cdots & \max(F_n^3(j)) \\ \max(F_1^4(j)) & \max(F_2^4(j)) & \cdots & \max(F_n^4(j)) \end{pmatrix} \tag{10.16}$$

Finally, matrix F'^k, whose generic element $F'^k(x, y)$ is the interval $[\min(F^k), \max(F^k)]$ is computed as follows:

$$F'^k = [\min(F^k)\ \max(F^k)] =$$

$$\begin{pmatrix} \min(F_1^1(j)) - \max(F_1^1(j)) & \cdots & \min(F_n^1(j)) - \max(F_n^1(j)) \\ \min(F_1^2(j)) - \max(F_1^2(j)) & \cdots & \min(F_n^2(j)) - \max(F_n^2(j)) \\ \min(F_1^3(j) - \max(F_1^3(j)) & \cdots & \min(F_n^3(j)) - \max(F_n^3(j)) \\ \min(F_1^4(j)) - \max(F_1^4(j)) & \cdots & \min(F_n^4(j)) - \max(F_n^4(j)) \end{pmatrix}. \tag{10.17}$$

The fuzzification step leads to treating each $F'^k(x, y)$ by means of a suitable function with good smoothing properties in $[0, 1]$. In this case, a sigmoid function can be chosen, but other ones can be taken into account as well. Eventually, the next equation shows the formulation of the fuzzification step. In this way, each range of possible values of feaures $F'^k(x, y)$ is "translated" into a fuzzy quantity located as an interval inside I^4, so obtaining four hyper-rectangles which represent each class of defect, respectively. The coordinates of each point are represented by:

$$[f(\min(F^k(x, y)),\ f(\max(F^k(x, y))] =$$

$$\left(\frac{1}{1 + \mathrm{e}^{-S^k(\min(F^k(x,y)) - H^k)}} \quad \frac{1}{1 + \mathrm{e}^{-S^k(\max(F^k(x,y)) - H^k)}} \right) \tag{10.18}$$

where x and y are a generic row and column of matrix F'^k, while S^k and H^k are suitable sigmoidal parameters (referred to each kth class of defects). In order to

[5] Here, x and y are the rows and columns of matrix F^k.

simplify the proposed procedure, each class of defect is represented inside I^4 by means of a class including its subclasses referred to superficial, middle, and deep defects respectively. In other words, for delamination and inclusion classes, the fuzzified ranges are computed as follows:

$$\left[\min_{subclasses}(f(\min(F^k(x,y)))),\ \max_{subclasses}(f(\min(F^k(x,y))))\right]. \tag{10.19}$$

If Y is a signal with an unknown defect that has to be classified, the vector $\boldsymbol{Y} = [F_{1x}, F_{2x}, \ldots, F_{nx}]$ must be fuzzified by a sigmoid function:

$$f(\boldsymbol{Y}(h)) = \frac{1}{1 + e^{-s(\boldsymbol{Y}(h))}} \tag{10.20}$$

with $h = 1, 2, 3, 4$. Note that $f(\boldsymbol{Y}(h))$ is a point inside I^4. Then,

$$FSO\left(f(\boldsymbol{Y}(h)),\ f\left(\max_{subclasses}(f(\min(F^k x, y)))\right)\right) \tag{10.21}$$

can be computed as explained in Eqs. (10.9), (10.10), and (10.11).

If:

$$FSO\left(f(\boldsymbol{Y}(h)),\ f\left(\max_{subclasses}(f(\min(F^k(x, y)))\right)\right) \tag{10.22}$$

is closer to the unity, the unknown signal, likely, belongs to the k class and, finally:

$$\min_k S\left(f(\boldsymbol{Y}(h)),\ f\left(\max_{subclasses}(f(\min(F^k(x, y)))\right)\right) \tag{10.23}$$

displays the pertaining class. In Fig. 10.1 the flowchart of the designed classification procedure is displayed.

10.3.1 An Example of Defects Detection and Classification in Carbon Fiber Reinforced Plastics by Fuzzy Subsethood Operator Approach

10.3.1.1 The Experimental Database

As showed in a previous chapter, top, middle, and bottom delamination/inclusion and defect-free areas were obtained on artificial panels (immersed into water) of CFRP [6] while another panel have been considered for porosity [7]. The analysis and investigations have been carried out at Laboratory of Electrical Engineering and NDT/NDE of

[6] $301 \times 355 \times 4$ mm

[7] $39 \times 40 \times 3$ mm

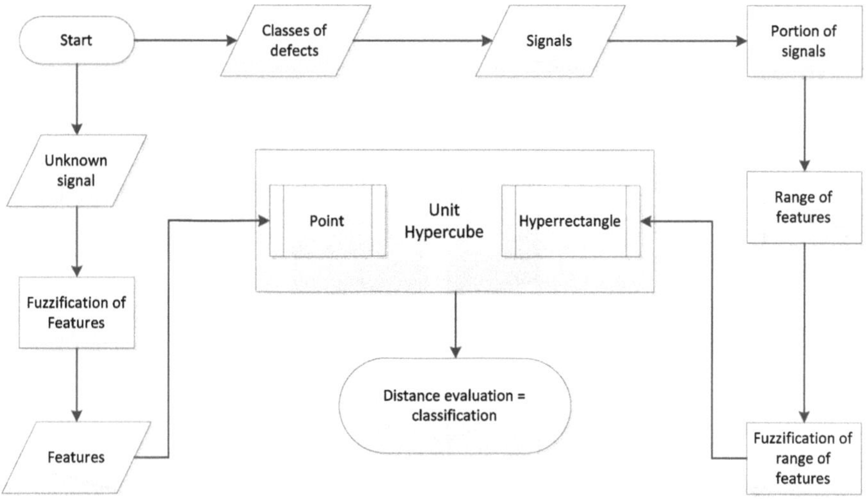

Fig. 10.1 Flow chart of designed classification procedure by means of FSO approach

Table 10.1 Distribution of signals in training and testing databases

	Training	Testing
Top delamination	42	21
Middle delamination	45	24
Bottom delamination	37	22
Top inclusion	35	28
Middle inclusion	38	25
Bottom inclusion	38	21
Porosity	40	25
No defects	45	30

University of Reggio Calabria. A 5 MHz ultrasonic transducer (UT) probe with 100 MHz sampling frequency and 3000 m/s speed of the ultrasonic wave on CFRP have been exploited. If an UT signal related to an undefected area is considered, a top-echo [8] and a bottom-echo [9] take place and any other peaks that appear in the signal highlight discontinuity in the specimen (defectiveness). In this case, qualitative inspection of the signals is not enough to classify defectiveness because different kinds of defects produce similar signals. So, our attention is focused on the pre-processing step in order to extract suitable features to make a classification procedure with a low computational complexity useful hardware device implementation. To the advantage of reading, we show Table 10.1 again where the training and testing sections of the experimental database (number of signals for each category of defect and class of signals without defects) are reported.

[8] Interaction between the exciting ultrasonic wave and the top surface
[9] Reflection of the wave on the bottom of the specimen

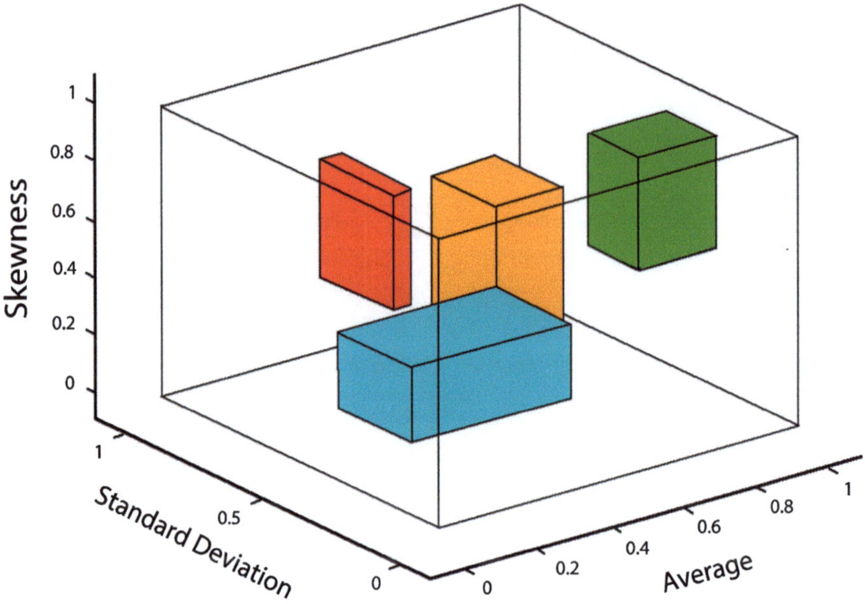

Fig. 10.2 Classes of defects inside I^n. Delamination, inclusion, porosity, and no-defects classes are depicted in different colors (*red* = class without defects; *green* = delamination; *yellow* = porosity and, *light blue* = inclusion)

Table 10.2 Comparison of the results from FSO by different formulations (Sanchez [7], Goguen [3], and Kosko [5])

	Detection (%)	Classification (%)
FSO_3	99.4	97.2
FSO_2	99.6	98.4
FSO_1	100	98.8

10.3.1.2 The Classification Procedure

The proposed algorithm has been applied to the above experimental dataset. As a result, fuzzified ranges for statistical quantities of each kind of defect have been obtained. Their Euclidean representation is displayed in Fig. 10.2 where color (red = class without defects; green = delamination; yellow = porosity; and light blue = inclusion). Finally, FSO carried out the detection and classification of possible defects (Kosko's formulation [5]). Defects have been detected in 100 % of cases, but their type have been rightly identified only on 98.8 % of classes (Table 10.2). Misclassification, occurring for a reduced number of inclusions and porosities, can be justified since fuzzy clusters are really close to each other inside I^4. On the other hand, misclassified signals could be seen as collected in a new fuzzy class, and provided again to the procedure in order to be correctly detected and classified, in a sort of back-propagating, self-improving paradigm [8]. Values of sigmoidal parameters, S^k, H^k, have been set by a sensitivity analysis, in order to have the best performance of classification as possible. Retrieved results could be improved by a better choice of FSO and metrics. Finally, a suitable reformulation of the problem into non-Euclidean spaces could provide further information about misclassified cases.

10.4 Conclusion

In this chapter, a new approach for the detection and classification of defects in CFRP has been proposed, based on the joint use of fuzzy and ultrasonic techniques. In particular, detection and classification are operated by means of FSO translating the classification problem in terms of the distance among points within the fuzzy unit hyper-cube thus providing a geometrical point of view particularly effective for the study of these problems. Furthermore, the low computational load (mere sequence of elementary arithmetic operations) makes the procedure particularly attractive for online applications and/or technology transfer. Obviously, the results can be improved both by careful choice of FSOs and metrics, with the transposition in non-Euclidean spaces able to provide additional information especially in all cases of doubtful classification.

References

1. Cacciola M, Calcagno S, La Foresta F, Versaci M (2012) A new fuzzy geometrical approach to classify defects in composite materials. Int J Meas Technol Instrum Eng 2(4):12–20
2. Chiu SL (1994) Fuzzy model identification based on cluster estimation. J Intell Fuzzy Syst 2(3):267–278
3. Goguen JA (1969) The logic of inexact concepts. Synthese 19(3):325–373
4. Hanss M (2005) Applied fuzzy arithmetic. Springer Berlin
5. Kosko B (1996) Fuzzy engineering. Prentice-Hall. New York
6. Kuncheva LI (2000) Fuzzy classifier design. Springer Berlin
7. Sanchez E (1979) Inverses of fuzzy relations. Application to possibility distributions and medical diagnosis. Fuzzy Sets Syst 2(1):75–86
8. Sander W (1989) On measures of fuzziness. Fuzzy Sets Syst 29(1):49–55
9. Zimmermann HJ (1992) Fuzzy set theory and its applications, 2nd revised edn. Springer Berlin

Chapter 11
Industrial Applications of Noncontact Ultrasonics Techniques

Luigi Battaglini, Sergio Callegari, Salvatore Caporale, Lee Andrew John Davis, Stefano Laureti, Luca Senni and David Arthur Hutchins

Abstract In many application environments, it is not possible to practice nondestructive evaluation (NDE) by physically contacting the material under test with probes. Examples include the evaluation of objects that are moving (e.g., on an assembly line), hot materials, materials that cannot risk contamination, and more. Lack of contact means that, typically, the coupling material is air. This poses specific requirements, due to the acoustic impedance mismatch that an air layer can provoke and the consequent energy losses. In this chapter, some examples of setups, transducer arrangements, and signal processing strategies capable of addressing such requirements are illustrated, showing how concepts presented in the previous chapters can be deployed in this particular context. Furthermore, some typical achievable results are illustrated, with respect to a few different testing scenarios, such as the

L. Battaglini (✉) · L. Senni
Department of Engineering, Polo Scientifico Didattico di Terni, University of Perugia, Terni, Italy
e-mail: luigi.battaglini@gmail.com

S. Callegari
Department of Electrical, Electronic and Information Engineering "G. Marconi" (DEI) and Advanced Research Center on Electronic Systems for Information and Communication Technologies "E. De Castro" (ARCES), University of Bologna, Bologna, Italy
e-mail: sergio.callegari@unibo.it

S. Caporale
Advanced Research Center on Electronic Systems for Information and Communication Technologies "E. De Castro" (ARCES), University of Bologna, Bologna, Italy
e-mail: salvatore.caporale@unibo.it

L. A. J. Davis · S. Laureti · D. A. Hutchins
School of Engineering, University of Warwick, Coventry, UK
e-mail: lee.davis@warwick.ac.uk

S. Laureti
Dipartimento di Ingegneria, Polo Scientifico Didattico di Terni, University of Perugia, Terni, Italy
e-mail: s.laureti@warwick.ac.uk

L. Senni
e-mail: luca.senni@unipg.it

D. A. Hutchins
e-mail: d.a.hutchins@warwick.ac.uk

P. Burrascano et al. (eds.), *Ultrasonic Nondestructive Evaluation Systems*,
DOI 10.1007/978-3-319-10566-6_11

testing of composite materials, the inspection of concrete, imaging of surfaces or thin materials, and food inspection.

11.1 Introduction

This chapter outlines the transducers and measurement techniques that use pulse-compression (PuC) techniques for certain types of nondestructive evaluation (NDE). In particular, details are given of measurements that do not need to contact the material, such as air-coupled ultrasound. Noncontact methods are useful for many inspection situations where conventional contacting techniques are difficult, for example, the inspection of materials which are moving, the testing of materials at elevated temperatures, or where contamination of the sample by a couplant (commonly used in contact ultrasound) may be an issue. The need for additional signal processing techniques such as pulse compression arises mainly from the fact that air has a much lower acoustic impedance than most materials that are to be inspected. Thus, in air-coupled ultrasound (the technique of most interest within this chapter), the resulting acoustic impedance mismatch between air and most solids is the reason that pulse compression techniques are considered, and much energy is lost by reflection at the air/solid interfaces. In addition, frequency-dependent ultrasonic attenuation in air limits the upper frequency of operation, and thus, there is a constraint on the type of waveforms and bandwidths that can be used.

Despite the frequency bandwidth limitations imposed by the attenuation properties of air, pulse compression remains an essential method for improving signal to noise ratios in air-coupled ultrasonic NDE. Thus, the aim is to optimize the measurement not only by designing transducers that operate over the appropriate bandwidth, but also by teaming these up with the most appropriate waveform for cross-correlation. A description of the two main transducer designs for air-coupled ultrasound is included below, namely the use of either piezoelectric devices or those based on electrostatic principles (capacitive transducers). The latter design is discussed in some detail, including a description of the expected radiated fields, and the ways in which the energy can be focused onto the surface of a material to improve lateral resolution. Some applications of air-coupled NDE are then described. Note that pulse compression is also used as a technique of interest with other types of noncontact transduction systems, including those which use electromagnetic-acoustic transducers (EMATs) and laser-generated ultrasound. Together with air-coupled ultrasound systems, EMATs are also discussed because of the presence of an SNR issue due to the relative inefficiencies of the transduction processes being used.

11.2 Noncontact Ultrasonic Transduction Techniques

11.2.1 Laser-Generated Ultrasound

The use of pulsed lasers to generate ultrasound is now an established technique, and typically uses some form of optical detection to give a totally noncontact transduction system. There are many illustrations of how these can be used together [50], and various applications have been described. However, this technique suffers from several limitations—signal levels are low for thermoelastic generation, which is preferred for minimal damage to the surface. Here, heating mechanisms dominate the generation process, and although ablation can dramatically increase signal levels, damage to the surface often results. Although this process can be managed somehow, for example, by using CO_2 lasers on fibre-reinforced composites to minimise damage [54], there is a limit to the generation efficiency. Interferometers are also limited in terms of surface condition and sensitivity [35]. However, in terms of signal processing methods that could be applied, laser techniques have a big advantage: the available bandwidth is extremely wide if pulsed lasers are used. Furthermore, the technique can be adapted for each specific application.

One approach for fixing the afore mentioned issues is to modify the surface. Arca et al. [3] described pulsed laser generation techniques which included different absorption patterns on the surface for focusing, etc., as shown in Fig. 11.1. Here, it can be seen that the spatial distribution of absorbing strips leads to efficient Rayleigh-wave generation in a particular direction, which can also be focused. In the case shown in the figure, a narrow bandwidth would result (which could be used with synchronous detection in a lock-in type arrangement to reduce noise, if desired). It is additionally worth noting that bulk waves would also be generated across a broad bandwidth, but that there would be an angular dependence of the frequency content, due to the periodic distribution of energy at the surface. Also note that a chirp could easily be generated by varying the spacing of the strips. This had been demonstrated previously by Murray et al. [36], who overcame some of the limitations mentioned above by generating linear frequency-modulated signals using a suitable mask geometry. This meant that the peak power density of the laser source could be kept below the threshold for damage. The chirped ultrasonic surface wave was detected, and pulse compression was used to produce a 15-fold enhancement in the signal to noise ratio.

The main disadvantage of this masking approach is the need to coat the surface of the material. Thus, a more satisfactory approach is to use optics to generate diffraction patterns and/or modulate the laser source. While much lower peak powers can be delivered in this way, there is much more flexibility in applying subsequent signal processing methods. For example, in [49] the efficient production of surface acoustic waves using spatial light modulators with only 40 mW average optical power on the sample surface is reported. This technique could easily be extended to produce waveforms suitable for pulse compression. As an alternative approach, pulse

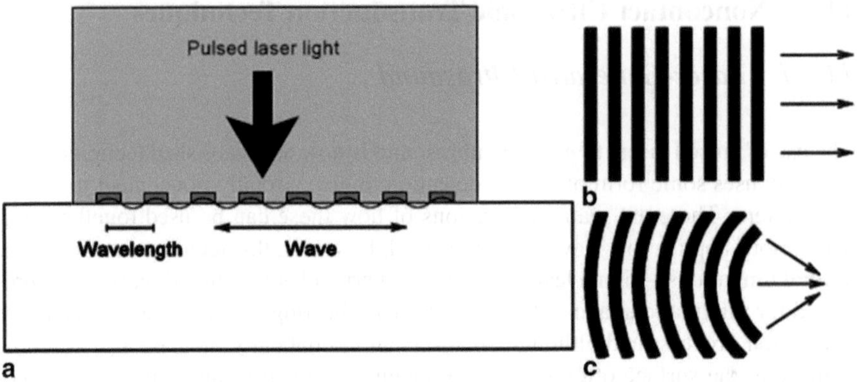

Fig. 11.1 The use of absorbing strips on the surface of a sample to enhance narrow bandwidth generation of ultrasound using pulsed lasers (**a**). Two types of strip geometries are shown in (**b**) and (**c**) (after [3])

compression techniques for laser-generated ultrasound have been described by Anastasi and Madaras [2], in which a laser diode was modulated with a pseudo-random m-sequence, and signals detected using a conventional piezoelectric detector. Cross-correlation was demonstrated, showing that the technique had a merit. Subsequently, Baloguna and Murray [4] used intensity-modulated laser sources as a frequency domain photoacoustic technique. This lock-in technique used an intensity-modulated continuous-wave laser for narrow bandwidth ultrasound generation, demonstrating that matched-filter techniques could be used. Despite these investigations, there has been relatively little material subsequently published on techniques that use pulse compression with laser generation, despite the obvious attractions.

11.2.2 Electromagnetic Acoustic Transducers (EMATs)

EMATs typically consist of two main components: an applied DC magnetic field, and a coil placed close to the surface. In ultrasonic generation, a current signal within the coil induces an eddy current within the material (which must have a sufficiently high conductivity), and this interacts with the applied magnetic field to generate forces. Ultrasonic detection is also possible using the same arrangement. A limitation of EMATs is that they have to be relatively close to the surface, but despite this, they are very useful in applications such as the inspection of hot, moving metal components [24]. The electronics needs to be designed carefully, so as to deal with the characteristics of the coil, but if this is done, EMATs are able to operate over a wide bandwidth. Thus, they are a natural choice for pulse compression techniques,especially, as their transduction efficiency is rather low when compared to contacting piezoelectric transducers. Such a system was described by Hirao and Ogi in [26], who used a chirp signal for pulse compression with an SH-Wave EMATs.

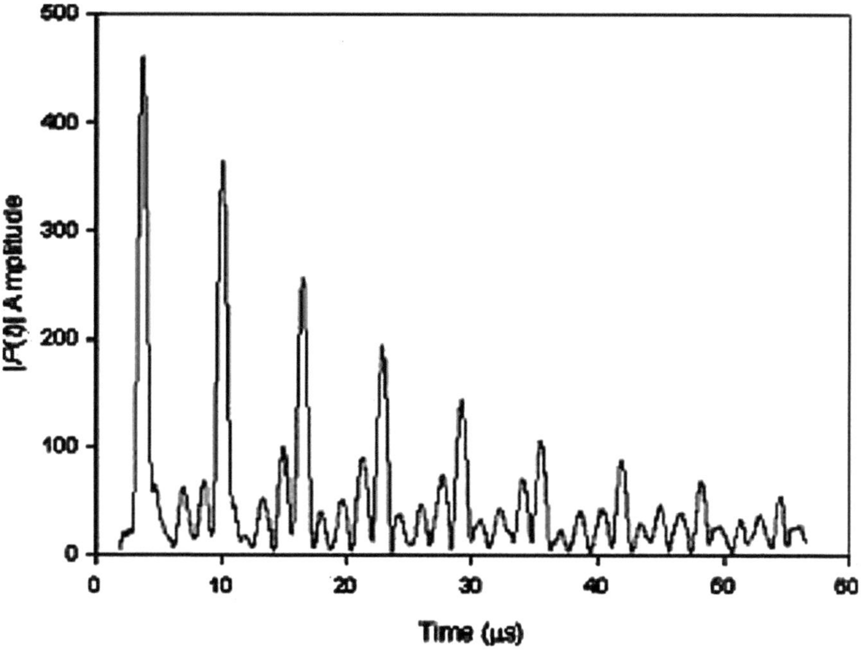

Fig. 11.2 Example of the pulse compression output signal obtained across 10 mm thick aluminium in through transmission mode, using a pancake-coil EMAT

Additional work by Ho et al. [27] used a chirp signal to generate ultrasonic waveforms within metal samples. Pulse compression allowed multiple echoes to be generated within steel and aluminium samples. A typical output is shown in Fig. 11.2, showing that the rectified and smoothed cross-correlation output gave the expected signals within a metal plate. It was demonstrated that the signal to noise ratio was much lower than that expected from a conventional transient impulse drive waveform.

11.2.3 Air-Coupled Ultrasound

As the name implies, this approach uses ultrasonic transducers operating in air, so that air becomes the medium by which signals are coupled into the sample being tested [9]. The need for techniques such as pulse compression arises from the large acoustic impedance mismatch between air and solid samples. In this section, some of the features that need to be addressed for successful implementation of the technique are discussed. This includes bandwidth limitations caused by attenuation in air, and the design of transducers that operate over the bandwidth defined by this phenomenon. Acoustic propagation in air has been widely investigated, due to the importance that

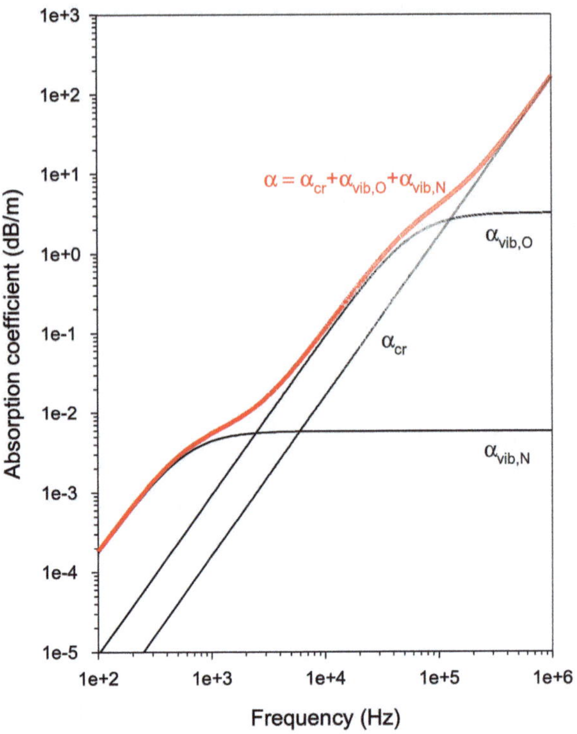

Fig. 11.3 The absorption coefficient (α) for acoustic signals in air as a function of frequency (at STP and for a relative humidity of 70 %)

it plays in general acoustics and noise control. In particular, atmospheric sound absorption in air has been studied extensively [5, 6, 17], as has the expected changes in acoustic velocity with changing gas composition. As an ultrasonic wave propagates through a homogenous gas such as air, absorption and scattering contribute to the overall attenuation level, although at ultrasonic frequencies in a controlled environment, absorption will be the main mechanism for attenuation. In a typical air-coupled NDE measurement, attenuation of the signal at high frequencies is a problem that will limit the bandwidth available. Attenuation is due to two basic mechanisms: classical losses (due to the change of kinetic energy of molecules into heat) and relaxation losses (where translational kinetic energy is changed into internal vibrational and rotational energy within the molecules themselves [17]). Both are functions of temperature and pressure for a given frequency. Air can be thought of as a mixture of nitrogen (N), oxygen (O), carbon dioxide (CO_2), and water vapour (H_2O). It is found that absorption due to nitrogen, oxygen, and their interaction with water vapour dominates, with the result that the total absorption coefficient α increases with frequency, as shown by the graph in Fig. 11.3. This also shows the relative contributions of vibrational absorption in oxygen ($\alpha_{vib,O}$) and nitrogen ($\alpha_{vib,N}$) and all other mechanisms (α_{tr}).

Note that the relative humidity (due to the presence of water vapour) is the most common contributor to increased ultrasonic attenuation in practical applications, and

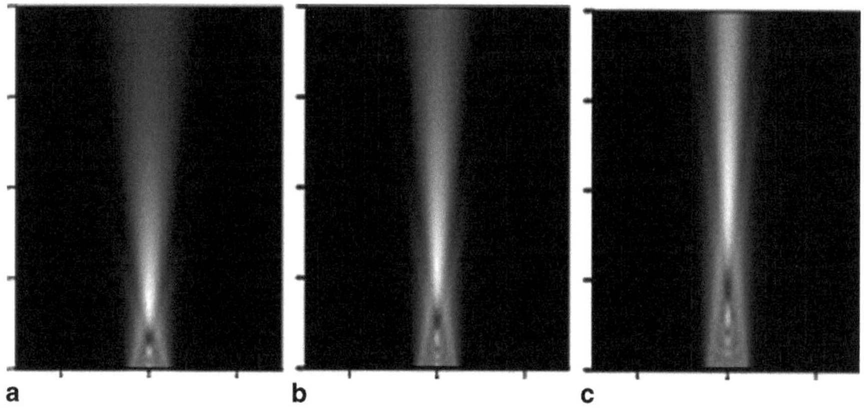

Fig. 11.4 Predicted radiated sound fields for a 10 mm diameter transducer in air, using tone-burst excitation at: **a** 200 kHz; **b** 300 kHz; and **c** 500 kHz. Both axes scaled at 20 mm per division

this effectively limits the frequencies that can be used for most ultrasonic work in air to a maximum of 1 MHz in many cases. This also effectively limits the bandwidth available for pulse compression processing.

A common method used for contact ultrasonic NDE is to use a piezoelectric transducer together with some form of intermediate impedance matching layer or a viscous couplant. Another common method of coupling is to use water immersion, where impedance matching layers are commonly used. There are, however, many materials and industrial manufacture processes where the presence of water will cause damage or inconvenience, and where air-coupled ultrasound may have a place [20]. In addition, many materials intended for the aerospace industry are not compatible with water [11]. Air-coupled ultrasound allows for rapid scanning without damage to the specimen material [53]. The relatively short wavelengths in air provide the potential for high-resolution imaging as the transducer is more directional than in either water or in a solid [47]. This is illustrated in Fig. 11.4 for a 10 mm diameter transducer in air [29].

Piezoelectric transducers are extensively used for the generation and detection of ultrasound in solids, but are inefficient when operated in air, due to the high impedance mismatch which occurs at the boundary between transducer and air, as shown in Table 11.1. The transmission coefficient (T_{air}) can also be calculated from the acoustic impedance (Z) on either side of the boundary, as shown.

It can be seen that transmission coefficient is much higher when the transducer is surrounded by water (T_{water}) as opposed to air, and this is because the impedance of water is much higher than air (1.5×10^6 compared to 414 kg m^{-2} s^{-1} respectively at 20 °C). Even for the relatively low impedance of piezoelectric polymer (PVDF), the transmission coefficient into air is still very low. There are various solutions to the problem of acoustic mismatch, but the basic approach is to try and match the impedance of the device to that of air more efficiently [1]. The most common approaches are:

Table 11.1 Transmission coefficients of some piezoelectric materials into air

Material	Density ρ (kg m^{-3})	Propagation velocity c_L(m s^{-1})	Acoustic impedance Z kg m^{-2} s^{-1}	T_{water}	T_{air}
Lead zirconate titanate (PZT-5 A)	7750	3880	30×10^6	0.181	5.5×10^{-5}
Piezoelectric polymer (PVDF)	1780	2260	4.6×10^6	0.742	36×10^{-5}

1. Reduce the acoustic impedance of the active element, for example with foam-like structures [7, 44], multilayer devices [16, 34], or the use of Lamb waves [15].
2. Use a damped resonant device (e.g., a piezoelectric element) and some form of impedance matching between the piezoelectric device and air (e.g., [32, 45]). The use of a single quarter-wavelength matching layer will reduce the emitted bandwidth [52], a problem if pulse compression is to be used. For this reason, multiple matching layers have been investigated [22]. Another approach is to use a piezocomposite element, which has a lower acoustic impedance than the piezoelectric active element alone. This can then be used with or without a matching layer [23].
3. Design a transducer with a thin membrane, which is inherently well-coupled to air. These are often referred to as capacitive devices. They have the advantage that they are of broad bandwidth [28, 48], and have thus been used extensively for ultrasonic noncontact NDE.

The capacitive transducer consists typically of a Mylar membrane, metalized on one side, which is placed against the face of a rigid conducting backplate, so that a capacitor is effectively formed. These transducers can be used in the ultrasonic region, provided that the membrane is sufficiently thin, and that the surface features of the backplate are well controlled. It is the improved impedance matching to air of the capacitive transducer when compared with the piezoelectric transducer, and the high bandwidth, which makes this type of transducer highly suitable for air-coupled applications involving pulse compression. Figure 11.5 shows a schematic diagram of such a device when used as a source. The polymer film is typically 2–5 μm thick, and the backplate is designed so that air is trapped between the metalised polymer membrane and the backplate, trapping air that then acts as a natural spring. Consequently, the device is intrinsically well suited to air-coupling. The conducting layer of the membrane is usually connected to a metal body surrounding the transducer, which is grounded to reduce noise. The behavior of a capacitance transducer, in terms of bandwidth, center frequency, and sensitivity, is critically dependent on the structure of the backplate [13]. There is also an effect caused by the DC bias voltage used, as this tends to decrease the air-gap as the membrane is attracted to the backplate. The overall response then is determined by the membrane thickness, the DC bias, and the surface features of the backplate. When a roughened

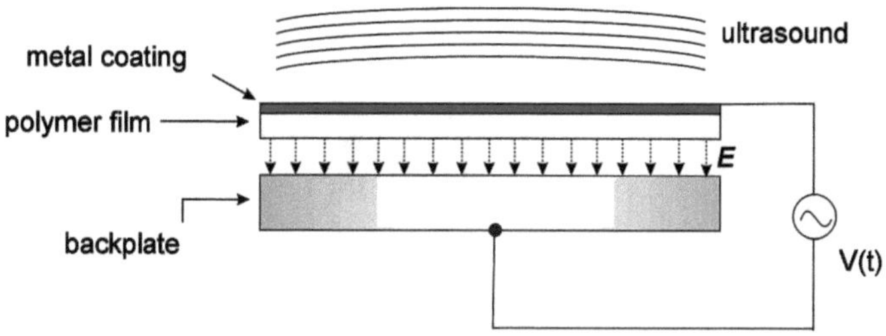

Fig. 11.5 A typical capacitive transducer acting as an ultrasonic source

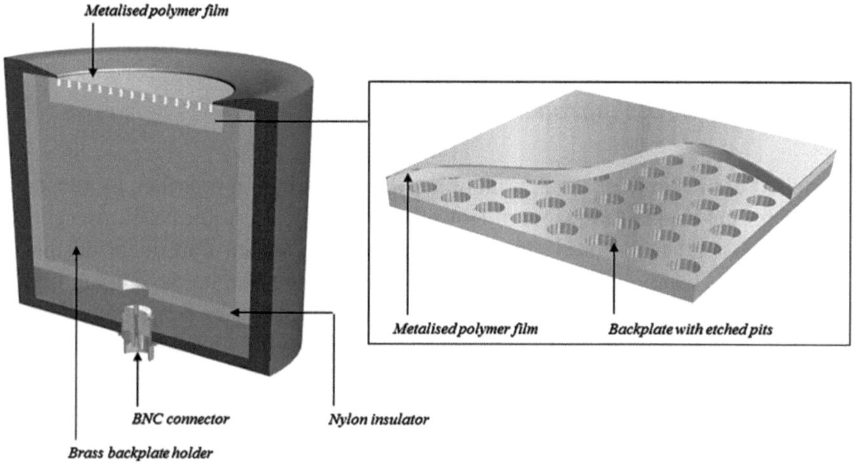

Fig. 11.6 Construction of a typical capacitive transducer for use in air, using a micro-machined silicon backplate

backplate is used it has been found that the frequency at which peak amplitude occurs is increased as roughness is decreased.

Good results have been obtained using micro-machined silicon backplates, in which 40 μm diameter pits are etched to a depth of 35 μm on a pitch of 80 μm [48]. These are illustrated in Fig. 11.6. In general, the upper frequency of operation increases with smaller amounts of trapped air. Treating the membrane as a piston constrained by an air spring, the approximate resonant frequency can be calculated using

$$f_0 = \frac{1}{2\pi}\sqrt{\frac{\gamma p_a}{\rho\, d_m d_a}} \tag{11.1}$$

where γ is the adiabatic constant of air, P_a is the atmospheric pressure, ρ is the membrane density, and d_m and $d_a$are the thicknesses of the membrane and air-gap respectively. Note that the membrane tension has little effect on the resonant frequency [25, 41].

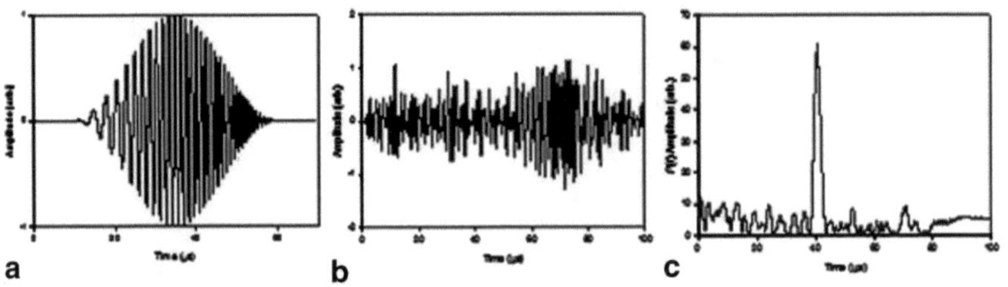

Fig. 11.7 Examples of pulse compression with air-coupled ultrasound chirp signals: **a** original chirp signal; **b** chirp buried in noise; and **c** the result following pulse compression

11.3 Pulse Compression and Air-Coupled Ultrasound

11.3.1 General Approach

As stated earlier in Sect. 11.2.3, there are three standard approaches to transducer design for air-coupled ultrasound. However, in practical use for NDE, two general approaches tend to be used. The first is to use a resonant transducer design, so that a narrow bandwidth waveform results with a high amplitude. This is usually a piezoelectric-based system, with either a matching layer or some other form of resonant structure attached to it to enhance efficiency of transduction into air. A pair of such devices can then be used in a through-transmission configuration, with the frequency of operation tuned to coincide with a through-thickness resonance of the structure under test. These are the conditions for maximum transmission of the signal through the solid structure. This approach has been tried successfully in inspection of composites, for example. In many cases, however, the sample thickness and structure may vary with location, and it is thus necessary to consider a broad bandwidth solution. In that case, pulse compression is by far the most common approach to establishing a workable SNR.

There are a range of commercial systems available for air-coupled ultrasound, and it is very common to see a chirp signal as the waveform used for pulse compression. The process of using such waveforms has been covered in Chap. 4 of this volume, but the chirp is a very good method for providing a robust cross-correlation over a well-defined bandwidth. An example for an air-coupled signal and a capacitive transducer is shown in Fig. 11.7, which demonstrates that a chirp signal, even when almost buried in noise, can be recovered after pulse compression. Note that the output from the cross-correlation has been rectified and smoothed to give an impulse-like display.

11.3.2 Choice of Excitation Waveform

It is found that the performance of pulse compression schemes is affected by the choice of both the excitation waveform and the subsequent processing [30]. Two

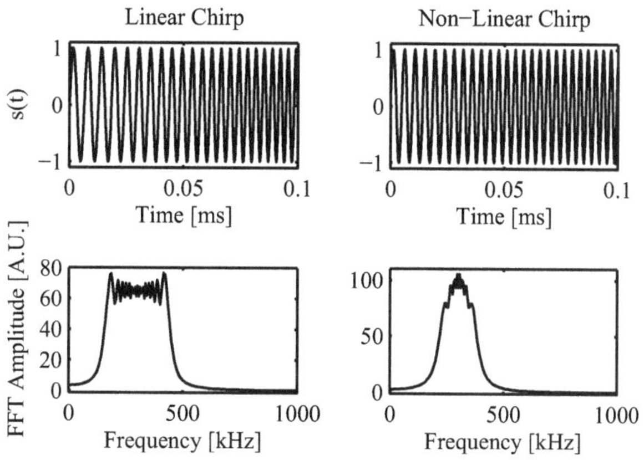

Fig. 11.8 Examples of chirp signals in the form of (*left*) linear and (*right*) nonlinear frequency sweeps. Waveforms are on the *top*, and corresponding frequency spectra are on the *bottom*

general types of coded waveform can be considered: chirps and binary coded sequences. Chirps using a linear frequency sweep or *linear chirps* (LChirps) have been most commonly used for pulse compression in air-coupled NDE in past studies. This is because they are both well understood and simple to implement in commercial systems. They are also very robust and easy to process. However, this does not mean that they are the optimum solution in all cases. A chirp signal can be defined in general terms as

$$S_{Chirp}\left(t\right) = A\ \sin\left(\Phi\left(t\right)\right) \tag{11.2}$$

where $\Phi(t)$ is a nonlinear phase function of t. The main characteristics of any given chirp signal are linked to the instantaneous frequency $f_{inst}(t)$ which is in turn related to the power spectral density (PSD) of the resulting function. It is expressed as

$$f_{\text{inst}}\left(t\right) = \frac{\pi}{2}\left(\frac{\mathrm{d}\Phi\left(t\right)}{\mathrm{d}t}\right). \tag{11.3}$$

The frequency can be swept in either a conventional linear fashion, or alternatively in a nonlinear way with respect to time. Two such signals sweeping within 150–450 kHz are shown in Fig. 11.8. The waveforms look similar, but their spectra indicate that the nonlinear chirp has a much different spectral range, more akin to that expected from a damped resonant transducer system. The nonlinear chirp may thus be a good choice if signal to noise ratio is to be maximised. It should be noted that chirps are well suited to systems with limited bandwidth, in that the duration and frequency range can be chosen in an attempt to match the characteristics of the transduction system—a major advantage in designing an experimental air-coupled system, for example.

Note that it is also conventional to impose a window function on the chirp signal. There are many such windows [39], which all have slightly different characteristics, but the aim in each case is to reduce side bands in the frequency domain. Note that this can also be achieved by windowing the spectrum itself. There are also other ways in which a chirp signal can be used. One is the frequency multiplication technique,

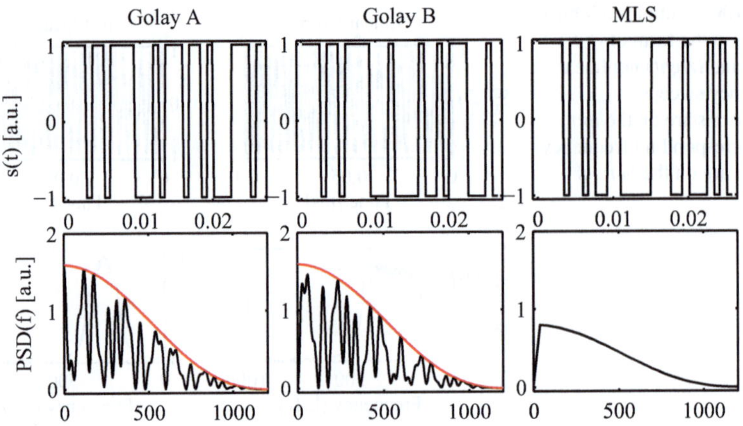

Fig. 11.9 Examples of waveforms useful for ultrasonic air-coupled signals derived from GCSs (A and B) and MLSs. At the *top* row, the "standard" waveforms, optimized for a central frequency of 500 kHz; at the *bottom* row, their respective PSDs. The *solid line* in the Golay cases represents the sum of the two PSDs (i.e., when both are used together). Vertical scales are arbitrary

described elsewhere in this volume, which has particular characteristics which suit some kinds of NDE measurement (e.g., where there is a high degree of attenuation or scattering).

The alternative approach is to use binary coded signals. Although many different types of coded signals exist, it is common to consider only those exhibiting a δ-like autocorrelation property in many ultrasonic applications, namely Golay complementary sequences (GCSs) and maximum length sequences (MLSs). Golay signals typically consist of two binary sequences, with the property that their out-of-phase aperiodic and periodic autocorrelation coefficients sum to zero. This makes them very well suited to air-coupled applications involving pulse compression, and indeed they are popular in many other ultrasonic measurements [19, 33, 55]. In practical applications, the two complementary sequences are both used as drive waveforms, and the matched filter is then applied separately to each output. The impulse response can then be reconstructed by adding the two contributions together. A further advantage of the GCSs approach is the existence of a fast algorithm to perform correlation-convolution between a generic signal and a Golay pair [12]. The properties arising from the use of MLSs techniques differ slightly from those of GCSs. Provided cyclic pulse compression is used (see 4.4.1), it can be shown that a perfect estimate of the impulse response can be retrieved. The immediate advantage is that only one sequence is necessary—a complementary pair of waveforms is not needed. In addition, a fast algorithm exists to perform cyclic correlation with MLSs [14].

In most air-coupled experiments, the transduction system has a limited bandwidth based around a center frequency. The wide bandwidth nature of binary sequences is thus not optimum in terms of energy efficiency. For this reason, schemes exist to perform spectral shaping of GCSs and MLS codes [38], leading to inverse repeated sequences (IRS) [51]. Examples of waveforms and spectra of GCS (complementary codes A and B) and MLSs are shown in Fig. 11.9 together with their equivalent IRS

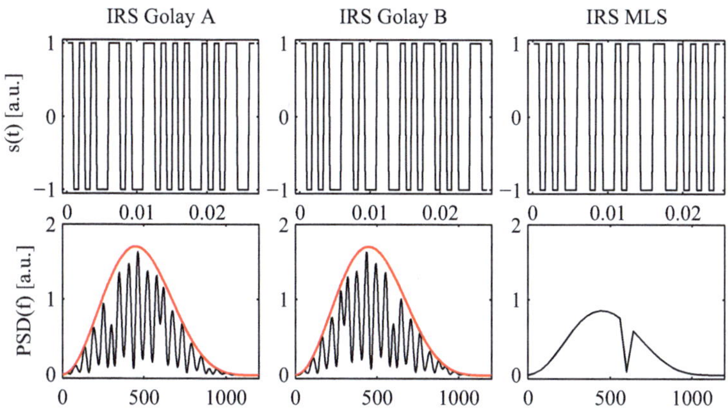

Fig. 11.10 In the *top* row, the equivalent IRS form for the waveforms of Fig. 11.9. In the *bottom* row, their PSDs. Vertical scales are arbitrary

Fig. 11.11 Impulse response for two different thicknesses of plexiglas: 1 mm (*top*) and 5 mm (*bottom*), for a linear Chirp (a) and an MLS signal (b). Vertical scales are arbitrary

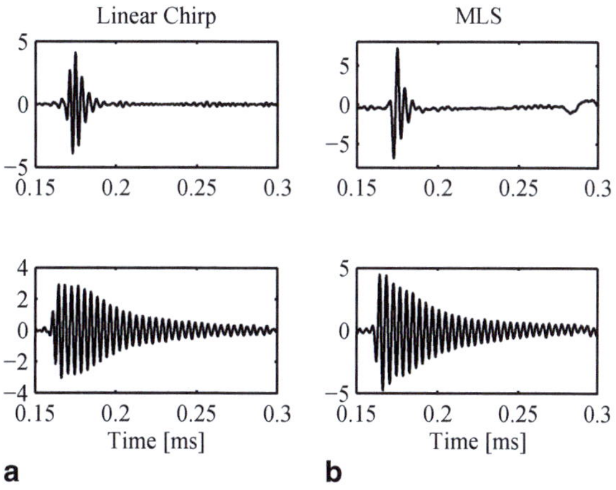

waveform. Figure 11.10 shows the equivalent frequency spectra, in terms of PSD, where it can be seen that the spectra of IRS signals are more suited to typical air-coupled transduction systems, which tend to have a particular center frequency. Note that as two complementary sequences are used for the Golay codes, the resultant PSD is double that of the MLS case.

The actual choice of waveform to use in a particular experiment depends on the type of measurement required [30]. As an example, consider an experiment which used air-coupled ultrasound to investigate ultrasonic transmission through a plexiglas plate of two different thicknesses. The resultant waveforms are shown in Fig. 11.11 for both sequences. It can be seen that the MLS result is of higher amplitude. Note also that the thickness resonance of the thinner plate is within the bandwidth of the measurement, and hence the waveform changes from an impulsive signal to a

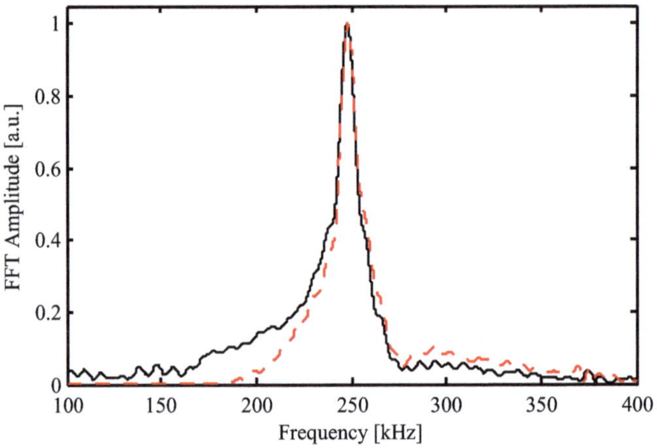

Fig. 11.12 Normalised frequency spectra obtained from Fig. 11.11, showing results for the MLS signal (*solid line*) and the linear chirp (*dotted line*)

decaying resonance, as expected. In addition, the impulse response is narrower for the MLS, as is evident in the waveforms of Fig. 11.11b.

The corresponding spectra of both signals types are normalized and plotted together for the thicker 5 mm plate in Fig. 11.12. In both cases, the narrow peak corresponding to the resonant frequency of the plate coincides. However, the larger received amplitude of the impulse response for the MLS sequences makes these codes preferable for thickness and time-of-flight (ToF) measurements.

There is, however, another issue of importance, namely that of side-lobe levels within the impulse response. It turns out that these are lower in linear chirp signals, and this could be an important feature in the detection of small features in NDE inspection.

In summary, binary and chirp signals achieve very similar results for wide bandwidth measurements, although binary codes appear to be the best choice. For a narrower available bandwidth, a chirp signal would be preferred, unless some spectral shaping is performed within the binary coding scheme. For time-of-flight and thickness measurement binary sequences give the highest received amplitude, and the shortest impulse response when used in a through-transmission measurement. For defect detection purposes, where very low side-bands levels are needed, linear windowed chirps appear to be the best solution for narrowband systems, at the cost of reducing the resultant spatial resolution.

11.4 Applications of Air-Coupled Ultrasound

11.4.1 Air-Coupled NDE of Composites

Consider the through-transmission arrangement shown in Fig. 11.13. Here, a pair of capacitive transducers of the type shown earlier in Fig. 11.6 are placed on both

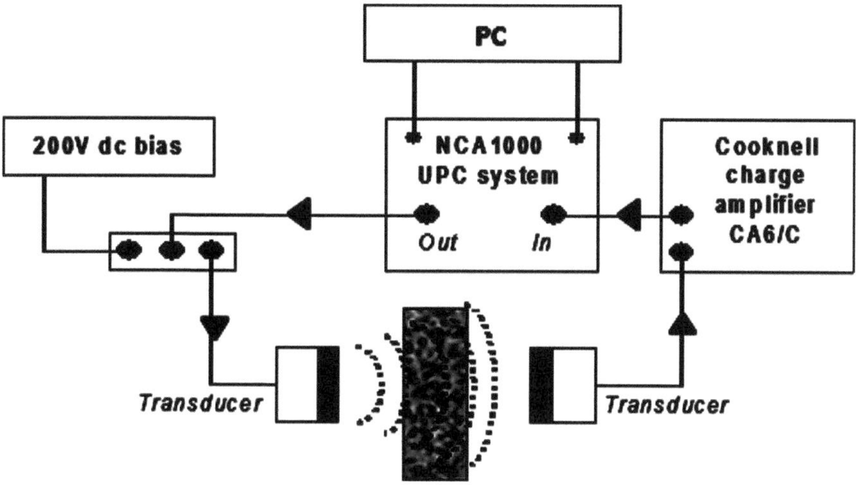

Fig. 11.13 Apparatus for the through-transmission air-coupled nondestructive evaluation (*NDE*) of solid samples

sides of the sample, and a signal is passed from one transducer to the other through the sample. The two transducers can be used as a pair for scanning the sample, with data collected by the receiver. In such cases, it is convenient to use an ultrasonic PuC system, as illustrated in Chap. 4 and reviewed in the previous sections, sending chirp-shaped voltage drive signals to the source and performing cross-correlations once the signals are recorded by the receiver. Note that most air-coupled experiments are performed in a through-transmission geometry. This is mainly because of the large reflection at the air/sample interface which would be detected in a pulse–echo system. In addition, alignment is much more difficult in pulse–echo systems, due to the large velocity change, and hence refraction, at the interface. Conversely, this problem is not so prevalent in through-transmission.

In Fig. 11.14, images obtained according to the illustrated procedure, that is, chirp excitation and pulse compression, are reported. In this experiment, an artificial square defect of 25.4 mm width was embedded within a carbon fiber plate. Images were produced using a chirp signal and cross-correlation and show variations across the sample in terms of changes in amplitude, illustrated in plot (a), and ToF (phase), as illustrated in (b). The defect was clearly identified.

This technique can also be employed to detect impact damage. In many cases, impact damage appears as a small indentation on one surface and would be barely visible, nevertheless there may be much more extensive damage within the structure. This is a particular problem with carbon fiber composites. Here, the advantage of air coupling is that the sample can be tested without a coupling liquid—an advantage for instance in testing ‘prepreg’ material (the layers that make up the composite before final curing under pressure). Figure 11.15 shows scans of impact-damaged samples that have been obtained using air-coupled ultrasound, and the results compared to

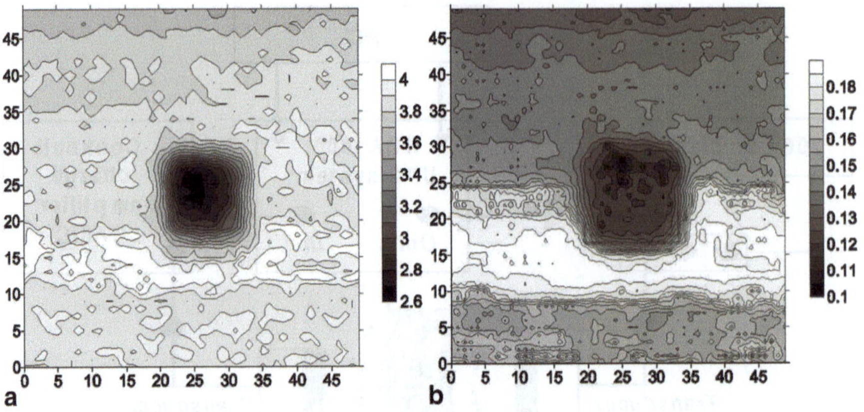

Fig. 11.14 Air-coupled ultrasonic images of apparatus for the through-transmission air-coupled NDE of solid samples

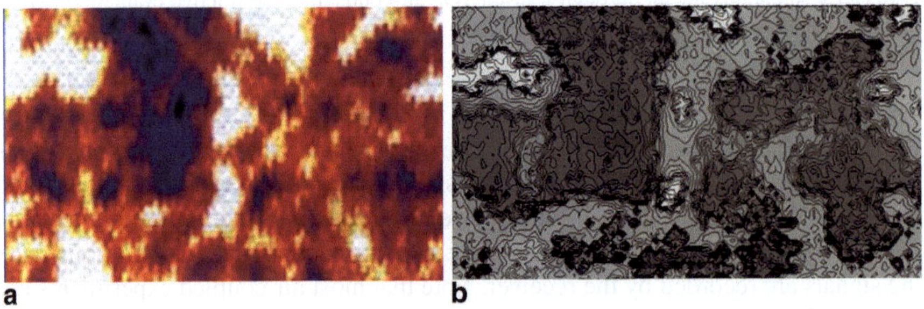

Fig. 11.15 Images of complex damage in a carbon fibre reinforced plate caused by a barely visible impact damage. Each image represents a 100 mm × 50 mm area of the composite, using: **a** a standard focused C-scan water immersion test at 10 MHz; and **b** an air-coupled focused scan at 500 kHz

those using C-scans in water immersion. It can be seen that both techniques give very similar results.

In addition, it is certainly relevant to show the imaging result achieved by using an LChirp and a *nonlinear chirp* (NLChirp) signal. The surface image of a carbon fiber composite for the two different coded signals is depicted in Fig. 11.16.

In this case, the NLChirp capability, which has been described in depth previously, emerges clearly.

11.4.2 Concrete Inspection

Despite the highly scattering nature of some industrial materials, air-coupled ultrasound can still be used, and in fact can result in some very interesting measurements.

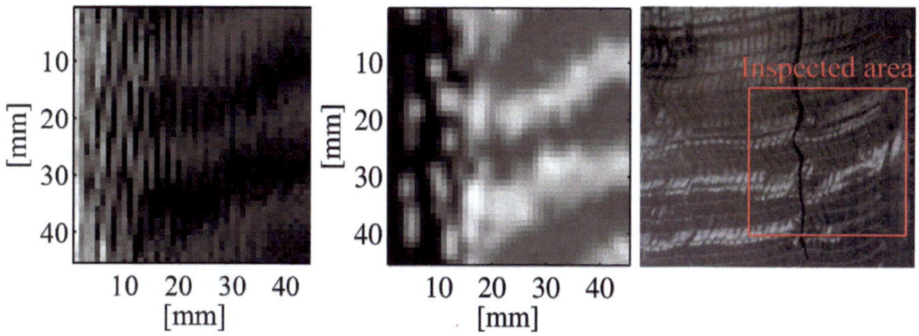

Fig. 11.16 Imaging results on a carbon fiber composite sample by using an LChirp (*left*) and NLChirp (*center*) excitation signal. A photo of the inspected area is shown beside (*right*)

Specifically, by a series of experiments, it has been found that a center frequency of 400 kHz, together with a chirp bandwidth of 500 kHz, gives a good result for many such industrial samples. An example is concrete, where the ultrasonic wavelength is approximately 10 mm at this center frequency. It is known that acoustic velocity (speed of sound) varies with the weight ratio between the added aggregate and the cement paste. Measurements of this are shown in Fig. 11.17, obtained using both a noncontacting air coupled system, and a more traditional measurement with contacting resonant transducers at 37 kHz. It is immediately evident that the two techniques measure different sound speeds, but that in both cases, the speed increases with a higher proportion of aggregate. The different sound speed values measured by air coupling could be due to the fact that it provides a more ready coupling with the cement paste phase. Yet, it may also be caused by dispersion, with the slower speeds seen for contact measurements being possibly introduced by the much lower frequency being used in this case (by approximately a factor of 10). The use of pulse compression allowed good sensitivity to be obtained at higher frequencies using air-coupled ultrasound, which is particularly useful for imaging applications (see below).

Note that the good coupling to the cement paste has been found to be a good way of measuring the cure of concrete over time, caused primarily by hydration of the cement [42]. The use of a noncontact technique also allows such measurements to be performed immediately after casting. It is also possible to use air-coupled ultrasound to study the effect of moisture content (expressed as relative humidity) on the measured sound speed in concrete. This has been done for five different mixes of concrete, with the proportions shown in Table 11.2.

The results are shown in Fig. 11.18. It can be seen again that the speed of sound is measured as a higher value when tests using contact transducers at 37 kHz are compared to those using air-coupled techniques. It is also evident that, for a given mix, the sound speed increases with the relative humidity, potentially giving a measurement of the latter if the initial mix is known.

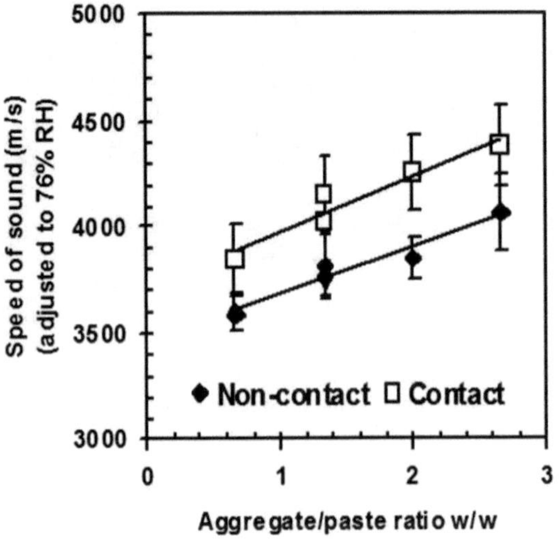

Fig. 11.17 Results of speed of sound vs. aggregate/paste ratio (ratio by weight or w/w) measurement in concrete, for both air-coupled and contact transducers. The results are adjusted to a constant relative humidity of 76 %, as the speed of sound varies with water content. Error bars are ±1 standard deviation

Table 11.2 Relative target ratios by weight for the concrete samples used in Fig. 11.17

Mix no.	Cement	Water	10 mm aggregate	Sand
1	1	0.5	1	0
2	1	0.5	2	0
3	1	0.5	0	2
4	1	0.5	1	2
5	1	0.5	2	2

It is also possible to image the internal structure of concrete using air-coupled ultrasound in through-transmission. An example is the detection of reinforcement bars. The complication here is that the aggregate within conventional concrete is highly scattering of ultrasound, a fact that conventionally limits imaging within this material. This is illustrated in Fig. 11.19, which shows the detection of the 10 mm diameter reinforcement bar as the vertical dark line in both images [8]. In Fig. 11.19a, there is no aggregate present, and the reinforcement bar is detected well. However, the presence of 10 mm stone aggregate, as in Fig. 11.19b, causes the signal to scatter to a much greater degree, as it has a size distribution which overlaps that of the reinforcement bar. This produces a greater masking effect.

11.4.3 Imaging of Surfaces and Thin Materials

It is possible to modify the field radiated from capacitive devices in air, using either a curved radiating surface or via an external mirror, to produce focusing and hence higher-resolution images. Curved surfaces have to be chosen for the easy application of the membrane. Two examples are shown in Fig. 11.20 [43]. The first one is a

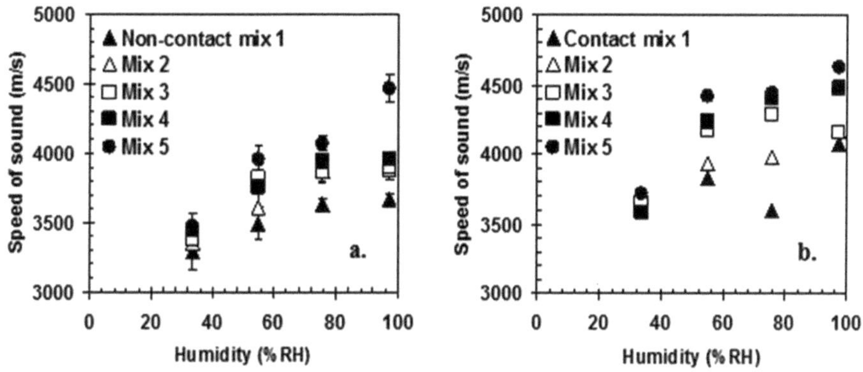

Fig. 11.18 Graphs of ultrasonic longitudinal wave speed vs. storage humidity, using: (*left*) the air-coupled system; and (*right*) a contact piezoelectric unit. Error bars are ±1 standard deviation

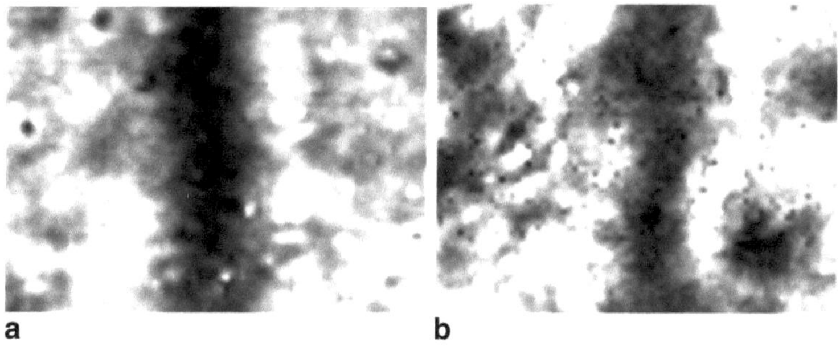

Fig. 11.19 Air-coupled ultrasonic images of embedded 10 mm diameter steel reinforcement bars. **a** Image for a cement paste sample (i.e., with sand but no stone aggregate) at a center frequency of 500 kHz. **b** The setup is as in (**a**), but now with the sample being concrete containing 10 mm stone aggregate. Scan areas are approximately 60 mm × 40 mm

cylindrical device focusing in one linear dimension with a limited depth of field, but not in the other, as in pane (a). Conversely, in the second example, a conical surface gives a much greater depth of field, as in pane (b), as seen in conventional medical devices for operation in water [31].

An alternative approach is to use a focusing external off-axis parabolic mirror [37, 46], as shown in Fig. 11.20c. Using such devices, spot sizes of less than 1 mm can be achieved, leading to the possibility of high resolution scanning. The use of such focusing devices allows thin materials to be inspected. This is because, in through transmission, the amplitude of the signal can be detected at closely-spaced intervals. In effect, the amplitude of transmission measures the local properties of the sample, so that scanning a pair of transducers over the surface allows these variations to be mapped over a finite area [18]. Here, a similar apparatus to that shown in Fig. 11.13 can be used to image thin paper samples, but with the source replaced by a focused device. An example of this is shown in Fig. 11.21, where the structure within a

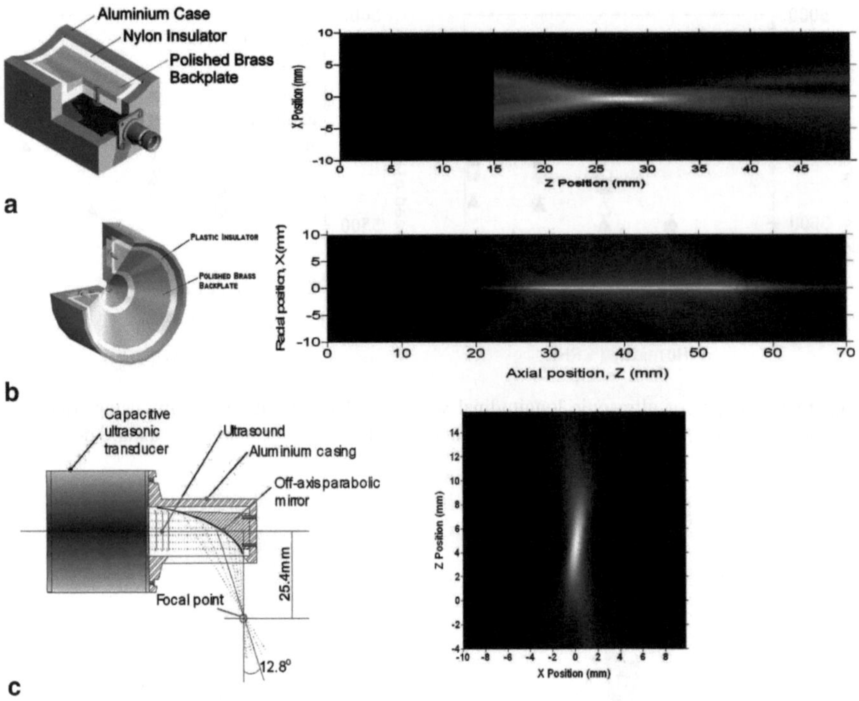

Fig. 11.20 Schematic diagrams of cylindrical (**a**) and conical (**b**) capacitive air-coupled transducers, each of which focuses in air with the amplitude distribution shown in gray scale on the right. **c** A capacitive transducer fitted with an off-axis parabolic mirror

sample of writing paper can be imaged using focused air-coupled ultrasound, to produce a good facsimile of the optical image seen under natural illumination. In effect, the ultrasound is measuring changes in density of the paper. This technology has application to the detection of counterfeit banknotes. The mirror arrangement can also be used in pulse–echo to image surfaces. An example of a coin is shown in Fig. 11.21c, indicating that good resolution is obtainable.

Good imaging result of a sample surface can be also achieved by using conventional unfocused transducer. As an interesting example, the imaging results on a carved wood sample gained by evaluating the ToF and the transmitted energy in through-transmission mode are here depicted in Fig. 11.22.

11.4.4 Food Quality Inspection

There are many types of material that can be inspected using air-coupled ultrasound. An interesting example is the detection of contaminants in food materials. Ultrasound is sensitive to small changes in density and/or acoustic properties; hence, it is

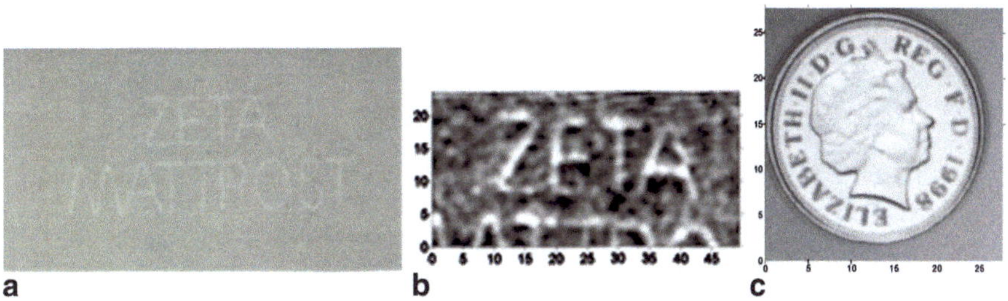

Fig. 11.21 Through transmission images of high quality writing paper, obtained using: **a** visible light; and **b** focused air-coupled ultrasound. **c** An image of a coin produced in pulse–echo mode in air

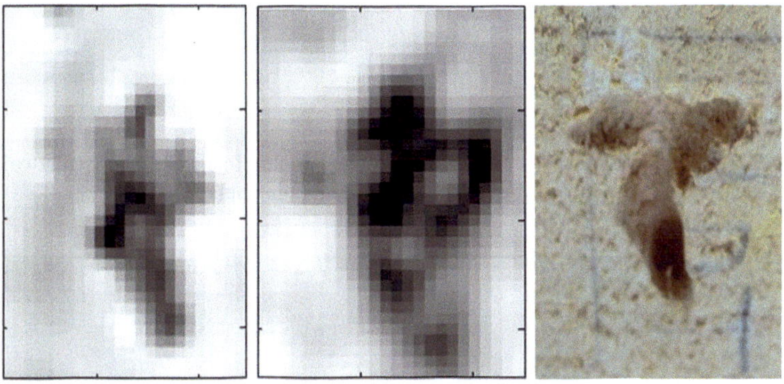

Fig. 11.22 Imaging result on a carved wood sample by evaluating the time of flight (*left*) and the transmitted energy (*center*). A photo of the inspected area is shown beside (*right*)

a candidate used to locate contaminants that would not be detectable in certain food materials using conventional X-ray and other imaging methods [10] and metal detector [21] techniques. These include unwanted pieces of glass, polymers, and wood, which are not electrically conducting and which do not easily show up under X-ray illumination. The detection of contaminants is simpler to undertake using ultrasound if the food material itself is fairly homogenous. In addition to such measurements, it is also possible to measure the consistency of food, and whether any deliberate additives are present [40]. Consider the case of chocolate. This is a homogenous material, which is a good transmitter of ultrasound at frequencies below 1 MHz. A scanning rig can be used for the air-coupled imaging of chocolate, using two 10 mm diameter capacitive transducers. These are scanned over a block of chocolate, and the internal structure is revealed. The results obtained for two different samples are shown in Fig. 11.23 for: (a) a solid chocolate sample; and (b) one containing nuts. In the uniform chocolate sample, the imaging system has been able to identify the individual squares of chocolate. In addition, while the resolution of these images is somewhat limited by the 10 mm aperture of the imaging system, it is clear that the added nuts have been successfully detected in the image to the right as the darker

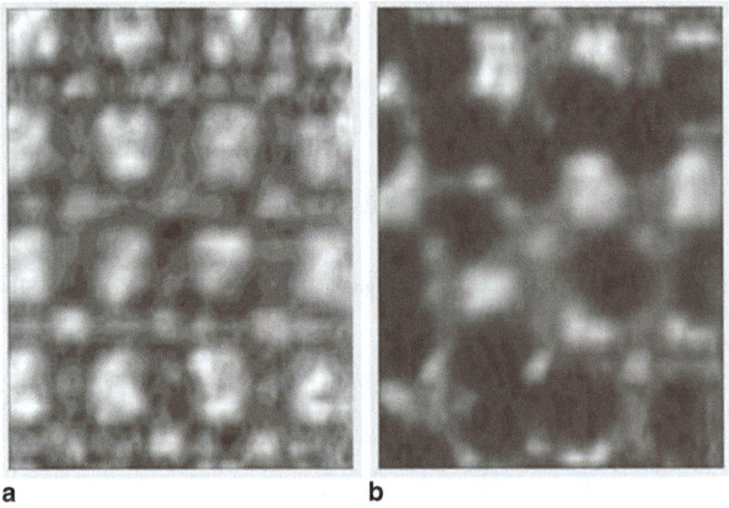

Fig. 11.23 Air-coupled ultrasonic imaging of chocolate samples. Images are shown for two different samples, namely: **a** chocolate only; and **b** chocolate containing nuts

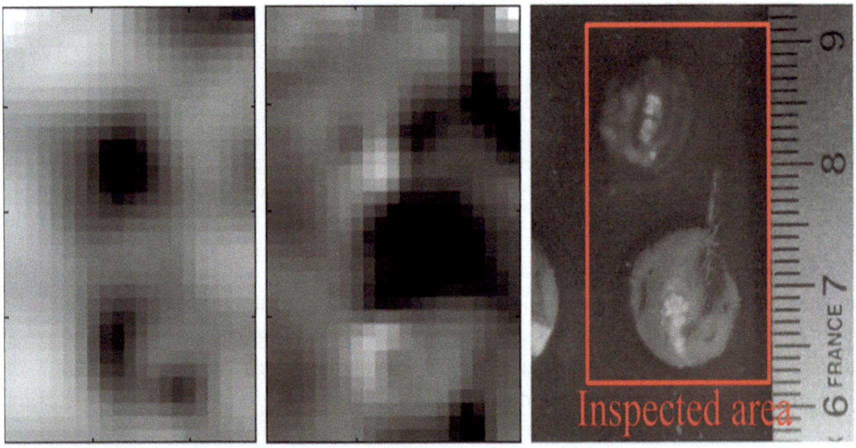

Fig. 11.24 Chocolate with hazelnuts—transmitted energy (*left*) and time of flight (*right*) imaging results. A photo of the inspected area is shown beside (*right*)

areas (indicating a lower transmission amplitude). This allows the spatial distribution of the nuts to be estimated—a valuable tool for process control.

Other example of ultrasonic imaging for chocolate samples are shown in Figs. 11.24 and 11.25.

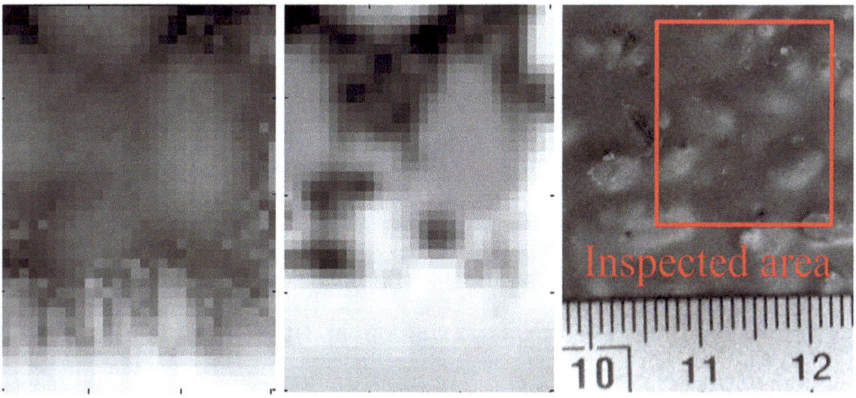

Fig. 11.25 Chocolate with rice—transmitted energy (*left*) and time of flight (*center*) imaging results. A photo of the inspected area is shown beside (*right*)

11.5 Conclusion

It has been shown that noncontact techniques are of interest in many NDE applications, ranging from imaging industrial samples to the quality inspection of food. Throughout, the signal to noise ratio is a challenge, and for these reasons modulation and pulse compression has been found to be a very valuable tool in processing signals and obtaining measurements.

References

1. Álvarez-Arenas TEG (2004) Acoustic impedance matching of piezoelectric transducers to the air. IEEE Trans Ultrason Ferroelectr Freq Control 51(5):624–633
2. Anastasi RF, Madaras EI (1999) Pulse compression techniques for laser generated ultrasound. In: Proceedings of the 1999 IEEE Ultrasonics Symposium, IEEE, vol 1, pp 813–817
3. Arca A, Aylott J, Marques L, Clark M, Somekh M, Smith R, Sharples S, Stratoudaki T, Chen X (2011) CHOTs optical transducers. Nondestruct Test Eval 26(3–4):353–366
4. Balogun O, Murray TW (2011) Frequency domain photoacoustics using intensity-modulated laser sources. Nondestruct Test Eval 26(3–4):335–351
5. Bass HE, Shields FD (1974) Vibrational relaxation and sound absorption in O_2/H_2O mixtures. J Acoust Soc Ama 56(3):856–859
6. Bass HE, Sutherland LC, Piercy J, Evans L (1984) Absorption of sound by the atmosphere. In: Physical acoustics: principles and methods, vol 17. Academic, Orlando, pp 145–232
7. Bauer S, Gerhard-Multhaupt R, Sessler GM (2004) Ferroelectrets: soft electroactive foams for transducers. Phys Today 57(2):37–43
8. Berriman JR, Hutchins DA, Neild A, Gan TH, Purnell P (2006) The application of time-frequency analysis to the air-coupled ultrasonic testing of concrete. IEEE Trans Ultrason Ferroelectr Freq Control 53(4):768–776

9. Blomme E, Bulcaen D, Declercq F, Deveugele J, Lust P (2008) Air-coupled ultrasonic pitch-catch method for material inspection. Emerging technologies in NDT. Taylor & Francis, London, p 137
10. Brosnan T, Sun DW (2004) Improving quality inspection of food products by computer vision — a review. J Food Eng 61(1):3–16
11. Buckley J (1998) Principles and applications of air-coupled ultrasonics. Insight 40(11):755–759
12. Budišin SZ (1991) Efficient pulse compressor for Golay complementary sequences. Electron Lett 27(3):219–220
13. Carr H, Wykes C (1993) Diagnostic measurements in capacitive transducers. Ultrasonics 31(1):13–20
14. Cohn M, Lempel A (1977) On fast M-sequence transforms (corresp.). IEEE Trans Inf Theory 23(1):135–137
15. Dabirikhah H, Turner CW (1994) Leaky plate wave airborne ultrasonic transducer. Electron Lett 30(18):1549–1550
16. Durris L, Goujon L, Pelourson A, Gonnard P, Brissaud M, Richard C (1996) Airborne ultrasonic transducer. Ultrasonics 34(2):153–158
17. Evans LB, Bass HE, Sutherland LC (1972) Atmospheric absorption of sound: theoretical predictions. J Acoust Soc Am 51(5B):1565–1575
18. Gan TH, Hutchins DA, Billson DR, Schindel DW (2003) High-resolution, air-coupled ultrasonic imaging of thin materials. IEEE Trans Ultrason Ferroelectr Freq Control 50(11):1516–1524
19. Garcia-Rodriguez M, Yañez Y, Garcia-Hernandez MJ, Salazar J, Turo A, Chavez JA (2010) Application of Golay codes to improve the dynamic range in ultrasonic lamb waves air-coupled systems. NDT&E Int 43(8):677–686
20. Grandia WA, Fortunko CM (1995) NDE applications of air-coupled ultrasonic transducers. In: Proceedings of the 1995 IEEE Ultrasonics Symposium, IEEE, vol 1, pp 697–709
21. Graves M, Smith A, Batchelor B (1998) Approaches to foreign body detection in foods. Trends Food Sci & Technol 9(1):21–27
22. Gudra T, Opieliński KJ (2002) Influence of acoustic impedance of multilayer acoustic systems on the transfer function of ultrasonic airborne transducers. Ultrasonics 40(1):457–463
23. Haller MI, Khuri-Yakub BT (1994) Micromachined 1-3 composites for ultrasonic air transducers. Rev Scientific Instrum 65(6):2095–2098
24. Hernandez-Valle F, Dixon S (2011) Pulsed electromagnet EMAT for ultrasonic measurements at elevated temperatures. Insight-Non-Destruct Test Cond Monit 53(2):96–99
25. Hietanen J, Stor-Pellinen J, Luukkala M, Mattila P, Tsuzuki F, Sasaki K (1993) A Helmholtz resonator model for an electrostatic ultrasonic air transducer with a V-grooved backplate. Sensors Actuators A Phys 39(2):129–132
26. Hirao M, Ogi H (2003) EMATs for science and industry: noncontacting ultrasonic measurements. Springer
27. Ho KS, Gan TH, Billson DR, Hutchins DA (2005) Application of pulse compression signal processing techniques to electromagnetic acoustic transducers for noncontact thickness measurements and imaging. Rev Scientific Instrum 76(5):054902
28. Hutchins DA, Schindel DW, Bashford AG, Wright WMD (1998) Advances in ultrasonic electrostatic transduction. Ultrasonics 36(1):1–6
29. Hutchins DA, McIntosh JS, Neild A, Billson DR, Noble RA (2003) Radiated fields of capacitive micromachined ultrasonic transducers in air. J Acoust Soc Am 114(3):1435–1449
30. Hutchins D, Burrascano P, Davis L, Laureti S, Ricci M (2014) Coded waveforms for optimised air-coupled ultrasonic nondestructive evaluation. Ultrasonics
31. Ju KC, Tseng LT, Chen YY, Lin WL (2006) Investigation of a scanned cylindrical ultrasound system for breast hyperthermia. Phys Med Biol 51(3):539
32. Krauss O, Gerlach R, Fricke J (1994) Experimental and theoretical investigations of SiO_2-aerogel matched piezo-transducers. Ultrasonics 32(3):217–222

33. Lee BB, Furgason ES (1983) High-speed digital Golay code flaw detection system. Ultrasonics 21(4):153–161
34. Manthey W, Kroemer N, Magori V (1992) Ultrasonic transducers and transducer arrays for applications in air. Meas Sci Technol 3(3):249
35. Monchalin JP (1985) Optical detection of ultrasound at a distance using a confocal Fabry–Perot interferometer. Appl Phys Lett 47(1):14–16
36. Murray TW, Baldwin KC, Wagner JW (1997) Laser ultrasonic chirp sources for low damage and high detectability without loss of temporal resolution. J Acoust Soc Am 102(5):2742–2746
37. Neild A, Hutchins DA, Robertson TJ, Schindel DW (2004) A model for the radiated field of a plane piston after reflection from a curved surface. J Acoust Soc Am 116(5):2793–2801
38. Nowicki A, Trots I, Lewin PA, Secomski W, Tymkiewicz R (2007) Influence of the ultrasound transducer bandwidth on selection of the complementary Golay bit code length. Ultrasonics 47(1):64–73
39. Pallav P, Gan TH, Hutchins DA (2007) Elliptical-Tukey chirp signal for high-resolution, air-coupled ultrasonic imaging. IEEE Trans Ultrason Ferroelectr Freq Control 54(8):1530–1540
40. Pallav P, Hutchins DA, Gan TH (2009) Air-coupled ultrasonic evaluation of food materials. Ultrasonics 49(2):244–253
41. Pizarro L, Certon D, Lethiecq M, Hosten B (1999) Airborne ultrasonic electrostatic transducers with conductive grooved backplates: tailoring their centre frequency, sensitivity and bandwidth. Ultrasonics 37(7):493–503
42. Purnell P, Hutchins D, Berriman J, Gan TH (2004) Testing without touching: advances in ultrasonic concrete interrogation. Concrete 38(5):10–11
43. Robertson TJ, Hutchins DA, Billson DR (2002) Capacitive air-coupled cylindrical transducers for ultrasonic imaging applications. Meas Sci Technol 13(5):758
44. Satyanarayan L, Vander Weide JM, Declercq NF, Berthelot Y (2009) Investigation of a novel polymer foam material for air coupled ultrasonic transducer applications. In: Ultrasonic wave propagation in non homogeneous media. Springer, Berlin, pp 329–340
45. Schiller S, Hsieh CK, Chou CH, Khuri-Yakub BT (1990) Novel high-frequency air transducers. In: Review of progress in quantitative nondestructive evaluation. Springer, Berlin, pp 795–798
46. Schindel DW (1998) Ultrasonic imaging of solid surfaces using a focussed air-coupled capacitance transducer. Ultrasonics 35(8):587–594
47. Schindel DW, Hutchins DA (1995) Applications of micromachined capacitance transducers in air-coupled ultrasonics and nondestructive evaluation. IEEE Trans Ultrason Ferroelectr Freq Control 42(1):51–58
48. Schindel DW, Hutchins DA, Zou L, Sayer M (1995) The design and characterization of micromachined air-coupled capacitance transducers. IEEE Trans Ultrason Ferroelectr Freq Control 42(1):42–50
49. Sharples SD, Clark M, Somekh M (2001) Efficient and flexible laser ultrasound generation using spatial light modulators. Electron Lett 37(18):1145–1146
50. Sharples SD, Clark M, Smith RJ, Ellwood RJ, Li W, Somekh MG (2011) Laser ultrasonic microscopy. Nondestruct Test Eval 26(3–4):367–384
51. Stan GB, Embrechts JJ, Archambeau D (2002) Comparison of different impulse response measurement techniques. J Audio Eng Soc 50(4):249–262
52. Stoessel R, Krohn N, Pfleiderer K, Busse G (2002) Air-coupled ultrasound inspection of various materials. Ultrasonics 40(1):159–163
53. Strycek JO, Loertscher HP (2000) High speed large area scanning using air-coupled ultrasound. In: Proceedings of the 15th World Conference on Nondestructive Testing (WCNDT), Rome
54. Taylor GS, Hutchins DA, Edwards C, Palmer SB (1990) TEA-CO_2 laser generation of ultrasound in non-metals. Ultrasonics 28(6):343–349
55. White JDH, Challis RE (1992) A Golay sequencer based NDT system for highly attenuating materials. In: IEE Colloquium on Non-Contacting and Remote NDT, IET, pp 1–7

Chapter 12
Ultrasonic Frequency Modulated Continuous Wave for Range Estimation

Luigi Battaglini, Pietro Burrascano, Marco Ricci and Luca Senni

Abstract As an alternative implementation of the pulse-compression (PuC) technique, the frequency modulated continuous wave (FMCW) approach is here presented as a method to perform time-of-flight (ToF) measurements by relaxing both hardware and software resources with respect to the standard PuC procedure based on the application of the *matched filter*. Precisely, by using linear chirps as excitations and by replacing the cross-correlation step between input and output signals with a mixing of the two waveforms, it is possible to retrieve the envelope $\mathrm{Env}[h](t)$ of the system impulse response by performing frequency analysis of the mixed signal. It will be shown that by exploiting this procedure, it is possible to attain a high precision in the measurement of ToF values also by using sampling intervals larger than the achievable temporal resolution. As a consequence, also the computational resources can be significantly reduced with respect to those requested by the standard cross-correlation approaches. Both numerical and experimental analyses are reported to illustrate this method and also some processing techniques are introduced and compared to further improve the ToF resolution.

12.1 Introduction

The evaluation of the time-of-flight (ToF) and signal amplitude attenuation in a nondestructive testing (NDT) application implies estimating the impulse response $h(t)$ of the sample under test (SUT). To this aim, the pulse-echo (PuE) method can be applied, which can be considered a direct application of the definition of impulse

L. Battaglini (✉) · P. Burrascano · M. Ricci · L. Senni
Department of Engineering, Polo Scientifico Didattico di Terni, University of Perugia, Terni, Italy
e-mail: luigi.battaglini@gmail.com

P. Burrascano
e-mail: pietro.burrascano@unipg.it

M. Ricci
e-mail: marco.ricci@unipg.it

L. Senni
e-mail: luca.senni@unipg.it

P. Burrascano et al. (eds.), *Ultrasonic Nondestructive Evaluation Systems*,
DOI 10.1007/978-3-319-10566-6_12

response: $h(t)$ is estimated by exciting the sample with a short pulse $\tilde{\delta}(t)$; the system response $\tilde{h}(t) = [\tilde{\delta} * h](t)$, where $*$ indicates convolution, is a good estimate of $h(t)$ if the exciting signal $\tilde{\delta}(t)$ is a good approximation of the Dirac delta.

Some intrinsic limitations of PuE method can be overcome by using the pulse-compression (PuC) method (see [10, 18] and Chap. 4 for a detailed review) where the system is excited by a wide-band coded signal $s(t)$ and $h(t)$ is estimated by convolving the corresponding output $y(t) = [s * h](t)$ with the so-called matched filter $\Psi(t)$ [4, 13, 14]. If $[s * \Psi](t)$ approximates the Dirac delta, $[s * \Psi](t) \approx \hat{\delta}(t)$ and then $\hat{h}(t) = [y * \Psi](t) = [s * h * \Psi](t) = [\hat{\delta} * h](t)$. The PuC method can be effectively exploited to estimate the impulse response in several ultrasonic NDT procedures, either in the cases requiring high SNR values or in those where high propagation attenuation may become a problem [3, 5, 19–21].

A further approach has been proposed in the technical literature to overcome the limitations of the PuE method [1]: the frequency modulated continuous wave (FMCW) approach. It makes use of self-interference [2], phase analysis, and application-tailored schemes [6]. In the present chapter, the possibility to pursue the FMCW approach in ultrasonic applications is analyzed and its efficiency and accuracy are compared with the standard PuE method. In the technical literature, a few applications of this methods to ultrasonic systems are reported, mainly for range measurement [12]; nevertheless in the past years, several schemes have been proposed that rely on this procedure both for diagnostic and imaging applications [8, 9, 11]. In the present chapter, the effectiveness of the FMCW is compared with the standard PuC scheme based on the matched filter with the aim of applying FMCW to the ToF measurement task. In the following, the basic theory underlying the method and the characteristic signals are reported. Numerical simulations, worked examples, and a comparison between FMCW and PuC are reported and also the use of some processing such as chirp z-Transform (CZT) can be exploited to further improve the technique.

12.2 Basic Principle, Processing Protocol, and Resolution Analysis

In an FMCW procedure, an ultrasonic transmitting transducer (TX) is driven with a sinusoidal chirp whose frequency, during a time period T, sweeps linearly between a start frequency f_0 and a stop frequency f_1. The ultrasonic generated field can propagate both in air (air-coupled ultrasound) or in solid materials (contact ultrasound); in the presence of discontinuities, defects, or isolated objects, the acoustic wave is reflected and turns back to a receiving transducer (RX). At the receiving stage, the output signal is mixed with the input one: this causes that, due to this frequency difference related to the propagation delay, signals with different instantaneous frequencies are multiplied, and a low frequency component becomes evident, due to the frequency difference. A schematic representation of the protocol is reported in Fig. 12.1.

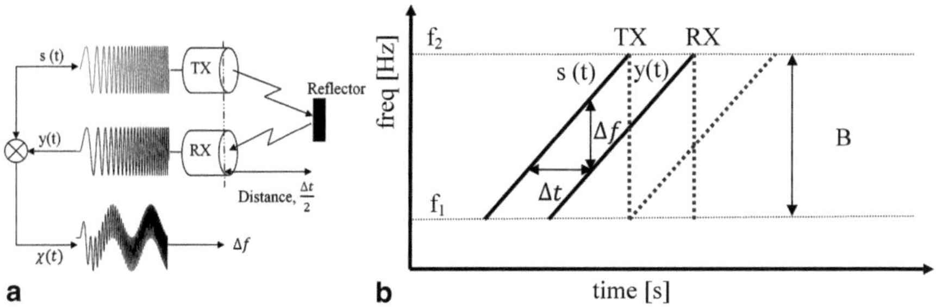

Fig. 12.1 Schematic representation of the FMCW ultrasonic radar working principle

To gain insight about the method, let us consider the easiest model of the FMCW: a simple time delay between input and output signals. The input signal $s(t)$ exhibits a periodic instantaneous frequency $f_{\text{inst}}(t)$ that changes linearly over time as:

$$f_{\text{inst}}(t) = f_0 + \frac{B}{T} \cdot t = f_0 + k_f \cdot t \qquad \text{for } 0 \leq t \leq T \tag{12.1}$$

where T is the time duration of the signal, $k_f(t) = \frac{B}{T}$, and B is the sweep bandwidth.

The ToF due to the propagation over a distance d can be evaluated from the geometrical characteristic as:

$$\Delta t = 2 \cdot \frac{d}{c}$$

where d is the distance between the source and the reflector and c is the ultrasonic wave velocity.

In this simplified model, the received output signal $y(t)$ can be assumed to be a time shifted replica of $s(t)$, and its instantaneous frequency $f_{\text{out}}(t)$ is equal to:

$$f_{\text{out}}(t) = f_0 + k_f \cdot (t - \Delta t) \qquad \text{for } \Delta t \leq t \leq T + \Delta t. \tag{12.2}$$

As illustrated in the time-frequency plot of Fig. 12.1, the difference between the two instantaneous frequencies, $\Delta f(t)$, is constant and equal to:

$$f(t) = f_{\text{out}}(t) - f_{\text{in}}(t) = k_f \cdot \Delta t = \frac{B}{T} \cdot 2 \cdot \frac{d}{c}\,. \tag{12.3}$$

Due to such frequency shift, multiplying the input and output signals leads to the presence of a low-frequency oscillation into the resultant product $\chi(t) = y(t) \cdot s(t)$; the frequency Δf of this low frequency oscillation is related to the ToF by the expression:

$$ToF = \frac{T}{B} \Delta f\,. \tag{12.4}$$

If we consider a more general model of the propagation process, the output signal can be regarded as the convolution between the input $s(t)$ and the system impulse response: $y(t) = [s * h](t)$. In this case, the mixed signal $\chi(t)$ is the superposition of low-frequency interference terms corresponding to the not vanishing values of $h(t)$. In this case, it can be found that the magnitude of the Fourier transform of $\chi(t)$ approximates the envelope of $h(t)$ as long as the chirp input signal $s(t)$ satisfies very general constraints.

It is worth highlighting that the great advantage of the technique is summarized in Eq. (12.4), which shows how it is possible to measure extremely short ToFs by measuring low-frequency signals. In several applications such as optics, radar, etc., the ToFs can be so short that no electronic device is able to directly measure them; in these cases, the FMCW approach is therefore an effective (indeed sometimes the best) strategy to retrieve information about the system impulse response, as for the swept optical coherence tomography. For ultrasonic nondestructive evaluation (NDE) applications, the FMCW technique can be exploited to develop high-frequency devices while relaxing the hardware requirements. For instance, the input signal $s(t)$ can be generated by a voltage controlled oscillator (VCO) or by other FM-generators; the $\chi(t)$ signal can be obtained by using an analog mixer; and then a low sampling-rate analog-to-digital converter (ADC) can be employed to represent the low frequency oscillation. At the same time, this technique can be used to relax the computational costs of standard PuC procedures, especially when a high throughput of data is required as in imaging applications. In this chapter, the application of FMCW to ultrasonic NDE is considered and numerical simulations are performed in order to evidence some fundamental aspects and to optimize the expected ToF resolution.

12.2.1 Chirp Signal, Expected Signal, and Resolution

In order to explicitly evaluate $\chi(t)$ and to demonstrate the efficiency of the method for ToF measurements, further mathematical details about the chirp signal must be given.

The use of chirp signals is motivated by a number of notable features: constant envelope, δ-like autocorrelation and easy bandwidth control. Mathematically, the most general chirp signal is defined as

$$s(t) = \mathrm{Re}\,[\mathrm{e}^{2\pi \mathrm{i}(f_c t + \frac{B}{2}\int_0^t x(\tau)\,\mathrm{d}\tau)}] = \mathrm{Re}\,[\mathrm{e}^{\mathrm{i}\Phi(t)}] \tag{12.5}$$

where: $f_c = \frac{f_0+f_1}{2}$ is the central frequency; B is the bandwidth; Re stands for the real part operator; $x(t)$ is a monotonically increasing, smooth modulating signal taking values in $[-1, 1]$ and $\Phi(t)$ is the accumulated phase [16]. The instantaneous frequency of the chirp signal is obtained by differentiating the accumulated phase as

$$f_{\mathrm{inst}}(t) = \frac{1}{2\pi}\frac{\mathrm{d}\Phi(t)}{\mathrm{d}t} \,. \tag{12.6}$$

Of course, depending on the behavior of $\Phi(t)$, the instantaneous frequency can be continuous or not. Furthermore, it can be linear or not, for instance taking a profile that is quadratic, logarithmic, etc. In this chapter, only the linear chirp case is considered for which the phase has a quadratic dependence on t:

$$\Phi(t) = 2\pi(f_c t + \frac{B}{2T}t^2 - \frac{B}{2}t) = 2\pi(f_0 t + \frac{B}{2T}t^2). \tag{12.7}$$

Therefore, the input and output signals are

$$\begin{aligned} s(t) &= \mathrm{Re}\left[\mathrm{e}^{\mathrm{i}2\pi(f_0 t + \frac{B}{2T}t^2)}\right] \\ y(t) &= \mathrm{Re}\left[\mathrm{e}^{\mathrm{i}2\pi[f_0(t-\Delta t) + \frac{B}{2T}(t-\Delta t)^2]}\right] \end{aligned} \tag{12.8}$$

respectively. With this, the mixed signal $\chi(t)$ is:

$$\begin{aligned} \chi(t) = y(t)\cdot s(t) = &\frac{1}{2}\mathrm{Re}\left\{\mathrm{e}^{-\mathrm{i}2\pi[(f_c-\frac{B}{2})\Delta t - \frac{B}{2T}\Delta t^2]}\cdot \mathrm{e}^{\mathrm{i}2\pi\frac{B}{T}t\Delta t}\right\} \\ &+\frac{1}{2}\mathrm{Re}\left\{\mathrm{e}^{\mathrm{i}4\pi[(f_c-\frac{B}{2})t+\frac{B}{2T}t^2]}\cdot \mathrm{e}^{\mathrm{i}2\pi[-(f_c-\frac{B}{2})\Delta t+\frac{B}{2T}\Delta t^2]}\cdot \mathrm{e}^{-\mathrm{i}2\pi\frac{B}{T}t\Delta t}\right\}. \end{aligned} \tag{12.9}$$

The mixed signal $\chi(t)$ consists of two terms, a sinusoidal low-frequency signal proportional to $\cos[2\pi(\frac{B}{T}\Delta t\cdot t+\phi_0)]$ and a chirp signal approximately proportional to $\mathrm{Re}[\mathrm{e}^{2\pi\mathrm{i}\{2[(f_c-\frac{B}{2}-\frac{B}{2T}\Delta t)t+\frac{B}{2T}t^2]\}+\phi_1}]$, where ϕ_0 and ϕ_1 are suitable initial phases. The frequency of the slow sinusoidal term is proportional to the ToF, while the chirped term exhibits a frequency range with a center frequency almost twice as high as that of the input signal.

The sinusoidal term and the chirp signal can be made nonoverlapping in the frequency domain by means of a suitable choice of the excitation chirp parameters, and in particular of f_0 and B; in this way, the ToF can be unambiguously retrieved from the Fourier analysis of $\chi(t)$. In the following, the constraints on the possible values of f_0 and BW are detailed both theoretically and with the aid of numerical simulations; moreover, the ToF measurement resolution of the FMCW ultrasound radar is compared with the resolution provided by the matched filter-based PuC approach.

Before this thorough analysis, it is worth noting that the signal processing is performed in the digital domain with a well-defined acquisition rate, that we denote by f_{ADC}. For the sake of clarity, it is convenient to take into account the digital nature of signals and spectra and to state the constraints by scaling all the quantities to f_{ADC}.

To accomplish this comparison, a virtual experiment to evaluate the results of measuring a time delay Δt with the two techniques is considered. If the output and the mixed signal are both sampled at a rate f_{ADC}, by using the PuC method, a peak in the $\hat{h}(t)$ appears at the time Δt corresponding to the sample $n_t = \Delta t \cdot f_{\mathrm{ADC}}$. In the same way, if we analyze the discrete Fourier transform (DFT) of $\chi(t)$, the peak corresponding to the low-frequency interference term happens to be at:

$$\Delta f = \frac{B}{T}\cdot\Delta t \quad\Rightarrow\quad \frac{n_f\cdot f_{\mathrm{ADC}}}{N} = \frac{B}{T}\cdot\frac{n_t}{f_{\mathrm{ADC}}} = \frac{B}{\frac{N}{f_{\mathrm{ADC}}}}\cdot\frac{n_t}{f_{\mathrm{ADC}}} \tag{12.10}$$

where n_f is the sample index at which the peak is located in the DFT output, $\frac{1}{T} = \frac{f_{\text{ADC}}}{N}$ is the resolution of the DFT, and N is the number of acquired samples on which the DFT is practiced.

It is straightforward to see that the following relation holds:

$$n_f = \frac{B}{f_{\text{ADC}}} \cdot n_t. \tag{12.11}$$

This observation makes evident that the two techniques, for the same delay time Δt, produce a maximum for two different sample numbers. Since these quantities are linearly proportional to Δt itself, a difference in the ToF resolution of the two techniques emerges.

In particular, since in the standard PuC procedure, it always happens that $\frac{B}{f_{\text{ADC}}} < 1$, the resolution in the frequency domain is poorer than the one obtainable in the time domain. This does not represent a limitation for two main reasons: the first one is that the actual aim of the FMCW is to reduce the sampling rate by introducing hardware processing. In this case, sampling rates lower than B can be used depending on the required inspection range, i.e., the expected ToF value. The second reason is that the resolution in the frequency domain can be enhanced by exploiting proper processing techniques; in particular, by replacing the DFT with the CZT or by applying interpolation on the DFT results, a significant increase of resolution can be obtained even with respect the matched filter approach [7, 15, 17].

A typical example of the signals involved in an FMCW procedure is illustrated in the following. Figure 12.2 shows an input signal $s(t)$ with: $R = 100$ MHz, $f_0 = 3$ MHz, $B = 4$ MHz, $T = 1$ms, $N = 100{,}000$ and an output signal $y(t) = s(t - 1\,\mu\text{s})$ for the first 10 μs while the resulting mixed signal $\chi(t) = s(t)y(t)$ is represented in Fig. 12.3, where the superposition of a sinusoidal carrier and a chirp signal is evident. Finally, Fig. 12.4 shows a detail of the plot in Fig. 12.3.

12.2.2 Cyclic Evaluation and Inspection Range

The signals used in practical implementations are periodic, that is $s(t) = s(t \pm nT)$ and $y(t) = y(t \pm nT)$. For each period, there are therefore two distinct regions $\Gamma_1 = \{t : \Delta t \leq t \leq T\}$ and $\Gamma_2 = \{t : 0 \leq t \leq \Delta t\}$, for which the difference in the instantaneous frequencies are, respectively, $\Delta f_1 = \Delta f = \frac{B}{T}\Delta t$ and $\Delta f_2 = B - \Delta f_1$, as shown in Fig. 12.5. In these regions, we have two different interference terms in $\chi(t)$, corresponding to two different frequencies in the DFT of $\chi(t)$, whose amplitudes depend on the temporal extension of the ith Γ region. Of course for $\Delta t \ll T$, the Γ_1 region is significantly larger than Γ_2. When $\Delta t \to \frac{T}{2}$, the two terms tend to have the same weight. Moreover, when $\Delta t \to \frac{T}{2}$ the two frequency shifts become closer and closer, i.e., $\lim_{\Delta t \to \frac{T}{2}} \Delta f_2 = \Delta f_1 = \frac{B}{2}$. For $\Delta t > \frac{T}{2}$, Δf_2 becomes smaller than Δf_1 and the Γ_2 region larger than Γ_1. This sort of *aliasing* leads to an ambiguity in the ToF measurement, which has to be avoided by limiting the range of inspection to $\Delta t < \frac{T}{2}$ or, more generally, assuming that the impulse response duration is shorter than the sweep period: $h(t) = 0$ for $t > \frac{T}{2}$.

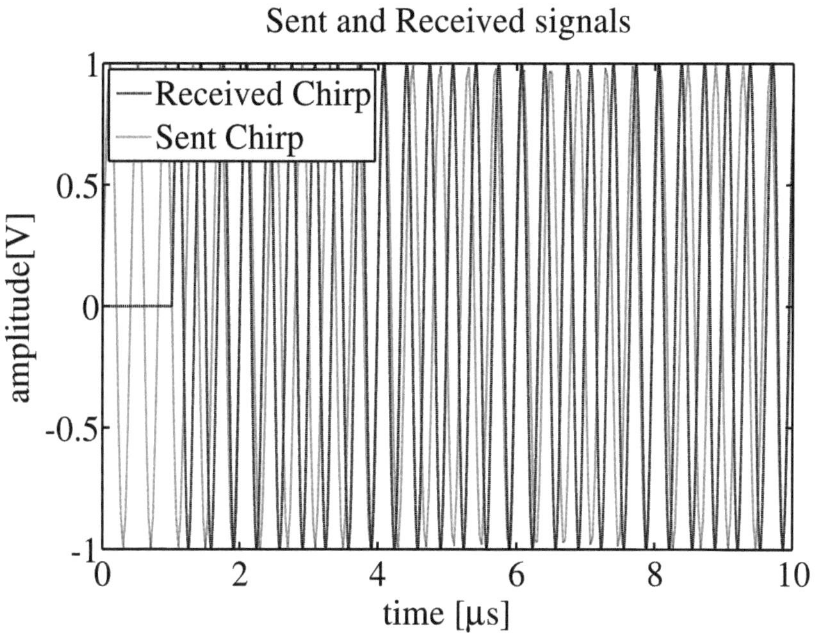

Fig. 12.2 Ideal representation of a chirp fed to the SUT (*thin trace*) and a received chirp (*thick trace*) with a 1 μs delay

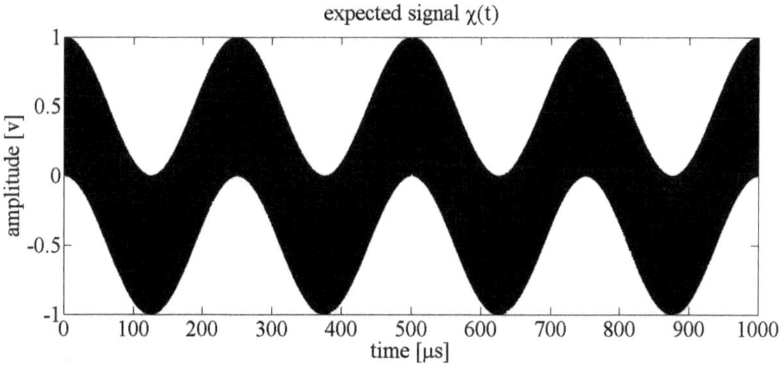

Fig. 12.3 Ideal representation of expected signal referred to sent and received signals in Fig. 12.2

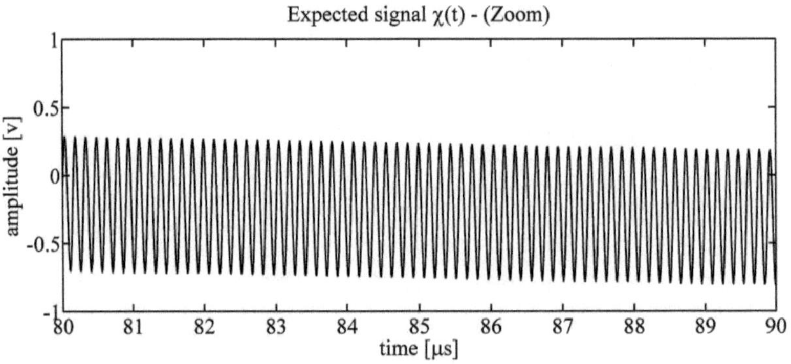

Fig. 12.4 Enlarged version of the signal in Fig. 12.3 representing the time interval [0.08–0.09] ms

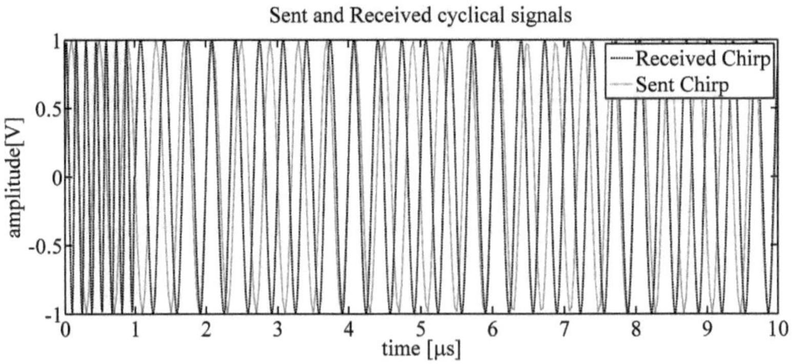

Fig. 12.5 Cyclic overlap of chirp signal fed to the SUT (*thin trace*) and acquired chirp (*thick trace*) in cyclic evaluation.

In the frequency domain, this condition implies:

$$\begin{aligned} \Delta f_{\max} &= (n_f)_{\max}\frac{f_{\mathrm{ADC}}}{N} = \frac{B}{2} \\ (n_f)_{\max} &= \frac{B}{f_{\mathrm{ADC}}}\cdot\frac{N}{2}. \end{aligned} \tag{12.12}$$

In order to illustrate the *aliasing-like* behavior of the interference frequency, the value of Δf_1 vs Δt is plotted in Fig. 12.6.

Another constraint that should be taken into account derives from Eq. (12.10): the expected signal is the composition of a low frequency component, and a chirp with starting frequency $2f_0$, where f_0 is the starting frequency of the chirp originally sent.

In order to avoid the superposition in the frequency domain of the low frequency component with the superimposed chirp, the following condition must be verified:

$$\Delta f_{\max} < 2f_0 \quad \Longrightarrow \quad (n_f)_{\max} \leq 2\cdot n_{f0}\,. \tag{12.13}$$

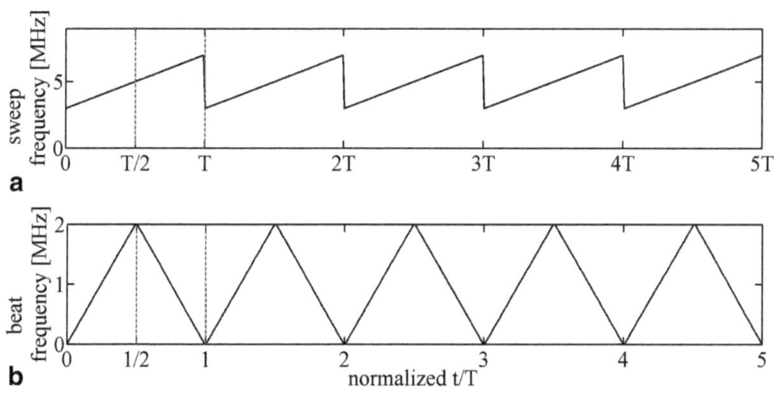

Fig. 12.6 **a** Plot of $f_s(t)$ versus t(a). **b** Plot of Δf_1 vs Δt. Both plots are obtained for an input signal $s(t)$ defined by $R = 100$MHz, $f_0 = 3$MHz, $B = 4$ MHz, $T = 1$ ms, $N = 100,000$.

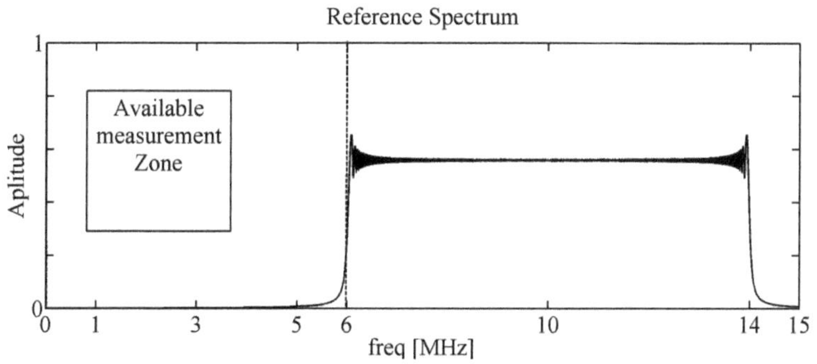

Fig. 12.7 Working zone in DFT spectrum for the same reference conditions used for the previous plots (Zoomed up to 15 MHz).

This condition is illustrated in Fig. 12.7.

In summary, in cyclic operation, the inspection range is limited independently by either the previous constraints. However, we note that, if $f_0 = \frac{B}{4}$, the two conditions coincide:

$$2f_0 = \frac{B}{2} = \Delta f_{\text{max}} \quad \Longrightarrow \quad 2 \cdot n_{f_0} = (n_f)_{\text{max}} = \frac{B}{f_{\text{ADC}}} \cdot \frac{N}{2} \,. \tag{12.14}$$

By adopting this choice of the parameter, it follows that:

$$f_1 = f_0 + B = \frac{5}{4} \cdot B \,. \tag{12.15}$$

With the previous limitations, it is possible to define a lower value for the chirp start frequency; however, we note that usually it is the devices that lead to the bandwidth choice.

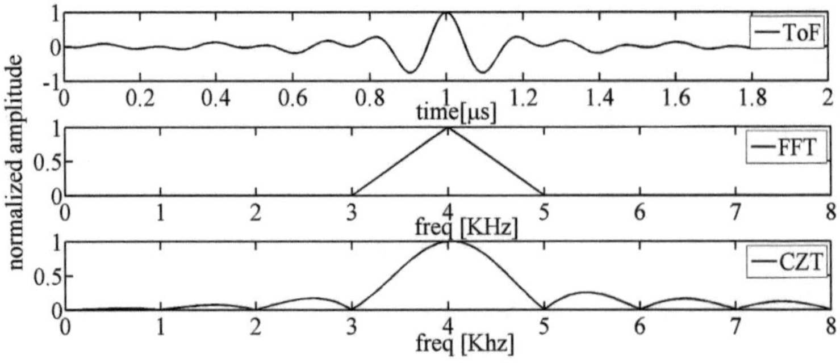

Fig. 12.8 Comparison among samples distribution in different domain after matched filter (*top*), DFT (*center*), CZT (*bottom*)

Moreover, in order to faithfully compare the matched filter results with the ones of the FMCW technique, henceforth, we consider $f_1 \le \frac{f_{\text{ADC}}}{2}$.

12.2.3 Discrete Fourier Transform vs. Chirp z-Transform

In digital signal processing, the number of acquired samples is a limiting factor for the standard procedures based on the use of transforms, such as the DFT. The ToF measurement resolution can be enhanced by using the CZT instead of the DFT. The CZT can compute the z-transform on an arbitrary spiral contour and in this case all the acquired samples are used to calculate the CZT of $\chi(t)$ on the unit circle between two frequency values f_A, f_B. The resolution of the FMCW can be, therefore, significantly improved and the amount of this increase in resolution can be easily computed.

For instance, in compliance with the constraints indicated above for the inspection range, by setting $f_A = 0, f_B = \frac{B}{2}$, the frequency resolution of the CZT is $\frac{2f_{\text{ADC}}}{B}$ times higher than the DFT one. By resuming the discussion of the previous section the resolution analysis of the various methods can be summarized as follows:

$$\frac{n_t}{n_f} = \frac{f_{\text{ADC}}}{B} \qquad \frac{n_{f_{\text{CZT}}}}{n_f} = \frac{2f_{\text{ADC}}}{B} \qquad \Longrightarrow \qquad \frac{n_{f_{\text{CZT}}}}{n_t} = 2\,. \tag{12.16}$$

The example shown in Fig. 12.8 illustrates these results by showing the signals calculated with the three methods in the case $\Delta t = 1$ μs, $R = 100$ MHz, $f_0 = 3$ MHz, $B = 4$ MHz, $T = 1$ ms.

It is worth noting that Eq. (12.16) has been obtained by considering the maximum obtainable inspection range, corresponding to $f_A = 0, f_B = \frac{B}{2}$. Of course, the CZT can also zoom in on a more reduced frequency range, in order to obtain higher resolution. This can be a limitation or not, depending on the specific application considered. If, for example, the FMCW technique is applied to precision thickness estimates, reducing the frequency range gives higher resolution but reduces the maximum measurable thickness. On the other hand, it is not a limitation if the goal is to

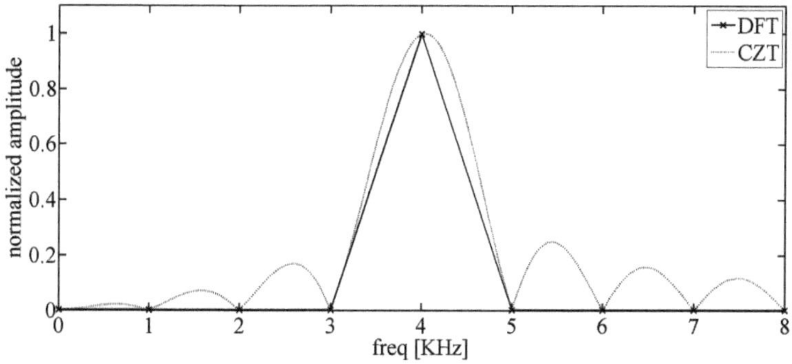

Fig. 12.9 Comparison between DFT and CZT resolution for $f_A = 2\text{kHz}$ and $f_B = 4\text{kHz}$. In the selected range, the DFT curve has five points, while the CZT curve has 100×10^3 points

measure a small thickness. An example of the CZT zoom capability is reported in Fig. 12.9.

12.3 Numerical Simulations

In order to evaluate the performance of the FMCW ultrasonic range finder, it is of utmost importance to verify the obtainable ToF resolution. Besides the theoretical resolution, it is also important to test the robustness of the method and evidence its limitations; this can be done by performing numerical simulations, as we will see they show that both DFT and CZT approaches suffer problems for small delays, as evidenced in Fig. 12.10; on the contrary the matched filter method, in the same context does not suffer any lack of resolution. Moreover, the simulations clearly show the coarse resolution obtainable with the DFT analysis for delays corresponding to a few samples.

To study the reasons of this lack of robustness of the FMCW procedure for small delays, we have to go in further detail. For very small Δt values, it may happen that $\Delta f = \frac{B}{T}\Delta t < \frac{1}{T} = f_{\text{REP}}$, this means that the low-frequency interference term does not accomplish any complete cycle within the signal repetition period T, thus hampering the correct measurement of the oscillation frequency, as shown in Figs. 12.11 and 12.12.

This limitation can be tackled by following appropriate strategies. A simple and effective solution consists in introducing a nonzero shift bias in the reference signal, i.e., instead of mixing the output signal $y(t)$ with a copy of the input signal $s(t)$, $y(t)$ is mixed with a shifted replica of $s(t)$ as in $\chi(t) = y(t) \cdot s(t + t_0)$. In this way, if $t_0 > \frac{1}{B}$ then $\Delta f > f_{\text{REP}} \forall \Delta t$ and at least a whole period of the interference term is completed in a period T.

By introducing this artificial shift, the CZT works correctly and small thicknesses can be well measured even if oscillations in the ToF estimation are yet visible, as shown in Fig. 12.13, since the number of cycles for period, $\frac{\Delta f}{f_{\text{REP}}}$, is very small. If the

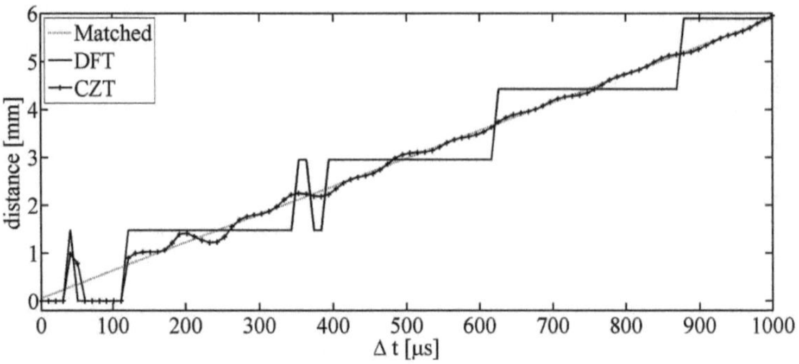

Fig. 12.10 Range evaluation with: matched filter (*light dashed curve*); DFT transform (*solid line*); and CZT (*line with marker*)

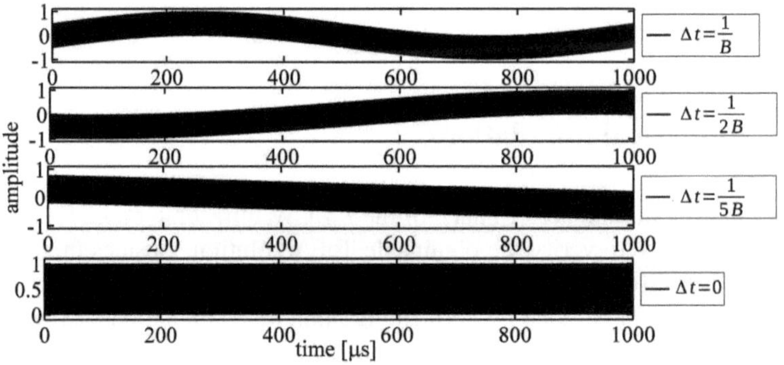

Fig. 12.11 Trends of the mixed signal $\chi(t)$ for small Δt values: from *top to bottom*: $\Delta t = \frac{1}{B}$, $\Delta t = \frac{1}{2B}$, $\Delta t = \frac{1}{5B}$, and $\Delta t = 0$

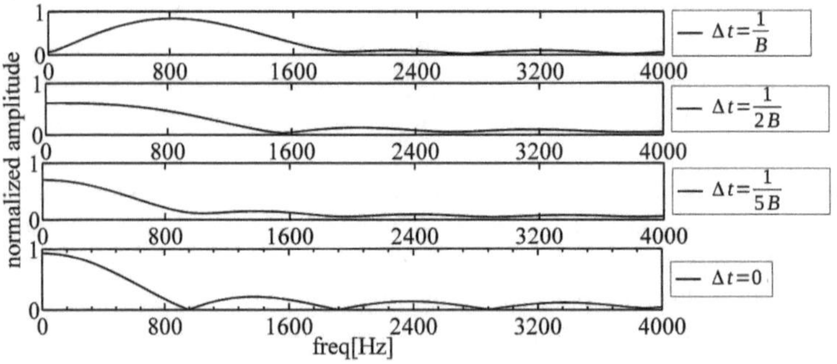

Fig. 12.12 CZT of $\chi(t)$ for small Δt values: from *top to bottom*: $\Delta t = \frac{1}{B}$, $\Delta t = \frac{1}{2B}$, $\Delta t = \frac{1}{5B}$ and $\Delta t = 0$

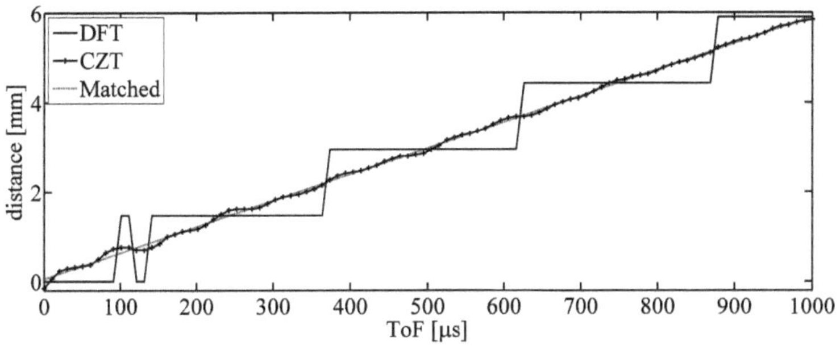

Fig. 12.13 Trends obtained introducing in processing the reference signal shifted by $\frac{f_{ADC}}{B}$ samples: matched filter (*thin dashed line*), DFT (*solid line*), CZT (*line with marker*).

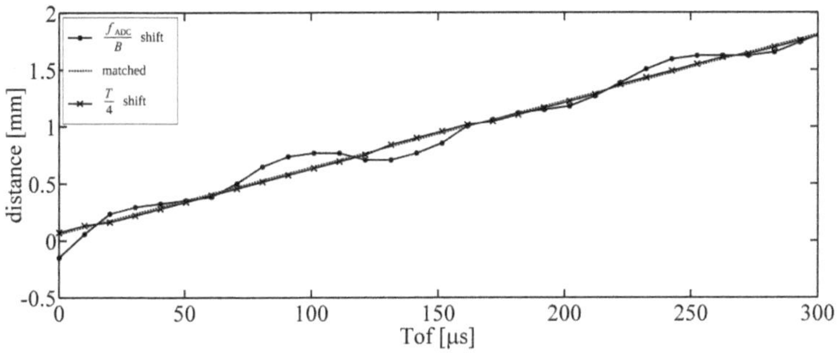

Fig. 12.14 Illustration of shift effect for small delays in CZT evaluation. Plot shows, reference behavior obtained with matched filter (*thin dashed line*), CZT with $\frac{f_{ADC}}{B}$ shift (*line with bullets*), and CZT with $\frac{T}{4}$ shift (*line with cross markers*).

optimal resolution and robustness are needed for very little delays, the bias shift can be further increased; in this way, the performances of the matched filter approach are recovered, as shown in Fig. 12.14.

The performance of FMCW technique has been also evaluated by means of numerical simulations in the presence of additive white Gaussian noise (AWGN). In order to accomplish this aim, a campaign of simulations was accomplished according to the experimental scheme represented in Fig. 12.15 and considering two different cases:

- CASE 1: after fixing the time delay Δt, corresponding to a fixed samples delay @ $f_{ADC} = 100$ MHz, and fixing the number of averages on periods of the chirp, the noise was grown up to reach an SNR set to $\frac{1}{60}$. A comparison was then performed by varying the the length T of the chirp, i.e., the factor $T \cdot B$. The procedure was repeated for the DFT, CZT and matched filter approaches.
- CASE 2: after fixing both the time delay Δt and the period T, the noise was grown up to reach an SNR value set at $\frac{1}{60}$. Then, a comparison was performed by increasing the number of averages. The procedure was repeated for the DFT, CZT, and matched filter approaches.

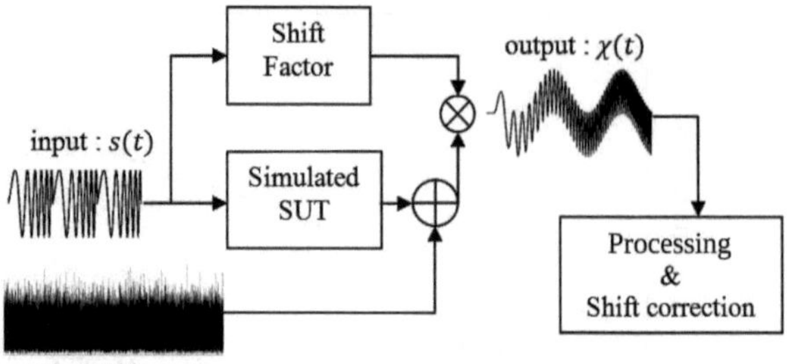

Fig. 12.15 Configuration for the introduction of noise into the system

The simulations show that the matched filter approach, as expected, is the most robust against AWGN noise; it also exhibits the highest resolution; indeed it allows the correct measurement of the ToF also for values of the SNR low enough to lead the other techniques to fail. Nonetheless, the DFT and the CZT analysis reach very similar performances with respect to the matched filter method in term of SNR enhancement, and the differences with the matched filter approach emerge only for high noise levels.

In conclusions, the simulations indicate that the FMCW can be a valid alternative to the matched filter for the measure of the ToF in most situations. In the following section, some analysis performed on experimental data are reported to corroborate the results of the numerical simulations.

12.4 Experimental Results

In order to test the procedure and the numerical simulations, a number of laboratory experiments was performed. The first set of tests was performed in order to have precision estimates of the thickness of three test pieces using contact ultrasonic probes: a forged cylinder 300 mm tall, a 10 mm thick aluminum plate, and a steel plate 0.8 mm thick.

The second set of tests was performed with noncontact ultrasonic probes applied to range measurements in air on a regular grid.

12.4.1 Frequency Modulated Continuous Wave for Thickness Estimate

The three different samples considered for the first set of tests were: a forged cylinder 300 mm tall, a 10 mm thick aluminum plate, and a steel plate 0.8 mm thick. The experimental procedure included an initial test with the probes in direct contact, so as to perform a calibration of the system: the acquired signal is plotted in Fig. 12.16. After this initial calibration, we started the measurements on the test

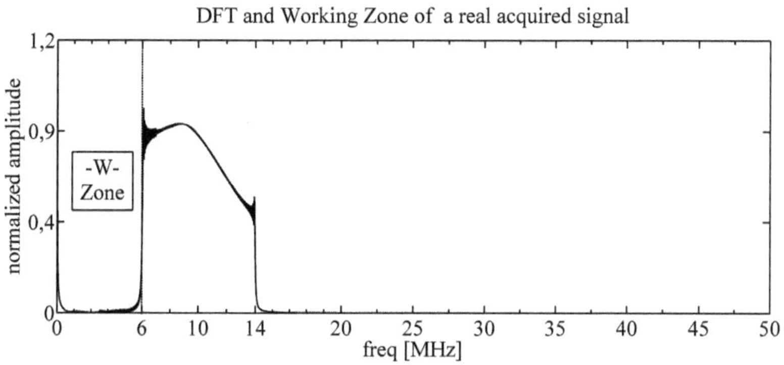

Fig. 12.16 DFT of a zero delayed signal acquired in laboratory with probes in contact

pieces. The forged component was analyzed in reflection mode, while the thickness of the two plates was measured in trough-transmission mode. The probes used are 0.5 in. Olympus Panametrics-NDT V109 broadband transducers with center frequency of 5 MHz and 75 % relative bandwidth. The parameters corresponding to the data acquisition and generation steps are: sampling rate equal to the generation rate and set at $f_{ADC} = 100$ MSa/s; chirp start frequency $f_0 = 3$ MHz; chirp bandwidth $B = 4$ MHz; chirp duration $T = 1$ ms implying $N = 100,000$. The same acquired data was processed by following the three different procedures, in order to compare the resolution obtainable with the DFT, CZT, and the matched filter techniques, the three Figs. 12.17, 12.18, and 12.19 report the results for the three examples. Each figure allows the comparison of the three techniques (DFT, CZT, and matched filter) applied to the same data set. The results evidence that, among the various techniques, the CZT guarantees the highest resolution, due to the possibility to concentrate all the available points in the frequency range of interest, i.e., in the corresponding spatial range of interest; on the contrary, the DFT presents the lowest resolution among the three methods.

Quite generally, the results show that the FMCW approach achieves results very close to those of the standard PuC procedure both in term of resolution and of SNR. In particular, the FMCW output almost coincided with the envelope of the impulse response. The FMCW approach can therefore replace the matched filter in that cases where a reduced computation costs is required. Indeed, with respect to the PuC algorithm, FMCW needs: (i) sample-wise signal multiplication; (ii) an optional moving average filter; (iii) a CZT operation that can be practiced on subsampled data. On the other hand, the PuC algorithm needs, when optimized, (i) a sample-wise signal multiplication; and (ii) two DFT operations. For large numbers of samples, the computational costs are nearly halved if no subsampling is performed and the saving can be further increased if subsampling is performed.

For completeness, the robustness of the technique was tested also evaluating the subsampling effects. Figure 12.20 shows that the ToF measurement remains stable and reliable even for large subsamplings; the CZT exhibits the best resolution while

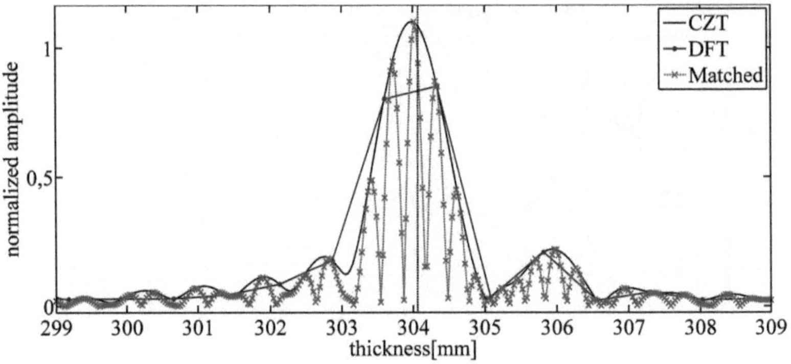

Fig. 12.17 Thickness measure on a forged steel component of approximately 300 mm: DFT (*line with round markers*), CZT (*solid line*), and matched filter (*dashed line with cross markers*)

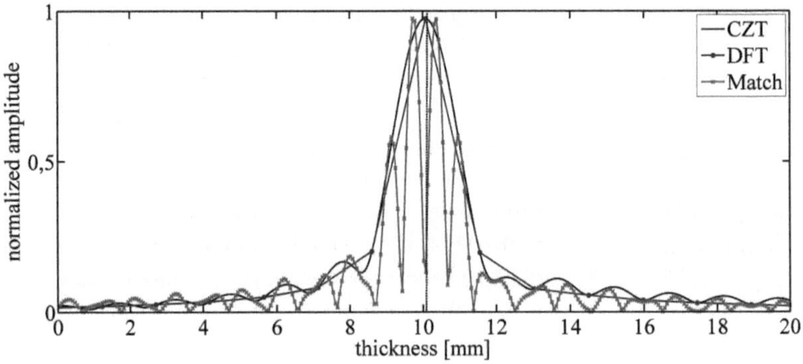

Fig. 12.18 Thickness measure on an aluminum plate 10 mm thick: DFT (*line with round markers*), CZT (*solid line*), and *matched filter* (*dashed line with cross markers*)

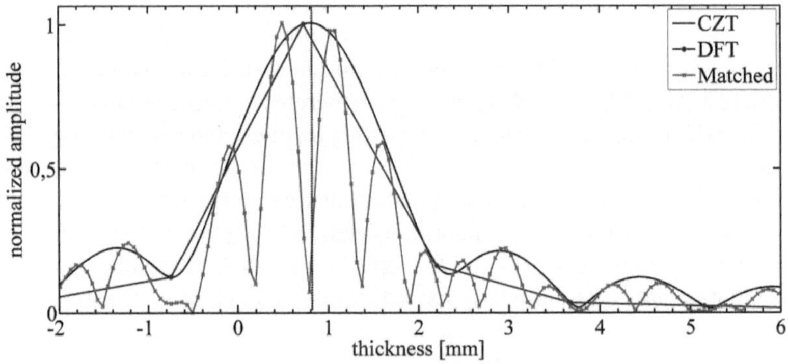

Fig. 12.19 Thickness measure on a steel plate 0.8 mm thick: DFT (*line with round markers*), CZT (*solid line*), and *matched filter* (*dashed line with cross markers*)

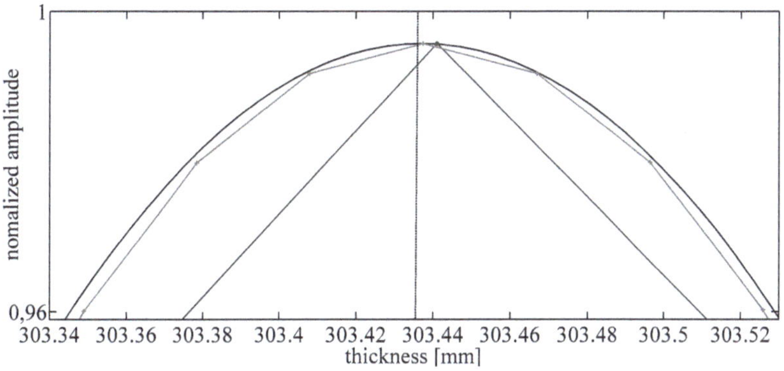

Fig. 12.20 Effect of subsampling in thickness evaluation: optimized CZT (*thick line*), matched filter envelope (*solid line*), signal sampled at $\frac{1}{25}$ of nominal rate with CZT processing (*thin line with star marker*)

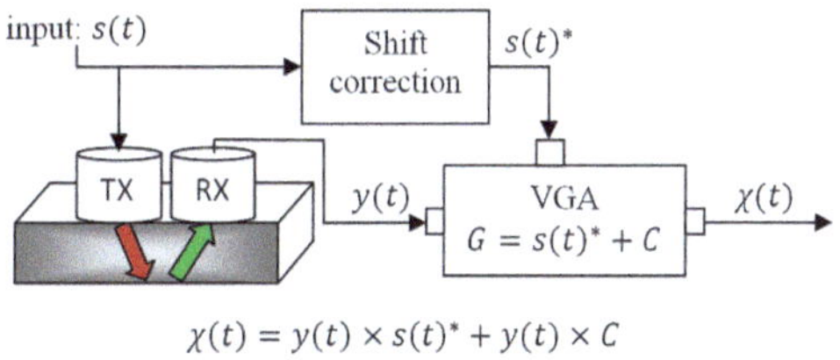

Fig. 12.21 VGA set-up configuration

the matched filter envelope almost coincides with a subsampled version of the CZT output. The subsampled signal was obtained with an acquisition rate of 4 MHz, whereas the chirp signal had a frequency range between 3 and 7 MHz.

The advantages of the FMCW approach are not only confined to the saving of the computation costs but, on the contrary, the key point is that the FMCW procedure can be entirely implemented in hardware by properly combining analog components such as VCOs, mixers, amplifiers, etc. By this, the CZT remains the sole step to be software implemented, in order to adjust the resolution depending on the setup parameters. With this, the whole procedure becomes at the same time very fast (in fact, so fast that it can be easily deployed for online inspections), and very economic, since it relaxes the hardware requirements. A number of tests can be performed to verify this aspect; as a preliminary test, we used a variable gain amplifier (VGA) to contextually implement the necessary amplification of the output signal and the mixing of the output signal $y(t)$ with the reference one $s(t + t_0)$. In particular, the output signal $y(t)$ was the input of the amplifier while the reference signal acted as the voltage controlling the gain. A schematic of the adopted setup is shown in Fig. 12.21.

It is straightforward to see that, in this configuration, the output of the *VGA* differs from the expected signal $\chi(t)$, since a constant bias component in the gain is present,

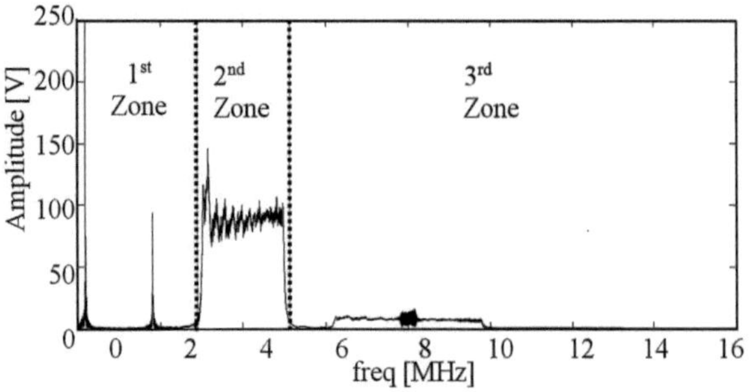

Fig. 12.22 Fast Fourier transform (FFT) of the signal processed with VGA

that leads to a slightly different output $\bar{\chi}$, that is expressed by:

$$\overline{\chi}(t) = y(t) \times (C \cdot s(t + t_0) + C) = C(\chi(t) + y(t)) \tag{12.17}$$

where $2C$ is the gain of the VGA.

By taking into account Eq. (12.12) and Fig. 12.6, the proposed setup introduces a term proportional to the original output signal, and this causes an extra region in the spectrum; a further constraint is thus present for the choice of f_0 and f_1. This effect was tested in laboratory using a chirp signal, with $R = 100$ MHz, $f_0 = 3$ MHz, $B = 2$ MHz, $T = 1$ ms, $N = 100{,}000$. The signal was sent to an aluminium plate 10 mm thick and processed with a VGA amplifier. The output is shown in Fig. 12.22.

In the plot, it is possible to identify three zones. The third one is due to the overlying chirp that comes from the multiplication effect ($C \cdot \chi(t)$). Zone 2 comes from the direct amplification of the probe's output signal ($C \cdot y(t)$). Finally, zone 1 is the working zone for measurements, as evidenced in the previous sections. This procedure, and the three related zones, introduce a limitation in the range of the measurable ToFs, and accordingly in the measurable thicknesses; nevertheless, it has the advantage to contextually implement both the necessary amplification of the output signal by using a low noise amplifier and the mixing of the amplified output with the reference signal.

By properly choosing the chirp frequency range, the three zones do not overlap and therefore the proposed scheme can be effectively used in real experiments.

12.4.2 Frequency Modulated Continuous Wave for Noncontact Range Estimate

The test consists in performing several measurements of distance with the help of a plotter scan that allows to move one of the probes on a regular grid in order to vary the distance between sending and acquiring probe. As a first step, also in this case the measurement system was characterized by transmitting the ultrasound signal in free air: ten measurements were taken in each position. The results allowed to extract a statistic of the estimates.

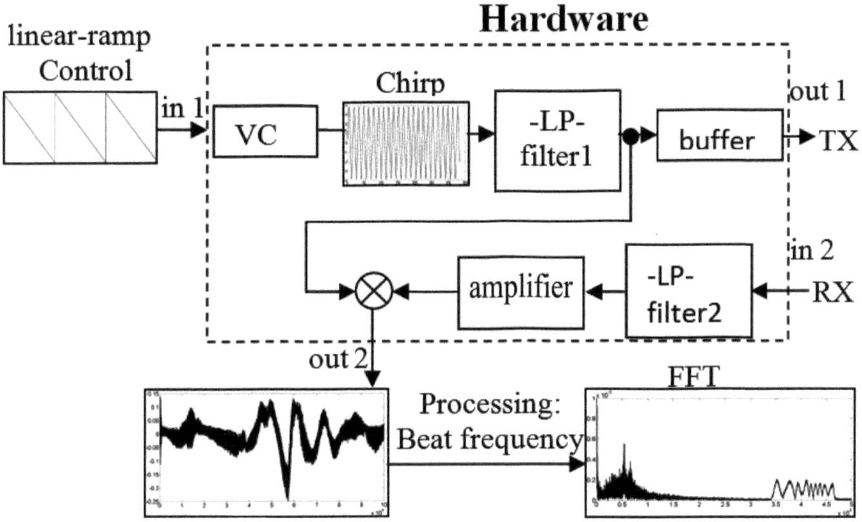

Fig. 12.23 Schematic layout of the hardware device adopted for the range measurements experiment

After this preliminary phase, the actual tests were performed; for each position, the system saved both the transmitted and the acquired signals, the mixer output, the acquisition rate, ToF statistics, signal spectra, and some external controls. The low frequency output was analyzed by using different techniques. By using the acquired data, it was also possible to perform an off line estimation of the ToF by using the matched filter techniques, in order to compare the two approaches.

Figure 12.23 shows the layout of the analog processing hardware adopted for the range measurements experiment, capable of generating an *linear chirp* (LChirp), transmitting the signal, receiving it, and mixing it with the transmitted chirp pulse train. In particular, the system includes a VCO, whose instantaneous frequency is controlled by an external voltage signal. Thus, by driving the VCO with a saw-tooth signal it outputs a periodic pseudo-chirp signal that is a rectangular pulse train with variable frequency. The pseudo-chirp signal is low-filtered (with cutoff at $f_L = 70$kHz) to eliminate the 3rd harmonic typical of the square wave, and any higher unwanted frequency component, in order to obtain a signal as close as possible to an ideal chirp. After that the signal is split, feeding both the TX element, after a buffering to prevent load effects, and one of the mixer inputs. Conversely, the received signal is band-pass filtered, removing unwanted spectral content outside the 10–100 kHz band, amplified (gain set to 1.875) and sent to the second mixer input.

The scanning setup, shown in Fig. 12.24, was used to generate and acquire several signals, with the following parameters:

- Measurement range: 50–560 mm
- Number of steps: 170 (resolution 3 mm)
- Propagation medium: air
- Acquisition rate: 500, 200, 100, 12.5 kHz

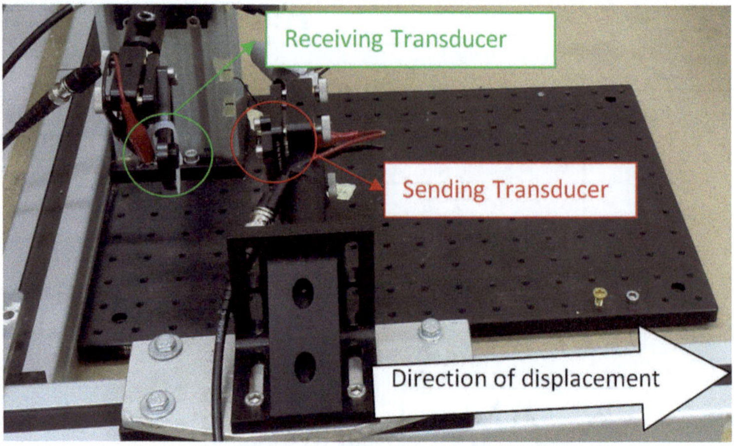

Fig. 12.24 The laboratory setup for the range measurements experiment

- Chirp bandwidth B: 12.5 kHz
- Chirp start frequency f_1: 35 kHz
- Chirp period T: 100 ms

Figure 12.25 shows: (a) the transmitted signal; (b) the acquired signal; and (c) the mixer output, obtained for a fixed distance between the transmitter and the receiver set at 300 mm. Furthermore, Fig. 12.25(d) shows the impulse response (IR), i.e., the correlation sequence obtained using matched filtering, while Fig. 12.25(e) shows the spectrum of the mixer output $\chi(t)$, obtained using FFT techniques. It can be seen that the peak frequency of the mixer output spectrum changes as well as the correlation peak, as a function of the distance d. The measurement resolution is limited by the width of a FFT frequency bin, and this can be a limitation, which can be reduced only by taking long acquisitions. Two different solutions were tested in this laboratory experiment to relax such requirement: (i) adopting the CZT; and (ii) making use of a parabolic analytical interpolation (INT) to define the beat frequency by imposing the parabola to pass through three points: the maximum revealed by FFT; the previous and the next sample [22]. Both approaches were implemented, and the resulting ToF estimations were compared with the traditional pulse compression method.

The low-frequency beat was acquired at different acquisition rates, using the proposed algorithms for various sampling approaches. In particular, Fig. 12.26 shows the estimated distance, obtained by converting the underlying ToF estimations, as a function of the actual distance d, for a sampling frequency of 500 kS/s. Figure 12.26(a) was obtained using the standard PuC (i.e., correlation techniques). Figure 12.26(b) was obtained using FFT evaluation. Figure 12.26(c) was obtained using the CZT transformation. Finally, Fig. 12.26(d) was obtained by applying parabolic interpolation to the FFT results.

Similarly, Fig. 12.27 was obtained following the same procedure but at a sampling rate of 12.5 kS/s. The comparison between the two measurements shows that the

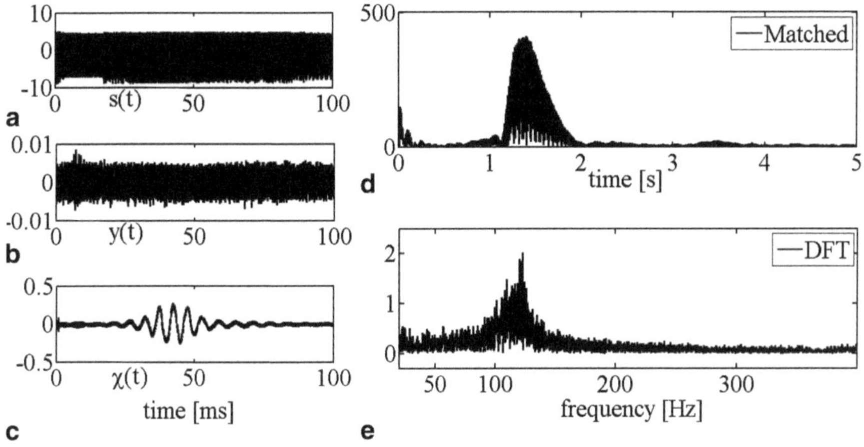

Fig. 12.25 Measured and processed signals to estimate the impulse response (*IR*)

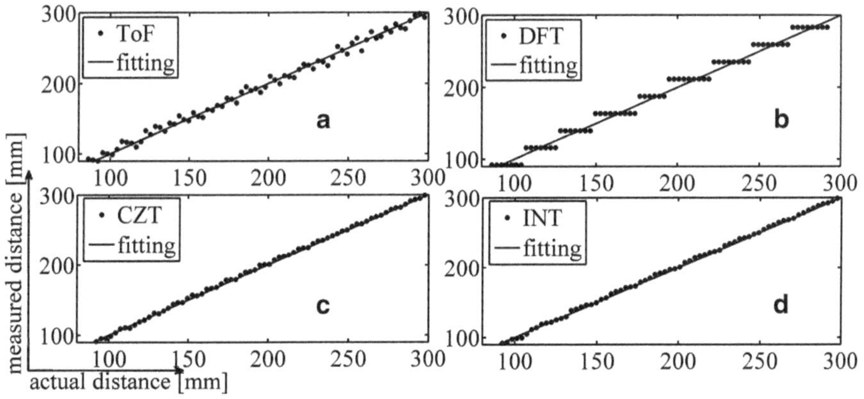

Fig. 12.26 Comparison among the results obtained for the estimated distance as a function of the actual distance d, for a sampling frequency of 500 kS/s

sample rates are the limit values for the chosen acquisition rates. For each technique, a best fit line was obtained, by applying the least mean square (LMS) method to the measured data. Then, the global error (GE), defined as the summation of the distance of each point from the best fit straight line, was evaluated for each algorithm, and used as indicator of the measurement precision. The LMS method also allows the computation of the experimental standard deviation σ of the range measurement, that, assuming an underlying Gaussian distribution, defines the confidence interval with a probability 68.3 % of confidence. Through these two parameters, it is possible to compare the different measurements and to assess their precision and accuracy. The analysis results are summarized in the following Table 12.1, that shows GE and σ for various sampling rates, ranging from 500 kS/s to 12.5 kS/s, well beyond the ultrasonic

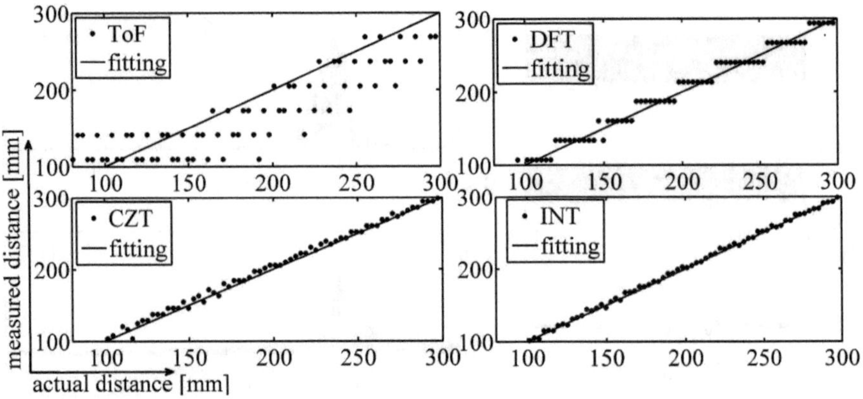

Fig. 12.27 Comparison among the results obtained for the estimated distance as a function of the actual distance d, for a sampling frequency of 12.5 kS/s

Table 12.1 Global Error (GE) and σ in measurement for different sampling rates; N is the number of acquired samples

	500 kS/s $N = 100{,}000$		200 kS/s $N = 40{,}000$		100 kS/s $N = 20{,}000$		12.5 kS/s $N = 2500$	
	GE	σ	GE	σ	GE	σ	GE	σ
ToF	13230	8.85	19032	10.61	19540	10.75	244380	38.02
FFT	11633	8.30	11327	8.19	11988	8.42	15427	9.55
CZT	1156	2.62	3133	4.30	8645	7.15	9050	7.31
INT	2458	3.81	2473	3.82	2696	4.00	3659	4.65

sensors central frequency and corresponding to an ultrasound wavelength in air of approximately 27 mm. It can be observed that the CZT and the FFT interpolation appear as the most promising approaches, offering both the minimum GE and σ. Also, they appear robust to reductions of sampling frequency and the record length N.

12.5 Conclusion

The FMCW ultrasonic measurement scheme performs very well in thickness and range measurements and its use in connection to CZT and/or interpolated-DFT ensures a ToF measurement resolution as high as the matched filter procedure and even better. The analysis in the presence of noise also showed that, although the matched filter approach is the most robust one, the results of the FMCW are quite similar and the SNR gain depends on the $T \cdot B$ product in the same manner.

Moreover, since the FMCW method relies on the measurement of low-frequency signals, a sampling rate significantly reduced can be used. The development of an

FMCW ultrasonic range-finder device performing most of the processing at an hardware level is therefore highly desirable. Such implementation could relax both the computational time and the ADC resources needed allowing for low-cost instrumentation, high-frequency apparatuses, online inspections, etc. Moreover, also the possibility to replace the linear chirp with other excitation signals, i.e., logarithmic chirps, should be investigated in order to further improve the ToF resolution and the robustness against noise.

The presented results show that the FMCW can be adopted to perform accurate distance measurements also when only a low sampling rate can be applied, due to constraints on components cost and performance. The CZT and the DFT parabolic interpolation seem to be the most accurate techniques, especially when the sampling frequencies and the record length assume low values.

References

1. Battaglini L, Ricci M, Senni L (2013) Frequency modulated continuous wave ultrasonic radar. In: 18th International Conference on Digital Signal Processing (DSP), IEEE, pp 1–8
2. Cai C, Regtien PPL (1993) Accurate digital time-of-flight measurement using self-interference. IEEE Trans Instrum Measure 42(6):990–994
3. Challis RE, Ivchenko VG (2011) Sub-threshold sampling in a correlation-based ultrasonic spectrometer. Measure Sci Technol 22(2):1–12
4. Chiao RY, Hao X (2003) Coded excitation for diagnostic ultrasound: a system developer's perspective. In: 2003 IEEE Symposium on Ultrasonics, IEEE, vol 1, pp 437–448
5. Gan TH, Hutchins DA, Billson DR, Schindel DW (2001) The use of broadband acoustic transducers and pulse-compression techniques for air-coupled ultrasonic imaging. Ultrasonics 39(3):181–194
6. Hymans AJ, Lait J (1960) Analysis of a frequency-modulated continuous-wave ranging system. Proceedings of the IEE-Part B: electronic and communication engineering 107(34):365–372
7. Jiang ZH, Huang-Fu K, Wan JW (2007) A chirp transform algorithm for processing squint mode FMCW SAR data. IEEE Geosci Remote Sensing Lett 4(3):377–381
8. Kunita M, Sudo M, Mochizuki T (2008) Range measurement using ultrasound FMCW signals. In: IEEE Ultrasonics Symposium, 2008 (IUS 2008), IEEE, pp 1366–1369
9. Kunita M, Sudo M, Inoue S, Akahane M (2010) A new method for blood velocity measurements using ultrasound FMCW signals. IEEE Trans Ultrasonics Ferroelectr Freq Control 57(5):1064–1076
10. Lam F, Szilard J (1976) Pulse compression techniques in ultrasonic non-destructive testing. Ultrasonics 14(3):111–114
11. Lee M, Singh RS, Culjat MO, Natarajan S, Cox BP, Brown ER, Grundfest WS, Lee H (2009) Waveform synthesis for the design and image reconstruction of step FMCW ultrasound imaging systems with conformal transducer arrays. In: SPIE Medical Imaging, International Society for Optics and Photonics, pp 72,650C–72,650C
12. Melcher RL, Bolef DI, Merry JB (1968) Frequency modulation CW technique for the measurement of very small changes in ultrasonic velocity. Rev Sci Instrum 39(11):1618–1620
13. Misaridis T, Jensen JA (2005) Use of modulated excitation signals in medical ultrasound. part I: basic concepts and expected benefits. IEEE Trans Ultrasonics Ferroelectr Freq Control 52(2):177–191
14. Misaridis T, Jensen JA (2005) Use of modulated excitation signals in medical ultrasound. part ii: Design and performance for medical imaging applications. Ultrasonics, Ferroelectrics and Frequency Control, IEEE Trans Ultrasonics Ferroelectr Freq Control 52(2):192–207

15. Oppenheim AV, Schafer RW (1989) Discrete-time signal processing. Prentice-Hall, Englewood Cliffs
16. Pollakowski M, Ermert H (1994) Chirp signal matching and signal power optimization in pulse-echo mode ultrasonic nondestructive testing. IEEE Trans Ultrasonics Ferroelectr Freq Control 41(5):655–659
17. Rabiner L, Schafer RW, Rader CM (1969) The chirp z-transform algorithm. IEEE Trans Audio Electroacoustics 17(2):86–92
18. Rao N, Mehra S (1993) Medical ultrasound imaging using pulse compression. Electron Lett 29(8):649–651
19. Ricci M, Callegari S, Caporale S, Monticelli M, Battaglini L, Eroli M, Senni L, Rovatti R, Setti G, Burrascano P (2012) Exploiting non-linear chirp and sparse deconvolution to enhance the performance of pulse-compression ultrasonic NDT. In: IEEE International Ultrasonics Symposium (IUS 2012), IEEE, pp 1489–1492
20. Ricci M, Senni L, Burrascano P (2012) Exploiting pseudorandom sequences to enhance noise immunity for air-coupled ultrasonic nondestructive testing. IEEE Trans Instrum Measure 61(11):2905–2915
21. Ricci M, Senni L, Burrascano P, Borgna R, Neri S, Calderini M (2012) Pulse-compression ultrasonic technique for the inspection of forged steel with high attenuation. Insight Non-Destruct Test Condit Monitor 54(2):91–95
22. Svilainis L, Lukoseviciute K, Dumbrava V, Chaziachmetovas A (2013) Subsample interpolation bias error in time of flight estimation by direct correlation in digital domain. Measurement 46(10):3950–3958

Index

P. Burrascano et al. (eds.), *Ultrasonic Nondestructive Evaluation Systems*,
DOI 10.1007/978-3-319-10566-6

MIX
Papier aus verantwortungsvollen Quellen
Paper from responsible sources
FSC® C105338

If you have any concerns about our products,
you can contact us on
ProductSafety@springernature.com

In case Publisher is established outside the EU,
the EU authorized representative is:
Springer Nature Customer Service Center GmbH
Europaplatz 3, 69115 Heidelberg, Germany

Printed by Libri Plureos GmbH
in Hamburg, Germany